EDITED BY
Dominic Emery and Keith Myers

BP Exploration, Stockley Park
Uxbridge, London

Sequence Stratigraphy

WITH CONTRIBUTIONS FROM
George Bertram
Cedric Griffiths
Nick Milton
Tony Reynolds
Marcus Richards
Simon Sturrock

Blackwell
Science

© 1996 by
Blackwell Science Ltd
Editorial Offices:
Osney Mead, Oxford OX2 0EL
25 John Street, London WC1N 2BL
23 Ainslie Place, Edinburgh EH3 6AJ
350 Main Street, Malden
  MA 02148 5018, USA
54 University Street, Carlton
  Victoria 3053, Australia

Other Editorial Offices:
10, rue Casimir Delavigne
75006 Paris, France

Blackwell Wissenschafts-Verlag GmbH
Kurfürstendamm 57
10707 Berlin, Germany

Blackwell Science KK
MG Kodenmacho Building
7–10 Kodenmacho Nihombashi
Chou-ku, Tokyo 104, Japan

All rights reserved. No part of
this publication may be reproduced,
stored in a retrieval system, or
transmitted in any form or by any
means, electronic, mechanical,
photocopying, recording or otherwise,
except as permitted by the UK
Copyright, Designs and Patents Act
1988, without the prior permission
of the copyright owner.

First published 1996
Reprinted 1997, 1998

Set by Setrite Typesetters, Hong Kong
Printed and bound in Great Britain
at the Alden Press Ltd,
Oxford and Northampton

The Blackwell Science Logo is a
trade mark of Blackwell Science Ltd,
registered at the United Kingdom
Trade Marks Registry

DISTRIBUTORS

Marston Book Services Ltd
PO Box 269
Abingdon Oxon OX14 4YN
(*Orders:* Tel: 01235 465500
         Fax: 01235 465555)

USA
Blackwell Science, Inc.
Commerce Place
350 Main Street
Malden, MA 02148 5018
(*Orders:* Tel: 800 759 6102
         617 388 8250
         Fax: 617 388 8255)

Canada
Copp Clark Professional
200 Adelaide Street, West, 3rd Floor
Toronto, Ontario M5H 1W7
(*Orders:* Tel: 416 597-1616
         800 815-9417
         Fax: 416 597-1617)

Australia
Blackwell Science Pty Ltd
54 University Street
Carlton, Victoria 3053
(*Orders:* Tel: 3 9347 0300
         Fax: 3 9347 5001)

A catalogue record for this title
is available from the British Library

ISBN 0-632-03706-7

Library of Congress
Cataloguing-in-Publication Data

Sequence stratigraphy/edited by Dominic Emery
and Keith Myers; with contributions from
George Bertram . . . [et al.].
        p.   cm.
     Includes bibliographical references
     and index.
     ISBN 0-632-03706-7
     1. Geology, Stratigraphic.
  I. Emery, Dominic
  II. Myers, Keith.
  III. Bertram, George T.
  QE651.S458 1996
  551.7 – dc20

Contents

List of Contributors, iv

Preface, v

Historical Perspective

1 D. EMERY: Historical Perspective, 3

Concepts and Principles

2 K.J. MYERS AND N.J. MILTON: Concepts and Principles of Sequence Stratigraphy, 11

Sequence Stratigraphic Tools

3 G.T. BERTRAM AND N.J. MILTON: Seismic Stratigraphy, 45

4 N.J. MILTON AND D. EMERY: Outcrop and Well Data, 61

5 N.J. MILTON: Chronostratigraphic Charts, 80

6 S.J. STURROCK: Biostratigraphy, 89

Applications to Depositional Systems

7 M.T. RICHARDS: Fluvial Systems, 111

8 A.D. REYNOLDS: Paralic Successions, 134

9 M.T. RICHARDS: Deep-marine Clastic Systems, 178

10 D. EMERY: Carbonate Systems, 211

11 K.J. MYERS: Organic-rich Facies and Hydrocarbon Source Rocks, 238

12 C.M. GRIFFITHS: Computer Modelling of Basin Fill, 258

References, 270

Index, 291

List of Contributors

GEORGE BERTRAM *Stratigraphic Research International, Braehead Avenue, Milngavie, Glasgow, UK*

DOMINIC EMERY *BP Exploration, Uxbridge One, 1 Harefield Road, Uxbridge, London, UK*

CEDRIC GRIFFITHS *NCPGG, Thebarton Campus, University of Adelaide, Adelaide, Australia*

NICK MILTON *BP Norge, Forus, Stavanger, Norway*

KEITH MYERS *BP Exploration, Uxbridge One, 1 Harefield Road, Uxbridge, London, UK*

TONY REYNOLDS *BP Exploration, Sunbury-on-Thames, London, UK*

MARCUS RICHARDS *BP Exploration, Anchorage, Alaska, USA*

SIMON STURROCK *56 Gloucester Court, Kew Road, Kew, London, UK*

Preface

In 1989, the chief geologist of BP Exploration and his senior colleagues recognized the need to expand the company's resource of sequence stratigraphers, and created the Stratigraphic Studies Group. This group initially included a few experts, but was composed chiefly of a mixture of willing geophysicists, sedimentologists and biostratigraphers, who were to train as sequence stratigraphers, but more importantly, were to bring the expertise from their own disciplines to bear on sequence stratigraphy. This merging of geological disciplines with sequence stratigraphic principles first saw the light of day as BP's 'Introduction to Sequence Stratigraphy' course. This course has been given internally since 1991, and has been presented in whole or in part to over a dozen national oil companies and at several international geological and geophysical conferences. The 'Introduction to Sequence Stratigraphy' course manuals formed the basis for this book, because, as we naively thought, it would not be too much trouble to recast the manuals in the form of a textbook, which could form the basis of university and professional courses. Inevitably, it was more difficult than we had imagined. Sequence stratigraphy is a rapidly evolving subject, new terminology was being added as we wrote, and the jargon we sought to demystify continues to grow. This book must thus be seen as a sequence stratigraphic synthesis for the mid-1990s, and not the final word on the subject. It is above all a practical guide and contains tools and techniques that the authors have found useful in their daily work.

We have arranged the book to cover four main themes; a brief history of sequence stratigraphy, concepts and principles, sequence stratigraphic tools and finally the application of sequence stratigraphy to different depositional systems. For the last theme, we have tried to emphasize the importance of seeing sequence stratigraphy in its sedimentological context, and it is recommended that the reader should have some familiarity with sedimentological processes before tackling the last five chapters. Otherwise, the book covers all the basics of sequence stratigraphy, and is intended to be a broad text suitable for undergraduate geologists of all years, MSc and PhD sedimentologists and stratigraphers, and for oil company geoscientists who wish to broaden their knowledge of the stratigraphic methods available for solving problems with which they are routinely faced. We hope you find it interesting and useful.

Acknowledgements

Bob Jones contributed several sections on biostratigraphy and Neil Parkinson contributed to Chapter 2. We are grateful to the following reviewers who have helped improve the book at various stages of its development: David Roberts, Henry Posamentier, Maurice Tucker, Dan Bosence, Mike Bowman, Andy Horbury and Andy Fleet.

We also acknowledge the support of BP Exploration, particularly David Roberts, Bob Rosenthal, John Wills and Peter Melville and we are grateful to BP's partners for permission to publish information on many of the areas and fields mentioned in the text. The following companies are also acknowledged for their assistance in providing seismic data; Lynx Information Systems, Trans-Asia Oil and Mineral Development Corp., Balabac Oil Exploration and Drilling Co. Inc., Crestone Energy Corp., Coplex (Palawan) Ltd., Oriental Petroleum and Minerals Corp., The Philodrill Corp., Seafront Resources Corp., Unioil and Gas Development Co. Inc., Vulcan Industrial and Mining Corp.

George Bertram, Dominic Emery,
Cedric Griffiths, Nick Milton,
Keith Myers, Tony Reynolds,
Marcus Richards and Simon Sturrock

BP Exploration
London, Sunbury-on-Thames,
Glasgow, Stavanger and Anchorage

Historical Perspective

CHAPTER ONE
Historical Perspective

1.1 What is sequence stratigraphy?

1.2 The evolution of sequence stratigraphy

1.1 What is sequence stratigraphy?

Sequence stratigraphy is a subdiscipline of stratigraphy, the latter being defined broadly as 'the historical geology of stratified rocks'. There have been many definitions of sequence stratigraphy over the years, but perhaps the simplest, and that preferred by the authors, is 'the subdivision of sedimentary basin fills into genetic packages bounded by unconformities and their correlative conformities'. Sequence stratigraphy is used to provide a chronostratigraphic framework for the correlation and mapping of sedimentary facies and for stratigraphic prediction.

Several geological disciplines contribute to the sequence stratigraphic approach, including seismic stratigraphy, biostratigraphy, chronostratigraphy and sedimentology. These are discussed in more detail in forthcoming chapters. Note that lithostratigraphy is not considered to contribute usefully to sequence stratigraphy. Lithostratigraphy is the correlation of similar lithologies, which are commonly diachronous and have no time-significance (Fig. 1.1). Lithostratigraphic correlation is useful provided the sequence stratigraphic boundaries enveloping the interval of interest are constrained.

1.2 The evolution of sequence stratigraphy

Sequence stratigraphy is often regarded as a relatively new science, evolving in the 1970s from seismic stratigraphy. In fact sequence stratigraphy has its roots in the centuries-old controversies over the origin of cyclic sedimentation and eustatic versus tectonic controls on sea-level. Much of this early debate has been summarized recently in a set of historical geological papers edited by Dott in 1992 (1992a), entitled 'Eustasy: the Ups and Downs of a Major Geological Concept', and the interested reader is referred to this volume for more detail. Other historically important collections of sequence stratigraphic papers include American Association of Petroleum Geologists (AAPG) Memoir 26, published in 1977, and Society of Economic Paleontologists and Mineralogists (SEPM) Special Publication 42, published in 1988.

Sacred theories

The Deluge and the story of Noah is the most well-known of the earliest references to sea-level change. To the early investigators of sea-level change, the veracity of the Deluge

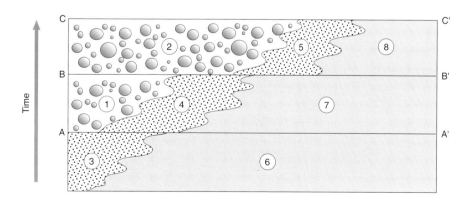

Fig. 1.1 The difference between sequence stratigraphy, which has a geological time significance, and lithostratigraphy, which correlates rocks of similar type. A lithostratigraphic correlation would correlate conglomerate units 1 and 2, sandstone units 3, 4 and 5 and mudstone units 6, 7 and 8. A sequence stratigraphic correlation would correlate time lines A–A', B–B' and C–C'

was not in question, but its origin was the subject of considerable debate by scientists and clergy alike. Perhaps the most popular of several theories were Burnet's *Sacred Theory of the Earth*, published in 1681, and the *Telliamed* of de Maillet, published in 1748 (and recently revisited by Carozzi in 1992). De Maillet proposed that following the formation of the Earth by the accretion of the ashes of burning suns over the cortex of an extinguished sun, a water envelope which developed around the planet gradually diminished in volume through time, and in so doing created the topography we see today. In effect, de Maillet interpreted sea-level changes on Earth as a 'single falling limb of a cosmic eustatic cycle' (Carozzi, 1992). This concept of a one-way sea-level fall was known as Neptunian theory. The erosion of primitive mountains by marine processes and the development of a series of offlapping sediment packages as implied by de Maillet and other Neptunists is illustrated schematically in Fig. 1.2.

The eighteenth century

The eighteenth century also saw the beginning of detailed stratigraphic analysis of rock units, and the recognition of unconformities as primary bounding surfaces. In 1788, Hutton first appreciated the significance of unconformities separating cycles of 'uplift, erosion and deposition', and unconformities were used by stratigraphers such as Sedgwick and Murchison in the following century to establish physical boundaries for geological periods (Sedgwick and Murchison, 1839). As the great stratigraphers continued with their practical approach, William Buckland (1823) proposed the concept of Diluvialism which was to

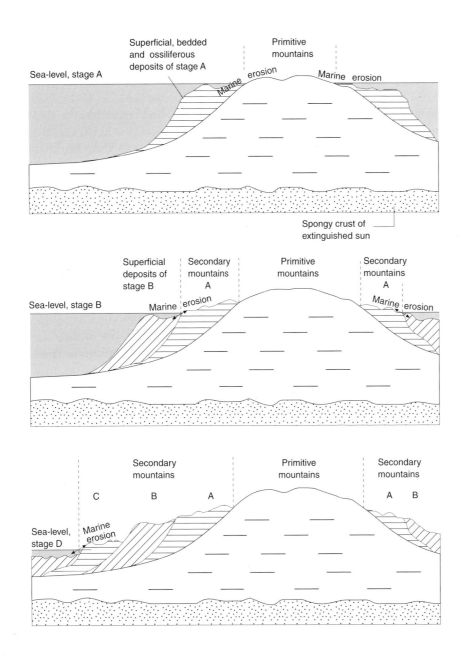

Fig. 1.2 Schematic illustrations of the development of sedimentary units by marine erosion and deposition during continuously falling sea-level, according to de Maillet (1748). After Carozzi (1992)

eclipse Neptunian theory. In diluvial theory, the geological products immediately preceding the flood were referred to as antediluvial, and those following the flood were referred to as post-diluvial or alluvial. However, the attraction of diluvial theory also soon waned as further geological evidence served to counter the simplistic notion of a single dramatic flooding event.

The nineteenth century

In the middle of the nineteenth century, the eustatic versus tectonic controls on sea-level change debate began in earnest with the glacial theories of Lyell and Agassiz. Lyell and others (including Celsius and Linnaeus) observed raised beaches along the coastline of Scandinavia and noted evidence of falling sea-levels from centuries-old marks on shoreline outcrops. Lyell concluded that the land was being slowly and differentially elevated (Lyell, 1835), a fact confirmed by Bravais in 1840 who had observed tilted beaches along fjords of the Scandinavian Arctic coast. At about the same time, Agassiz (1840) was developing his theories of glaciation, and MacLaren, on reviewing Agassiz's glacial theory in 1842, saw the potential of melting ice-caps as a major control on global sea-level. Unfortunately, neither Agassiz nor MacLaren received acceptance for their ideas for at least two more decades, until Croll (1864), in a forerunner of Milankovitch theory (1920), published the concept of orbitally forced glaciations.

The early twentieth century

By the late nineteenth century, glacial theory was thus able to explain eustatic sea-level change and isostatic uplift. However, it was to be several decades before glacial eustasy was resurrected as a control on sedimentary rhythmicity; other explanations of global eustasy took precedence, notably the work of Eduard Suess. Suess first coined the term eustasy in 1906, when he attributed the patterns of onlap and offlap of sedimentary units to global sea-level changes. Suess favoured a mechanism whereby sea-level was lowered by subsidence of the sea-floor, and raised by the displacement of seawater by oceanic sedimentation. He refused to believe the evidence for differential land uplift from Scandinavia, concluding that the Baltic was 'gradually emptying' (Suess, 1888). However, the majority of geologists in the early twentieth century still held the Lyellian view that the major control on sea-level at any point along the coast was the movement of the land. Despite the general lack of support for Suess' ideas, a number of American geologists began to develop concepts of global controls on unconformity development. Foremost amongst these was Chamberlin, who in 1898 and 1909 published his theory on the 'diastrophic control of stratigraphy by world-wide sea level changes'. Three diagrams from his first paper show this to be a precursor of modern sequence stratigraphic concepts (Fig. 1.3).

Chamberlin's ideas were developed by several American

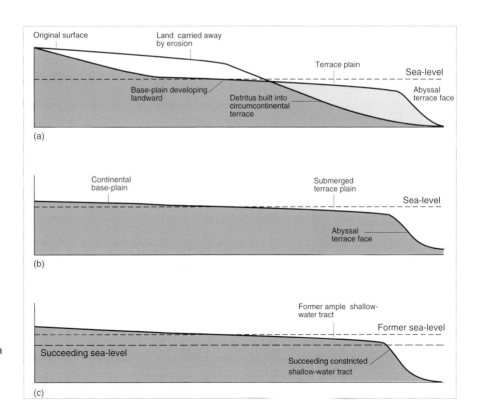

Fig. 1.3 Reproduction of figures from Chamberlin (1898) on diastrophism, unconformities and geological time divisions. From Dott (1992b)

geologists in the following decades, particularly in Palaeozoic systems of the Mid-west. Most notable amongst these were Ulrich and Schuchert, the latter using early palaeogeographic concepts and facies theory to re-create past environments bounded by global unconformities. However, the single most important publication from the 'eustatic' school was that of Grabau, a contemporary of Ulrich and Schuchert, whose 'pulsation theory' postulated rhythmic transgressions and regressions caused by changing heat flow from inside the Earth. The resulting 'pulse beat of the Earth', published in Grabau's 'The Rhythm of the Ages' in 1940, had a periodicity of about 30 million years and caused the development of global unconformities, which could be used to divide the stratigraphic record. Prior to Grabau's work, European geologists, notably Stille (1924) had begun to develop ideas about global unconformities caused by global tectonism with resulting eustatic effects, akin to modern low-order eustatic cycles.

On a smaller scale, sedimentary rhythms were being observed on a scale of metres in coal-bearing Carboniferous (Pennsylvanian) strata in Illinois and Kansas. In 1935, following further studies on Pleistocene glacio-eustatic changes, Wanless and Shepard proposed a control on the development of these Pennsylvanian 'cyclothems' by the accumulation and melting of Gondwana glaciers. This study and others like it resurrected the glacial-eustatic control on sedimentation developed by Croll many decades earlier.

The case for periodicity at a variety of scales in the stratigraphic record was thus becoming compelling. However, as with so many scientific bandwagons, it eventually ran out of steam. In a keynote address to the geological community in 1949, Gilluly argued for orogenesis as a continuous, rather than episodic process, and as a result the concept of rhythmicity of low order (tens of millions of years) gradually lost credibility. The Carboniferous cyclothems were then reinterpreted as autocyclic products, resulting from delta lobe switching and the internal re-organization of sedimentary systems. This latter point also emphasizes the ascendancy of process sedimentology in the early 1960s. Dott (1992a) amusingly points out that at that time many stratigraphers preferred to call themselves sedimentologists!

The late twentieth century

Ironically, Sloss, Krumbein and Dapples (1949) first outlined the concept of stratigraphic sequences at the same meeting that Gilluly proposed his ideas on orogenic continuum. Sloss, Krumbein and Dapples defined sequences as 'assemblages of strata and formations' bounded by prominent interregional unconformities. Despite the negative reaction to these ideas, Sloss (1963) published his major sequences correlateable across the North American Craton, the Indian Tribal names of which still appear as 'super sequences' on the Haq *et al.* (1987) chart. Sloss's ideas were developed further by his graduate students at Northwestern University, one of whom was Peter Vail. Also published at this time was Harry Wheeler's classic 1958 paper on time-stratigraphy which contains many of the concepts in use today, as well as an early attempt to introduce sequence stratigraphic terminology.

Seismic stratigraphy

The next major breakthrough in sequence stratigraphy was in the 1960s and 1970s, when the development of digitally recorded and processed multichannel seismic data made large scale two-dimensional images through basins available. Vail *et al.* (1977a) in AAPG Memoir 26 is perhaps the most referenced work on sequence stratigraphy to date. It summarizes work carried out by Vail and his co-workers, first in the Carter Oil Company and subsequently at the Exxon Production Research Corporation, through the 1960s and early 1970s (Vail and Wilbur, 1966; Mitchum *et al.*, 1976). This period of time marks a break where industry took the lead from academia in the development of sequence stratigraphy. Further papers on seismic sequence stratigraphy followed, and the ideas were gradually extended to incorporate both borehole and outcrop data (Vail *et al.*, 1984). In this work, eustatic sea-level was emphasized as the controlling mechanism for sequence development. In 1985 AAPG Memoir 39 appeared, in which Hubbard *et al.* proposed a tectonic mechanism for the subdivision of basin fill into 'megasequences', driven by changes in tectonic process. The tectonic versus eustatic debate was beginning afresh, although for many at this time seismic stratigraphy was synonymous with eustatic sea-level change, possibly because of its appeal as a global predictive tool for hydrocarbon exploration. In 1987, the Haq *et al.* global sea-level cycle chart was published. This is possibly the single most contentious of all the 'Exxon school' publications, chiefly because the supporting evidence for the curves has not been released. It remains unclear whether local corrections for tectonic uplift or subsidence have been applied, and the dating of unconformities to the accuracy implied by the chart has been challenged (Miall, 1991).

The sequence stratigraphy bandwagon rolls

Special Publication 42 of SEPM, *Sea Level Changes — an Integrated Approach*, was published in 1988 and introduced new concepts such as accommodation space and parasequences, and many of the concepts and principles described in Chapter 2 of this book. Special Publication 42 was important because it opened up the subject to a broader geological community beyond industrial seismic interpreters. In the late 1980s and in this decade, many sequence stratigraphic publications have appeared, some of which uncritically apply the tools and techniques, and some of which are strongly critical. Many question the validity of the interbasinal correlations upon which the Haq *et al.* (1987) curve is based, and others have questioned the validity of certain aspects of the sequence stratigraphic models presented in SEPM 42, such as Miall (1991) and Schlager (1992). Galloway (1989) presented an alternative model for the development of depositional units or 'genetic stratigraphic units' bounded by major flooding surfaces, rather than unconformities. Pitman (1978) has suggested that the origin of sequences and onlap patterns can be explained by variations in subsidence at continental margins, whereas Cloetingh (1988) and Kooi and Cloetingh (1991) proposed that relative sea-level changes and the formation of sequences of millions of years duration can be explained by intraplate stresses rather than eustatic sea-level changes.

The most recent developments in sequence stratigraphy have been in the area of high-resolution subseismic-scale sequence stratigraphy and computer modelling of sedimentary fill. Van Wagoner *et al.* (1990) led the way with the publication of a colourful text on high-resolution sequence stratigraphy from outcrops, logs and core. This stimulated excellent work in superbly exposed marine and marginal marine settings, such as the Jurassic of the Yorkshire Coast and the Cretaceous of the Western Interior Seaway, USA (see also Posamentier and Weimer, 1993, for review). High-resolution sequence stratigraphy also has been combined with work on metre-scale rhythmic successions, particularly bedded platform carbonates and mixed siliciclastic carbonate units (Hardie *et al.*, 1986; Goldhammer *et al.*, 1991). Milankovitch theory of orbital forcing has been revived by sequence stratigraphers in order to explain the origin of high-frequency subsequence-scale cycles. Computer modelling packages have been developed to analyse and replicate the sedimentary fill of basins, at scales from a few metres to entire basins. Basin-wide models include those developed by Royal Dutch/Shell, and published by Aigner *et al.* (1990), and the SEDPAK program developed at the University of South Carolina. Smaller scale cyclicity has been modelled by software such as Mr Sediment (Goldhammer *et al.*, 1989) and by Bosence and Waltham (1990).

The future

The future direction of sequence stratigraphy is difficult to predict, given the turbulent history of the sea-level change debate. At least in the short-term, carbonate systems require further case studies to demonstrate the importance (or otherwise) of controls other than sea-level change. Posamentier and Weimer (1993) have also emphasized the need for further work on the applicability of the concepts to non-marine and deep-marine settings, and further validation (or otherwise) of the sea-level cycle chart from outcrop and subsurface data. Schlager (1992) and others also argue for a more sedimentological approach to sequence stratigraphy, accounting for the autocyclicity of sedimentary processes within the sequence stratigraphic framework. At the very least we can expect considerable debate and further critiques of the subject. This level of activity and debate is all a far cry from the early 1960s when stratigraphy was unfashionable, before Peter Vail and others rescued the subject from its decline.

Concepts and Principles

CHAPTER TWO

Concepts and Principles of Sequence Stratigraphy

2.1 Introduction
   2.1.1 Basin forming processes
   2.1.2 Basin-margin concepts

2.2 Relative sea-level, tectonics and eustasy
   2.2.1 Definitions of sea-level
   2.2.2 Accommodation
   2.2.3 Accommodation through time
   2.2.4 Orders of cyclicity and global correlation

2.3 **Sediment supply**
   2.3.1 Principles of clastic sediment supply
   2.3.2 Filling of accommodation
   2.3.3 Basin architecture

2.4 **Sequences and systems tracts**
   2.4.1 Sequences and sequence boundaries
   2.4.2 Systems tract definition
   2.4.3 Lowstand systems tract
   2.4.4 Transgressive systems tract
   2.4.5 Highstand systems tract
   2.4.6 Type 2 sequence boundary and the shelf-margin systems tract
   2.4.7 Lowstand systems tracts on a ramp margin
   2.4.8 Controls on systems tract boundaries
   2.4.9 Other possible systems tracts within a relative sea-level cycle
   2.4.10 Composite (second and third order) sequences and systems tracts
   2.4.11 Genetic stratigraphic sequences

2.5 **High-resolution sequence stratigraphy and parasequences**
   2.5.1 Introduction
   2.5.2 Parasequences and their continental equivalents
   2.5.3 Parasequence sets
   2.5.4 Parasequence thickness trends
   2.5.5 Sequence boundaries
   2.5.6 Maximum flooding surfaces
   2.5.7 Ravinement surfaces
   2.5.8 Problems and pitfalls of high-resolution sequence stratigraphy

2.1 Introduction

The stratigraphic signatures and stratal patterns in the sedimentary rock record are a result of the interaction of tectonics, eustasy and climate. Tectonics and eustasy control the amount of space available for sediment to accumulate (accommodation), and tectonics, eustasy and climate interact to control sediment supply and how much of the accommodation is filled. Autocyclic sedimentary processes control the detailed facies architecture as accommodation is filled. The purpose of this chapter is to introduce the principles that govern the creation, filling and destruction of accommodation. It then shows how these principles are used to divide the rock record into sequences and 'systems tracts', which describe the distribution of rocks in space and time.

The chapter uses siliciclastic systems to introduce the concepts and principles of sequence stratigraphy. Carbonate systems differ from clastic systems in their ability to produce sediment '*in situ*', and they respond in a different manner to accommodation changes. Carbonates are therefore discussed separately in Chapter 10.

2.1.1 Basin forming processes

Tectonism represents the primary control on the creation and destruction of accommodation. Without tectonic subsidence there is no sedimentary basin. It also influences the rate of sediment supply to basins. Tectonic subsidence results from two principle mechanisms, either extension or flexural loading of the lithosphere. Figure 2.1 illustrates theoretical tectonic subsidence rates in extensional, foreland and strike-slip basins. These curves in effect govern how much sediment can accumulate in the basin, modified by the effects of sediment loading, compaction and eustasy.

Extensional basins form in a variety of plate tectonic settings, but are most common on constructive plate margins. In extensional basins, tectonic subsidence rates vary systematically through time, with an initial period of very rapid subsidence caused by isostatic adjustment to lithosphere stretching, followed by a gradual (60–100 million years) and decreasing thermal subsidence phase as the asthenosphere cools. This systematic change in tectonic subsidence rate has a strong influence on the geometry of the basin-fill, such that it may be possible to divide the stratigraphy into pre-, post- and syn-rift phases (these phases have been termed *megasequences*; Hubbard, 1988). In the simple *syn-rift megasequence* model the sediments are deposited in the active fault-controlled depocentres of the evolving rift and can show roll-over and growth into the active faults. Differential subsidence across the extensional faults may exert a strong control on facies distributions. In the *post-rift megasequence*, any remaining rift-related topography is gradually buried beneath sediments that fill the subsiding basin and onlap the basin margin, creating the typical 'steers head' geometry (McKenzie,

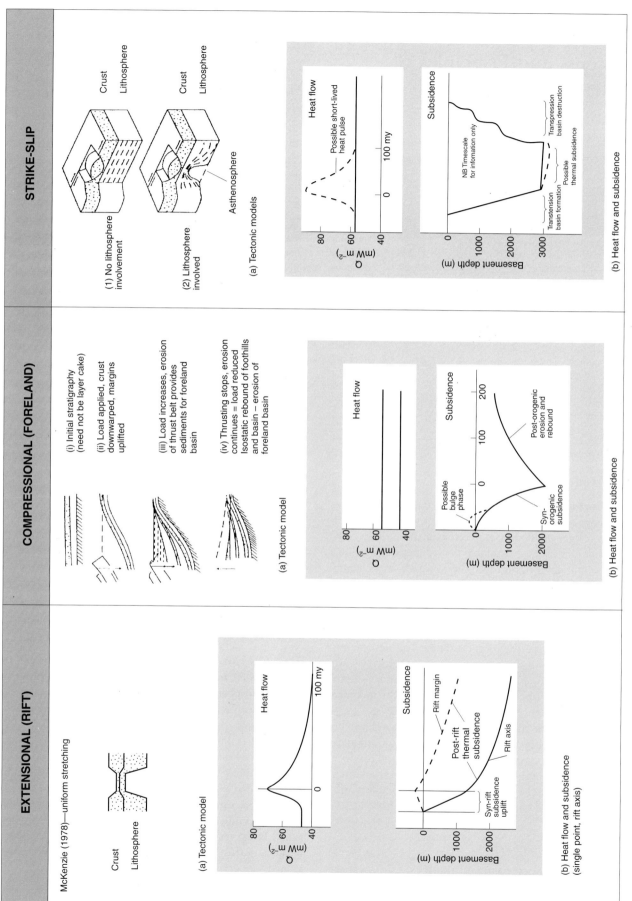

Fig. 2.1 Tectonic subsidence histories in rift, foreland and strike-slip basins

1978). The syn-rift and post-rift megasequences in a marine rift will contain sequences in which development is controlled by higher frequency changes in relative sea-level.

Foreland basins develop in response to loading of the lithosphere below thrust belts. The lithosphere bends in response to loading as the thrust sheets are emplaced, and creates a depression that is accentuated towards the load. The sedimentary fill to this foreland basin has a characteristic wedge shape, thickening towards the thrust front and forming a foreland basin megasequence. The width of the basin is proportional to the rigidity of the underlying lithosphere, and the depth is proportional to the size of the load. Foreland basins formed adjacent to growing mountain belts are characterized by large, and initially rapidly increasing, rates of sediment supply. Cessation of thrusting and continued erosion of the mountain belt leads to an eventual decrease in load, and many foreland basins become uplifted.

Strike-slip basins do not have a characteristic subsidence pattern, although in general, rates of subsidence (and uplift) are extremely rapid.

Tectonic subsidence curves provide a fundamental control on sediment accommodation, upon which higher frequency controls, such as eustasy, fault movement and diapirism, are superimposed. Figure 2.2 shows calculated tectonic subsidence curves for two real basins. In the Llanos Basin, Colombia, sediment supply has exceeded tectonic subsidence. The basin has remained full to base level, with excess sediment bypassed northwards to the sea. The subsidence curve shows slow subsidence through the late Cretaceous and early Tertiary, linked to thermal subsidence in a back-arc basin setting. Two distinct increases in subsidence rate occur in the mid–late Eocene and mid-Miocene, corresponding to two phases of mountain building in the Andes.

In the South Viking Graben example (Fig. 2.2), typical of a number of rifts, sedimentation has not always kept pace with true tectonic subsidence. This led to periods in the Cretaceous where water depths increased and sediment starvation occurred. In the Tertiary, uplift of the Scottish mainland and adjacent North Sea Basin resulted in increased sediment input to the basin (Milton *et al.*, 1990), which locally filled to base level. The remainder of the basin subsequently filled with sediment, resulting in the present-day shallow sea. Separation of the syn-rift and post-rift in this basin is difficult, because the transition occurred during a period of sediment starvation (Milton, 1993).

During periods of rapid basin subsidence, sequence boundaries generated by higher frequency eustatic sea-level falls will be obscured. In times of slow tectonic subsidence or basin uplift, sequence boundaries will be enhanced.

2.1.2 Basin-margin concepts

Many of the concepts and principles of sequence stratigraphy are based on the observation from seismic data that prograding basin-margin systems often have a consistent

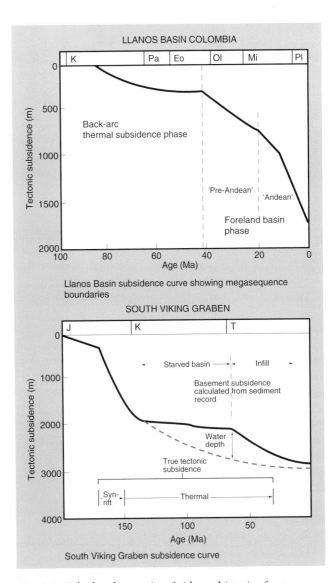

Fig. 2.2 Calculated tectonic subsidence histories for two sedimentary basins (reproduced by permission of BP Exploration Ltd)

depositional geometry (Fig. 2.3). *Topset* is a term used to describe the proximal portion of the basin-margin profile characterized by low gradients (< 0.1°). Topsets effectively appear flat on seismic data and generally contain alluvial, deltaic and shallow-marine depositional systems.

The *shoreline* can be located at any point within the topset. It can coincide with the offlap break or may occur hundreds of kilometres landward. The proximal termination of the topset is usually termed the point of *coastal onlap*, referring to the up-dip limit of coastal-plain or paralic facies. *Clinoform* is used to describe the more steeply dipping portion of the basin-margin profile (commonly > 1°) developed basinward of the topset. Clinoforms generally contain deeper water depositional systems characteristic of the slope. The slope of the clinoform generally

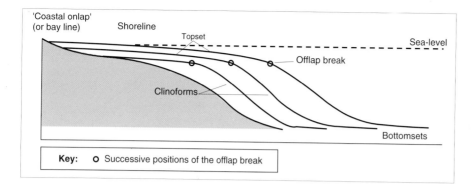

Fig. 2.3 Typical profile of a prograding basin-margin unit, comprising topsets and clinoforms separated by a break in slope; the offlap break. Bottomsets may also be present

can be resolved on seismic data. *Bottomset* is a term sometimes used to describe the portion of the basin-margin profile at the base of the clinoform characterized by low gradients and containing deep-water depositional systems.

The main break in slope in the depositional profile occurs between topset and clinoform and is called the *offlap break* (Vail et al., 1991). The offlap break previously has been termed the *shelf edge* (Vail and Todd 1981; Vail et al., 1984), leading to a confusion with the *shelf break*, i.e. the edge of the modern continental shelf, which is usually a relict feature rather than depositional feature. The term *depositional shoreline break* (Van Wagoner et al., 1988) also has been used, but this implies that the main break in slope in a depositional profile coincides with the shoreline. The term offlap break is preferred here as it does not imply coincidence of the main break in slope with the shoreline.

The topset–clinoform profile results from the interplay between sediment supply and wave, storm and tidal energy in the basin. Sediment enters the proximal end of the profile through river systems and is distributed across the topset area by wave- and/or current-related processes. These may include fluvial currents, tidal currents, storm currents, etc. However, these topset transport processes are effective only at relatively shallow depths of up to a few tens of metres, and to move sediment into deeper water a slope must develop in order to allow sediment transportation by gravity processes. The clinoforms build to the angle needed to transport sediment at the required rate. Slope angle is strongly influenced by sediment calibre. Coarse-grained sediment, with a higher angle of rest, will build up steeper slopes than fine-grained sediment (Kenter, 1990). Also, carbonate systems generally can build steeper depositional slopes (up to 35°) than fine-grained clastic systems (0.5–3°) owing to their greater shear strength. Steeper slopes in clastic systems generally are either made of coarser grade material or are zones of erosion and sedimentary bypass.

The importance of the offlap break on the depositional systems is most apparent during relative sea-level fall (see 2.2.1). When relative sea-level fall exposes the offlap break, rivers commonly incise in order to re-equilibrate to lowered *base level*, with the result that the river becomes entrenched at its mouth (discussed in 2.4.3). The response of the depositional systems to this fall in relative sea-level depends on the nature of the basin margin (Fig. 2.4).

Shelf-break margins are those with well developed depositional clinoforms. Fluvial entrenchment during sea-level fall may result in focusing of the sediment load to discrete locations on the clinoform slope. Failure of the sediment mass has the capacity for forming large turbidity currents and submarine fan deposits. Shelf-break margins are typical of passive continental margins at times of slow rise of relative sea-level, when the delta systems can easily prograde to the shelf edge.

Ramp margins are characterized by relatively shallow water depths, where storms and current processes can operate over much of the area of deposition. Depositional angles are generally less than 1° and seismic clinoforms (if resolved) are shingled with a dip of around half a degree. The offlap break on a ramp margin is likely to be at the shoreline, where fluvial gradients pass into slightly steeper shelf or delta-front gradients. The response of the depositional systems in a ramp setting to relative sea-level change is therefore different from the shelf-break margin. In particular, deep-water turbidite deposition during lowstand may be absent, or of only minor significance. Depositional systems will, instead, be translated basinward without significant slope bypass or basinal deposition. Any turbidites found on a siliciclastic ramp margin are likely to be delta-front turbidites, rather than detached submarine fans (Van Wagoner et al., 1990).

Many modern delta systems can be considered to form ramp margins, as generally they are shelf deltas prograding on to the drowned topsets of a previous shelf-break margin (Fig. 2.4). Frazier (1974) has shown that deposition on the continental shelf of the Gulf of Mexico is confined to the Mississippi delta, which is prograding into about 100 m of water. The rest of the shelf is effectively an area of non-deposition. The Mississippi delta presently forms a ramp margin, although very little extra progradation is needed for the delta to reach the shelf edge, and for the margin to become a shelf-break margin.

Rift margins characterize basins undergoing active crustal extension. Extensional faults have a strong influence on both palaeogeography and sediment influx rates. The spatial

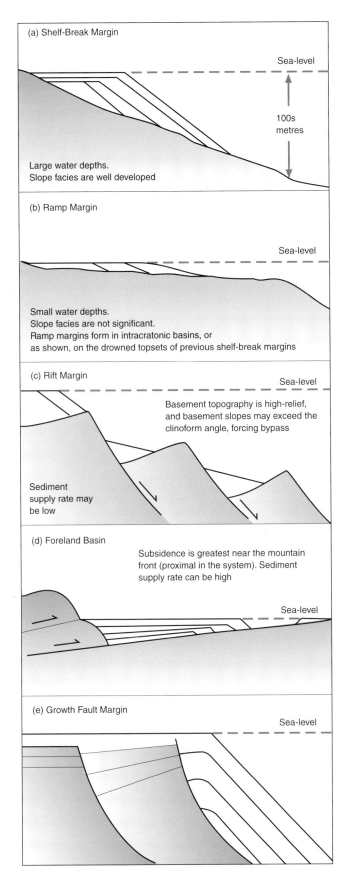

Fig. 2.4 Basin-margin types: (a) shelf-break margin; (b) ramp margin; (c) rift margin; (d) foreland basin, (e) growth-fault margin

distribution of sediment accommodation within the rift is controlled largely by tectonics. Subsidence rates generally will increase from the margins to the centre of the rift, although each individual fault block will have its own pattern of accommodation. The foot-wall crest will see the least subsidence and may experience uplift and erosion, whereas the hanging wall will experience progressively greater subsidence rates towards the controlling fault. The depositional systems that develop will depend on whether the rift is marine or continental. Transfer zones in the rift margin may control sediment entry points. Rift margins may be characterized by high topographic relief and relative sediment starvation, because sediment is bypassed towards the rift centre. Basin-margin systems may build out into deep water with long clinoform slopes and relatively minor topsets (Fig. 2.4). There is little potential for trapping coarse material in the topsets, and much may be bypassed to the basin.

Foreland-basin margins vary depending on whether sediment is being fed axially along the foreland basin or directly into the foreland basin from the thrust belt. In the latter case, the rate of tectonic subsidence increases towards the foreland thrust belt, i.e. the sediment source area. In other words, sediment accommodation may be relatively high in proximal areas compared with the basin centre. This has a marked affect on stratal geometries and may result in the aggradation of thick topset deposits, with little opportunity for seismic-scale clinoforms to develop (Posamentier and Allen, 1993).

Growth-fault margins are characterized by gravity driven syn-sedimentary extensional faults. The rate of subsidence is considerably greater on the hanging-wall side of the growth fault, resulting in an expanded sedimentary succession. The effect of the growth fault in the depositional systems developed will depend on whether the fault had a topographic expression on the sea-bed. At times when the hanging wall was a topographic low relative to the foot wall, facies differentiation occurs across the fault, with thick, deeper water clastic systems on the downthrown side. Growth-fault margins are discussed further in section 9.3.3.

2.2 Relative sea-level, tectonics and eustasy

2.2.1 Definitions of sea-level

In order to understand the controls on sequence development, it is first necessary to define what is meant by eustasy, relative sea-level and water depth (Fig. 2.5, from Jervey, 1988).

Global eustasy

Eustasy is measured between the sea-surface and a fixed datum, usually the centre of the Earth. Eustasy can vary by changing ocean-basin volume (e.g. by varying ocean-ridge volume) or by varying ocean-water volume (e.g. by glacio-

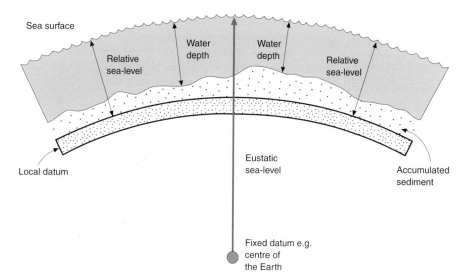

Fig. 2.5 Definitions of sea-level (after Jervey, 1988). 'Eustatic sea-level', base level for fluvial systems entering the ocean

eustasy). The interpretation of eustatic changes from the rock record is a complex and controversial topic, which will be discussed in detail in section 2.2.4. For the moment it is important only to emphasize that it can rise or fall, thus varying the base level for erosion on a global scale, where *base level* is defined as the level above which deposition is temporary and erosion occurs (see 2.2.2).

Relative sea-level

Relative sea-level is measured between the sea-surface and a local moving datum, such as basement or a surface within the sediment pile (Posamentier *et al.*, 1988). Tectonic subsidence or uplift of a basement datum, sediment compaction involving subsidence of a datum within the sediment pile, and vertical eustatic movements of the sea-surface all contribute to relative sea-level change. Relative sea-level 'rises' due to subsidence, compaction and/or eustatic sea-level rise, and 'falls' due to tectonic uplift and/or eustatic sea-level fall. Relative sea-level should not be confused with water depth, which is measured between the sea-surface and the sea-bed in any given geographic location at a point in time. The term *equilibrium point* is sometimes used to distinguish the point on a depositional profile where the rate of relative sea-level change is zero. The equilibrium point will separate, at any given time, the zone at the basin margin where relative sea-level is falling from the zone where relative sea-level is rising.

2.2.2 Accommodation

Eustasy and subsidence rate together control the amount of space available for sediment accumulation — this is conventionally termed accommodation. *Accommodation* is defined as the space available for sediment to accumulate at any point in time (Jervey, 1988). Accommodation is controlled by *base level* because in order for sediments to accumulate, there must be space available below base level. The base-level datum varies according to depositional setting (Fig. 2.6). In alluvial environments base level is controlled by the graded stream profile, which is graded to sea-level or lake-level at its distal end (Mackin, 1948; and Chapter 7). In deltaic and shoreline systems base level is effectively equivalent to sea-level. In shallow marine environments base level is ultimately also sea-level, although fairweather wave base can form a temporary base level in the form of a 'graded shelf profile'. Sediment supply fills the accommodation created and controls water depth:

$$\Delta\text{accommodation} = \Delta\text{eustasy} + \Delta\text{subsidence} + \Delta\text{compaction}$$

Sediment supply fills available accommodation. If the rate of sediment supply exceeds the rate of creation of accommodation at a given point, water depths will decrease:

$$\Delta\text{water depth} = \Delta\text{eustasy} + \Delta\text{subsidence} + \Delta\text{compaction} - \text{sediment deposited}$$

A series of cartoons in Fig. 2.7 illustrates the relationship between accommodation, relative sea-level and water depth in shoreline–shelf depositional systems. In these examples relative sea-level change and new sediment accommodation added are the same because base level is taken at the sea-surface.

In the discussion that follows we examine the relationship between relative sea-level and accommodation in shoreline–shelf depositional systems. Fluvial systems and the controls on the graded stream profile are discussed in Chapter 7, paralic systems are discussed in Chapter 8, submarine fans in Chapter 9 and carbonates in Chapter 10.

2.2.3 Accommodation through time

In order to understand how accommodation varies through time it is useful to consider how different rates of tectonic

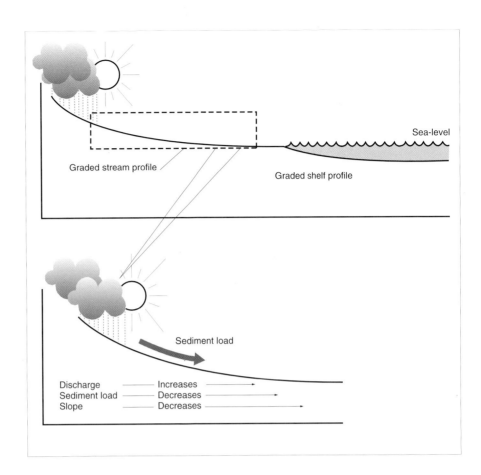

Fig. 2.6 Definitions of base level in fluvial, shoreline and shelf environments

subsidence and, in this case, the same sinusoidal eustatic sea-level curve combine to give different rates of addition and destruction of accommodation (rise and fall of relative sea-level) (Fig. 2.8; from Jervey, 1988).

In Fig. 2.8 subsidence is represented as a straight line, the gradient of which indicates the rate of subsidence at each point. The different gradients can be thought of as representing positions in a basin with increasing subsidence rates or changes in subsidence rate through time. Eustasy is represented by the same smooth curve in each case. The change in relative sea-level through time is found simply by the addition of the two curves. Relative sea-level is, in this case, equivalent to accommodation because the curves begin at zero water depth.

Where slow subsidence occurs, maximum accommodation is developed near the eustatic maximum. When eustasy falls to its original position, accommodation falls to a value representing that created only by subsidence. With increased rates of subsidence, the time of maximum accommodation is progressively later. Points in the basin where subsidence rates are very high experience no decrease in accommodation even though eustatic fall may be occurring. Note that the same curves could be produced theoretically by adding variable rates of tectonic subsidence and uplift to a flat eustatic curve.

2.2.4 Orders of cyclicity and global correlation

A depositional sequence represents a complete cycle of deposition bounded above and below by erosional unconformities. The sequence has a maximum duration, which is measured between the correlative conformities to the bounding unconformities. Thus, the duration of the sequence will be determined by the event controlling the creation and destruction of accommodation, i.e. tectonic subsidence and/or eustasy. Tectonic cycles of subsidence and uplift and eustatic cycles of rising and falling sea-level can operate over different time periods, and it is useful to classify sequences in terms of their order of duration, commonly termed first, second, third, fourth order, etc. (Fig. 2.9). A basin-fill can then be divided into a hierarchy of sequences, each representing the product of a particular order of tectonic or eustatic cycle.

In Fig. 2.9, from Duval *et al.* (1992), four orders of stratigraphic cycle are depicted. The continental encroachment cycle is defined by the very largest scale (> 50 million years) cycles of sedimentary onlap and offlap of the supercontinents. There are only two such cycles in the Phanerozoic, according to the Haq *et al.* (1987) sea-level curve. First-order continental encroachment cycles are considered to be controlled by tectono-eustasy, i.e. changes in ocean basin volume related to plate tectonic cycles (Pitman, 1978).

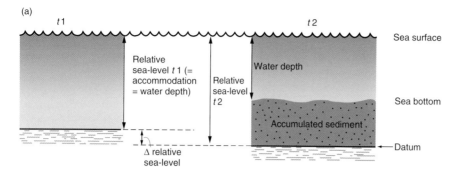

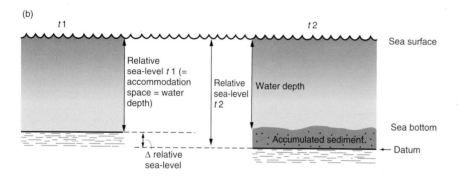

Fig. 2.7 This series of diagrams illustrates how eustatic sea-level rise/fall and subsidence/uplift can create/destroy accommodation. The rate at which sediment fills the space created controls water depth and whether *progradation* or *retrogradation* of facies belts is observed. In (a) relative sea-level rises and accommodation increases from time 1 to time 2 owing to subsidence, but the rate of sediment accumulation at this point is greater than the rate of relative sea-level rise and so water depth decreases from time 1 to time 2. In the depositional record, the interaction of these parameters would result in an overall regressive character to the vertical facies succession. (b) Relative sea-level rises and accommodation increases from time 1 to time 2 owing to subsidence, but the rate of sediment accumulation is less than the rate of relative sea-level rise and so water depths increase from time 1 to time 2. In the depositional record this may be apparent as a transgressive vertical facies succession. The same patterns would occur for a relative sea-level rise due to a eustatic rise of the same rate. (c) (*opposite*)

Second-order (3–50 million years) cycles are the building blocks of the first-order sequences and represent particular stages in the evolution of a basin. They may be caused by changes in the rate of tectonic subsidence in the basin or rate of uplift in the sediment source terrane.

Third-order (0.5–3 million years) sequence cycles are the foundation of sequence stratigraphy because they are often of a scale well-resolved by seismic data. They are identified by the recognition of individual cycles of accommodation creation and destruction. These cycles are considered by Vail *et al.* (1991) to be controlled by glacio-eustasy, although other tectonic mechanisms are possible (Cloetingh, 1988).

Composite sequence is a term sometimes used to describe second- or third-order sequences made up of higher order sequences (Mitchum and Van Wagoner 1991; and see 2.4.10).

Fourth-order (0.1–0.5 million years) 'parasequence' cycles represent individual shallowing upward facies cycles bounded by surfaces of abrupt deepening. These may be related in part to autocyclic processes within the sedimentary system.

The theory of eustatic control on deposition is a unifying stratigraphic concept that has attracted geologists for many generations (see Chapter 1 and papers in Dott, 1992a,b). If it were true that a global eustatic signature was overprinted on all stratigraphic successions, then it would be possible to date a stratigraphic section from the pattern of sequences and systems tracts, and to predict stratigraphy in unsampled areas from a knowledge of the global standard. A proposed global sea-level chart was first published by Vail *et al.* (1977a), and updated by Haq *et al.* (1987), based on measurements from basins around the globe. This chart is taken to support the theory that third-order relative sea-level variations are mostly eustatic in origin. Sceptics would argue that the chart is based on that theory rather than proof of it.

More discussion and controversy have been caused by

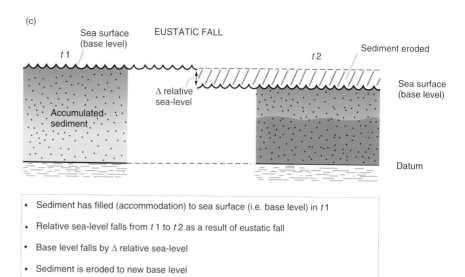

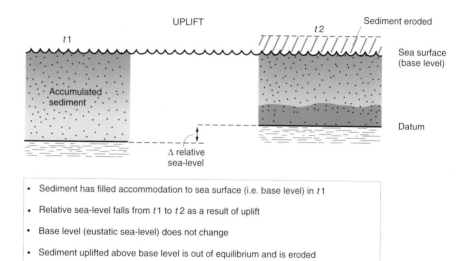

Fig. 2.7(c) Illustrates time when accommodation space is destroyed. This can occur by two mechanisms — tectonic uplift (and more locally salt or mud diapirism) and/or eustatic sea-level fall. The effect of relative sea-level fall is manifest at the basin margin (where accommodation space is limited) by *erosion*, and at the basin centre (where accommodation space is not limited) by *an increase in sediment supply*. The increase in sediment supply to the more basinal areas is due to both the erosion of previously deposited sediment and bypass of areas where accommodation space is now filled

the concept of a global eustatic signal than any other aspect of sequence stratigraphy. It is beyond the scope of this work to cover this discussion in detail and the following is a brief summary only.

There are a number of aspects of the Haq *et al.* (1987) chart that have excited comment:

1 The data to support the Haq chart have never been published fully, in particular the evidence for the global correlatability of the sequence boundaries. Miall (1986, 1992) has been a consistent critic of the Haq *et al.* curve, stating:

> The basic premise of the Exxon cycle chart, that there exists a globally correlatable suite of third order eustatic cycles, remains unproven... There are some specific cases where global synchrony is suggested by detailed stratigraphic documentation (e.g. 4th- and 5th-order glacioeustatic cycles in the Neogene and, possibly the late Palaeozoic; 1st- and 2nd-order cycles

related to changing rates of global sea floor spreading), but for the greater part of the Phanerozoic column no such proof is available. (Miall, 1991)

Miall also points out that it is arguable whether global biostratigraphic control is accurate enough to correlate third-order relative sea-levels unambiguously. Thus, for the moment, the global synchroneity of eustatic cycles has to remain to a degree an article of scientific faith rather than scientific fact.

2 The mechanism for generating third-order-scale eustatic sea-level changes is problematic in certain periods of geological time. Increases in the global ice volume during glaciations provide a mechanism for eustatic falls in sea-level in the late Cenozoic and late Palaeozoic, but no such mechanism exists for the (presumed) ice-free Cretaceous and Jurassic. Cloetingh *et al.* (1985) have proposed intra-plate stress as a tectonic mechanism for generating third-order plate-wide relative sea-level cycles. Finally, it is not

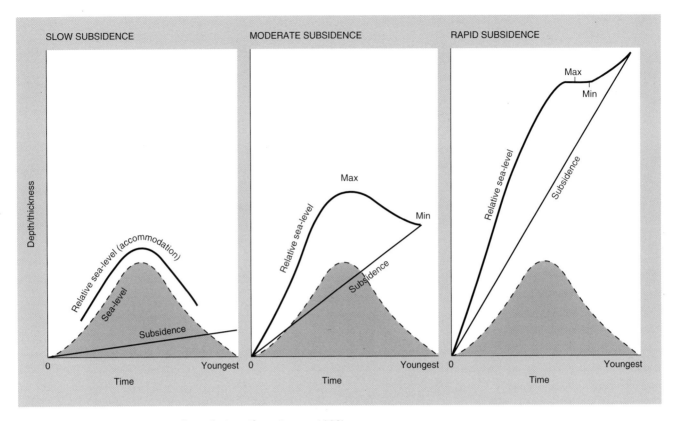

Fig. 2.8 Accommodation space through time (from Jervey, 1988)

accepted universally that the eustatic signal will be visible in all basins, and several stratigraphers believe that it may be obscured by a tectonic signal (e.g. Hubbard, 1988).

However, work is continuing to date basin-margin unconformities more accurately and to correlate these with the Neogene oxygen isotope record and hence directly to ice volume changes (e.g. Miller et al., 1991, 1993). Also, various projects are underway to accurately date and correlate sequence boundaries on a regional scale in Europe, e.g. De Graciansky et al. (1993).

2.3 Sediment supply

The rate of sediment supply controls both how much and where accommodation is filled. The balance between sediment supply and relative sea-level rise controls whether facies belts prograde basinward or retrograde landward, and the calibre of sediment supplied to the basin has a strong influence on sedimentary facies. The first part of this section considers the principles controlling siliciclastic sediment supply to the basin margin and how sediment supply may vary through time. The second part considers how accommodation is filled in locations with high, moderate and low rates of sediment supply. The principles of carbonate sediment production and supply are discussed in Chapter 10.

2.3.1 Principles of clastic sediment supply

River transport is the principle means of transporting material from the continental interior to the depositional basin. The volume and grade of sediment delivered to the basin margin is a complex function of hinterland physiogeography, tectonics and climate. Studies of modern rivers show huge variations in the rate of sediment supplied to the continental margins (Fig. 2.10). Around 70% of the total load is supplied from only 10% of the land area, and just three rivers, the Ganges, the Brahmaputra and the Huang He (Yellow) supply 20% of the total fluvial load (Summerfield, 1991).

The amount of sediment supplied to the basin margin is a function of both the fluvial drainage basin area and the mechanical denudation (erosion) rate. Tectonism at both local and regional scale affects fluvial drainage basin shape, size and relief and also the geology of the provenance area, and will control the calibre of sediment eroded. The rate of fluvial denudation is a complex function of relief within the drainage area and climate. Climate influences not only the erosive power of the river by controlling discharge but also the erodibility of soils in the drainage basin, and the presence or absence of stabilizing vegetation. Present-day mechanical denudation rates vary from less than 1 mm per 1000 years in the St Lawrence river drainage basin to 640 mm per 1000 years in the Brahmaputra drainage basin.

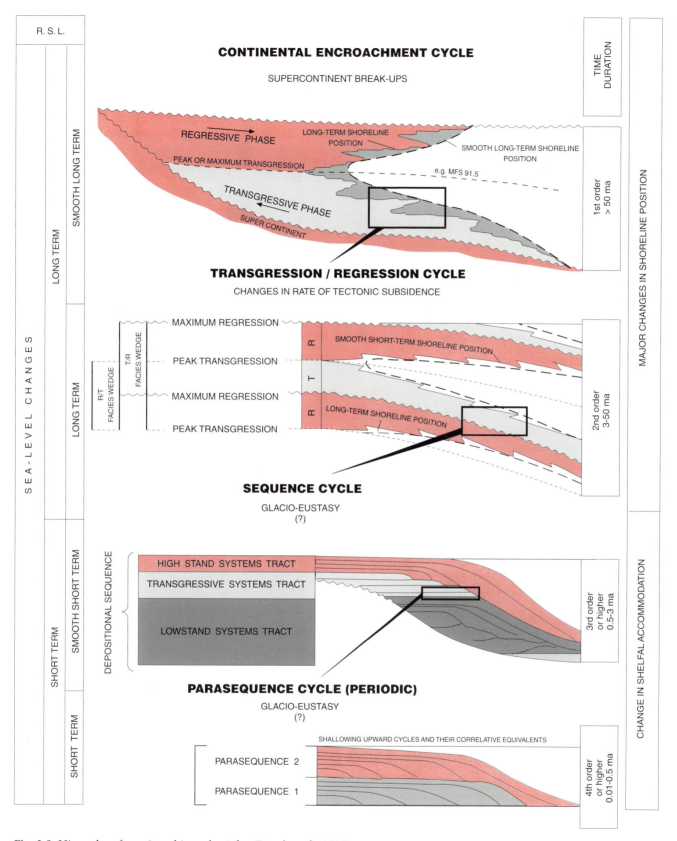

Fig. 2.9 Hierarchy of stratigraphic cycles (after Duval *et al.*, 1992)

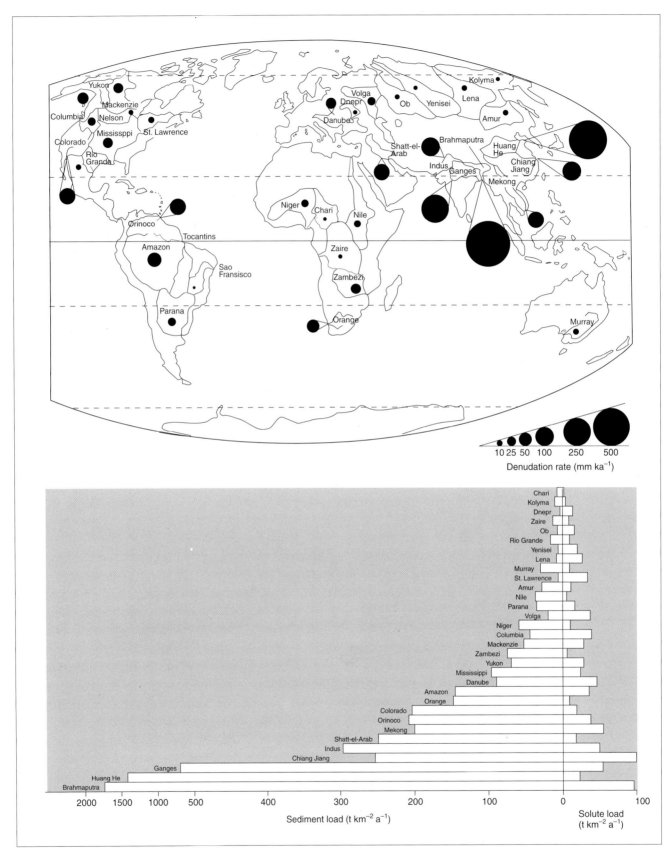

Fig. 2.10 (a) Denudation rates for the world's 35 largest drainage basins based on solid and solute data. Allowance has been made for the non-denudational component of solute loads. Source rock density is assumed to be $2700 \, \text{kg m}^{-3}$. (b) Sediment and solute loads for the world's largest drainage basins. Both figures from Summerfield (1991)

A tributary of the Huang He river in China is an extreme example, with a denudation rate of 19 800 mm per 1000 years, because it drains over 3000 km² of loess-covered terrain in a semi-arid region of sparse vegetation (Summerfield, 1991).

It clearly would be unwise to think of sediment supply to the basin margin as being either spatially or temporally constant. Local sediment supply will depend on the proximity to a fluvial entry point to the basin margin. There also may be a linkage between glacio-eustatically controlled relative sea-level cycles and climate in the fluvial drainage basin (Blum, 1990). This will mean that sediment supply may vary through a sea-level cycle in a fashion characteristic of the drainage basin.

2.3.2 Filling of accommodation

The amount of sediment supplied to locations in the basin is a function of both the general rate of sediment supply to the basin and the proximity to sediment entry points to the basin. Figure 2.11 (from Jervey, 1988) considers the relationship of facies, relative sea-level and rates of sediment accumulation at three fixed points in the basin with identical relative sea-level curves but with differing rates of sediment supply. These could represent points along a continental margin at varying distances from a point source. Each model begins at time 0 with zero water depth, i.e. the shoreline is exactly located at the model location. For the purpose of illustration, Jervey distinguishes marine 'mud-prone' from coastal plain 'sand-prone' facies in Fig. 2.11. Sediment grain-size is clearly a function both of sediment supply 'type' as well as sediment supply 'rate'. The sediment supply rate is held constant through the relative sea-level cycle in this simple model.

At the location with low rates of sediment influx, accommodation always exceeds sediment accumulation, the coastline migrates landward and transgression ensues, with considerable water depths being developed. Mud-prone marine facies may be expected to accumulate at some distance from a coastline located marginward of the figured depositional site. The rate of accumulation in this case reflects rate of supply of sediment to this point in the basin.

With a moderate rate of sediment influx, the sea-floor can aggrade to sea-level (base level). The rate of increase of accommodation initially exceeds the ability of sediment supply to maintain the sediment surface at sea-level and a transgression ensues. During the transgression water depth increases at this location and marine shales are deposited. As the rate of relative sea-level rise diminishes, regression of the shoreline commences. Regression of the shoreline

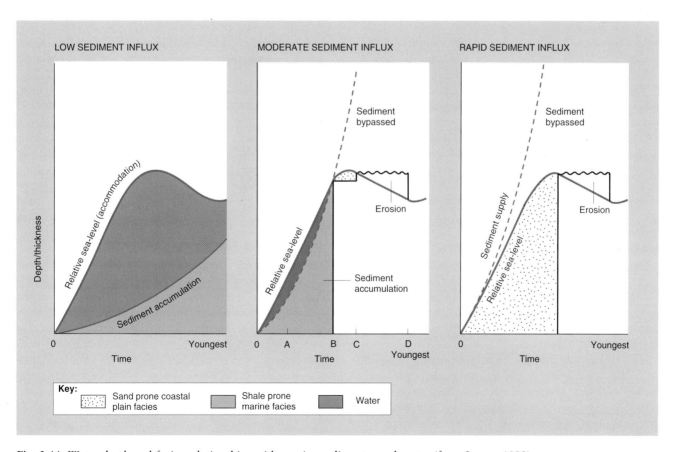

Fig. 2.11 Water depth and facies relationships with varying sediment supply rates (from Jervey, 1988)

continues until marine facies have aggraded to sea-level and the coastline again reaches the location shown. Thereafter, sediment supply rate exceeds the rate of creation of accommodation, the sediment surface is maintained at sea-level and coastal-plain facies accumulate. Excess sediment which cannot be accommodated in the coastal plain is transported basinwards. When accommodation decreases (relative sea-level falls) there is the potential for erosion of sediment deposited previously.

Where sediment influx is rapid, sediment supply rate always exceeds the rate of creation of accommodation and coastal/delta-plain sediments will accumulate. Regression of the shoreline will be continuous through the sea-level cycle. The rate of accumulation at this point in the basin is limited by the rate of accommodation increase. Erosion is likely when accommodation is removed during relative sea-level fall.

2.3.3 Basin architecture

In order to understand the behaviour of a topset/clinoform margin through time it is necessary to consider the balance between the rate of sediment supply and the rate of creation of *topset accommodation volume* (sometimes termed 'shelfal accommodation volume'). The rate of change of accommodation volume is a function of the magnitude of the sea-level rise multiplied by the topset area (Milton and Bertram, in press). If, during the same interval, the basin margin is supplied with a greater volume of sediment, then topset accommodation volume will be completely filled, and sediment will deposit on the clinoforms allowing the offlap break to prograde basinwards (Fig. 2.12).

Progradational geometries therefore occur when sediment supply exceeds the rate of creation of topset accommodation volume and facies belts migrate basinward. On seismic data progradation is expressed as *clinoforms* that show the basinward migration of the offlap break. *Regression* is a term that will be used here to refer specifically to basinward movement of the shoreline.

Aggradational geometries occur when sediment supply and rate of creation of topset accommodation volume are roughly balanced. Facies belts stack vertically and the offlap break does not migrate landward or basinward.

Retrogradational geometries occur when sediment supply is less than the rate of creation of topset accommodation volume. Facies belts migrate landward and the former depositional offlap break becomes a relict feature. *Transgression* is used here to refer specifically to the landward movement of the shoreline.

These phases of progradation, aggradation and retrogradation are not continuous but are made up of smaller (subseismic) scale progradational units called *parasequences* (see 2.5). Parasequences stack together in *parasequence sets* to make up the depositional geometries observable on seismic data.

The next section will show that the principles of cyclic changes in accommodation through time can be used to divide the sedimentary record into packages deposited during characteristic phases of the sea-level cycle.

2.4 Sequences and systems tracts

2.4.1 Sequences and sequence boundaries

The term 'sequence', as applied in sequence stratigraphy, was defined originally by Mitchum *et al.* (1977a) as:

> A stratigraphic unit composed of a relatively conformable succession of genetically related strata bounded at its top and base by unconformities or their correlative conformities.

This generalized definition does not specify the scale or duration of the sequence, nor does it imply any particular mechanism for causing the unconformities. The term 'unconformity' in this definition was an initial cause of confusion, because the precise usage of the term can vary. Mitchum *et al.* (1977a) initially included marine hiatuses and condensed intervals in the term 'unconformity', but as models of cyclic deposition driven by relative sea-level variations developed, it became clear that basin-margin subaerial unconformities needed to be distinguished from basin-centre marine hiatuses. For the purpose of defining sequences, the term 'unconformity' is now restricted to a much narrower definition, namely 'a surface separating younger from older strata along which there is evidence of subaerial erosion and truncation (and in some areas correlative submarine erosion) and subaerial exposure and along which a significant hiatus is indicated' (Van Wagoner *et al.*, 1988).

Thus sequences are units bounded by significant subaerial erosion surfaces. Units bounded by marine condensed surfaces, surfaces of transgression, or marine onlap surfaces, are not sequences by this definition. It is interesting to note that the Exxon workers 'seriously considered using the term "synthem" instead of "sequence"' (Mitchum *et al.*, 1977a). In retrospect this might have avoided a lot of confusion with the sedimentological use of the term 'sequence', and with other cycle-defined 'sequences' (such as the genetic depositional sequences of Galloway, 1989). However 'synthem stratigraphy' does not have the same ring to it!

At first sight the definition quoted above seems simple enough. However, in practice, it is not so simple to apply. It is difficult to demonstrate non-marine exposure from a well log or from a seismic data set and tracing an erosion surface offshore into its 'correlative conformity' is often problematic. The term 'significant' in the clarification of Van Wagoner *et al.* (1988), quoted above, is not particularly helpful because it gives no indication of what magnitude of discontinuity is significant. *Composite sequences* (discussed in 2.4.10) are allowed to contain unconformities, provided these unconformities are of a higher 'order' than the ones which bound the sequence, and are therefore not significant.

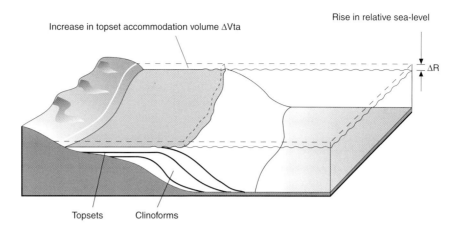

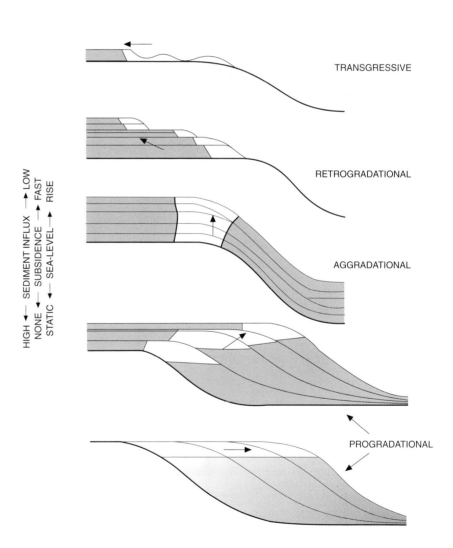

Fig. 2.12 Depositional architecture as a function of accommodation volume and sediment supply (after Galloway, 1989)

The stratigrapher must define 'significant' on the scale of the particular study.

Nevertheless the restricted principle is straightforward, and a sequence represents one cycle of deposition bounded by non-marine erosion, deposited during one 'significant' (on the scale of the study) cycle of fall and rise of base level. In the majority of basins, base level is controlled by sea-level, and a sequence is the product of a cycle of fall and rise of relative sea-level. An idealized sequence resulting from one cycle of base level change is shown in Fig. 2.13, after figures in Van Wagoner *et al.* (1988). This is a *type 1 sequence*, where the fall in relative sea-level is sufficiently large that the first topsets within the sequence onlap the clinoforms of the previous sequence, implying a fall in relative sea-level at the position of the offlap break. *Type 2 sequences* are described below in section 2.4.6.

According to Van Wagoner *et al.* (1988), a type 1 sequence boundary is characterized by subaerial exposure and concurrent subaerial erosion associated with stream rejuvenation, a basinward shift in facies, a downward shift in coastal onlap, and onlap of overlying strata. Coastal onlap is a term used to describe the onlap point on topset strata at the basin margin (see Chapter 3). As a result of the basinward shift in facies, non-marine or marginal marine rocks, such as braided-stream or estuarine sandstones, may directly overlie shallow-marine rocks, such as lower shoreface sandstones or shelf mudstones, across a sequence boundary with no intervening rocks deposited in intermediate depositional environments. This facies superposition is termed a *facies dislocation*. A type 1 sequence boundary is interpreted by Van Wagoner *et al.* (1988) to form when the rate of eustatic fall exceeds the rate of basin subsidence at the offlap break, producing a fall in relative sea-level at that position.

2.4.2 Systems tract definition

The idealized type 1 sequence shown on Fig. 2.13 is representative of a shelf-break margin. It can be seen to be comprised of a number of distinct depositional packages. It was observed in the early days of seismic stratigraphy that deposition in a basin was not uniform and continuous but occurred in a series of discrete 'packets' bounded by seismic reflection terminations (see Chapter 3). Workers in Exxon found that these packages generally were arranged in a predictable fashion in the majority of sequences they observed on seismic data. These packages are known as *systems tracts*.

This term systems tract was first defined by Brown and Fisher (1977) as a linkage of contemporaneous depositional systems, where a depositional system is a three-dimensional assemblage of lithofacies, genetically linked by active (modern) or inferred (ancient) processes and environments (after Fisher and McGowen, 1967).

A systems tract is therefore a three-dimensional unit of deposition, and the boundaries of a systems tract are depositional boundaries of onlap, downlap, etc. The seismic expression of a systems tract is a unit of conformable reflections bounded by surfaces of reflection termination ('seismic-stratigraphic units' of Brown and Fisher, 1977; 'seismic sequences' of Mitchum *et al.*, 1977a; referred to as 'seismic packages' in Chapter 5).

Systems tracts are recognized and defined by the nature of their boundaries and by their internal geometry. Within any one relative sea-level cycle, three main systems tracts that characterize different parts of the relative sea-level cycle are frequently developed (Fig. 2.13).

It is easy to become lost in the complexities of systems tract terminology, and it is always worth remembering the purpose of stratigraphic division into systems tracts. The systems tract represents the fundamental mapping unit for stratigraphic prediction, because it contains a set of depositional systems with consistent palaeogeography and depositional polarity, and for which a single palaeogeographic map can be drawn.

2.4.3 Lowstand systems tract

The basal (stratigraphically oldest) systems tract in a type 1 depositional sequence is called the *lowstand systems tract*.

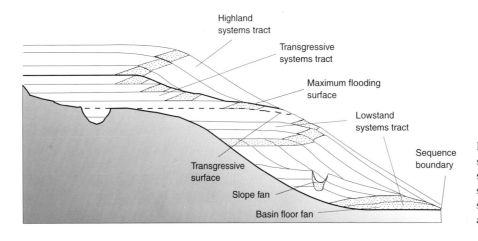

Fig. 2.13 Stratal geometries in a type 1 sequence on a shelf-break margin. Five separate sedimentary packages are shown, traditionally assigned to three systems tracts; lowstand, transgressive and highstand

The lowstand systems tract is deposited during an interval of relative sea-level fall at the offlap break, and subsequent slow relative sea-level rise.

Falling relative sea-level at the offlap break of a shelf-break margin will have an extreme effect on the river systems. Prior to the fall in relative sea-level, the rivers will have more-or-less maintained a graded river profile with an erosional upper portion and a depositional lower portion (alluvial plain and coastal plain). The rivers will have been free to avulse, responding to rises in relative sea-level over this lower portion. When relative sea-level falls at the offlap break, the river profile must adjust to the lowered base level (see Chapter 7). The river incises into the previously deposited topsets; the alluvial plain, coastal plain and/or shelf deposits of the previous sequence. These reworked sediments, and the fluvial load from the hinterland, are delivered directly on to the previous highstand clinoform slope. Because the river is not free to avulse, the sediment is focused towards the same point on the slope. This is an inherently unstable situation, and sedimentation processes are dominated by large-scale slope failure resulting in bypass of the slope and deposition of submarine fans in the basin. These processes continue to dominate the sedimentary record while relative sea-level is falling and the river system is forced to incise.

At the relative sea-level low point the river profile stabilizes again, and a prograding topset–clinoform system can then be established. The first topset of this system will onlap below the level of the previous offlap break. This is known as a *downward shift in coastal onlap below the level of the offlap break*, and is indicative of a type 1 sequence boundary. The rate of rise of relative sea-level is initially low, and together with the limited topset area of the prograding system, this results in a low rate of creation of topset accommodation (see Fig. 2.15). This will be outpaced by sediment supply, and so the system will prograde. However the accelerating rate of creation of accommodation volume eventually may outpace sediment supply, resulting in a change from progradation to aggradation and retrogradation, and the onset of the next (transgressive) systems tract.

The lowstand systems tract therefore consists of two parts; a unit of submarine fans deposited during falling relative sea-level, and a topset/clinoform system, initially progradational but becoming aggradational, deposited during a slow rise of relative sea-level. These can be treated as separate and distinct systems tracts, because the fans and the topset–clinoforms need never have been in depositional continuity. They are traditionally both placed in a single lowstand systems tract, on the basis that the boundary between the two may be gradational rather than distinct, with submarine fans forming much of the slope portion of the lowstand wedge (Posamentier and Vail, 1988).

Lowstand submarine fans

Frequently two distinct fan units can be recognized within the lowstand submarine fans; an initial *basin floor fan* unit, detached from the foot of the slope, and a subsequent *slope fan* unit, abutting the slope, occasionally referred to in older literature as 'slope front fill' (see Fig. 2.14). Van Wagoner *et al.* (1988) describe the basin-floor fan as being characterized by submarine fan deposits on the lower slope or basin floor. Fan formation is associated with the erosion of canyons into the slope and the incision of fluvial valleys into the shelf. Siliciclastic sediment bypasses the shelf and slope through the valleys and the canyons to feed the basin-floor fan. The base of the basin-floor fan (coincident with

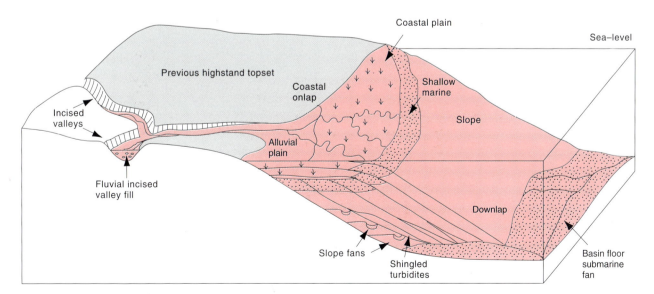

Fig. 2.14 Components of the lowstand systems tract on a shelf-break margin. These include a basin-floor fan and a slope fan, but the diagram also shows the active systems of the lowstand wedge; namely valley fill, alluvial and coastal plain topsets, a shallow marine belt and an active slope system, which in its early stages may contain shingled turbidites

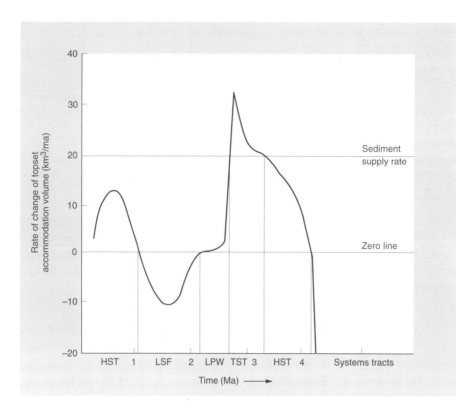

Fig. 2.15 The relationship between relative sea-level, topset accommodation volume, and systems tracts, in a simple numerical model with sinusoidal relative sea-level and constant sediment supply. The large change in topset accommodation volume between the lowstand and transgressive systems tracts is caused by flooding back over the highstand topsets. HST, highstand systems tract; LSF, lowstand submarine fan; LPW, lowstand prograding wedge; TST, transgressive systems tract

the base of the lowstand systems tract) is correlated with the type 1 sequence boundary and the top of the fan may be a downlap surface to the subsequent lowstand prograding wedge, if it prograded far enough, or may be a downlap surface for any overlying slope fans. Basin-floor fan deposition, canyon formation, and incised-valley erosion are interpreted to occur during a fall in relative sea-level over the entire topset area.

Slope fans are described by Van Wagoner *et al.* (1988) as characterized by turbidite and debris-flow deposition on the middle or the base of the slope. Slope-fan deposition can be coeval with the basin-floor fan or with the early portion of the lowstand wedge. The top of the slope fan may be a downlap surface for the middle and upper portions of the lowstand wedge. Slope fans are typically described as being composed of channel–levee complexes (see Chapter 9).

It is not clear whether two distinct fan units will be visible in all sequences in all basins, and the interpreter should beware of force-fitting a twofold subdivision on a submarine fan succession if the data do not warrant it.

Lowstand prograding wedge

As described above, the lowstand prograding wedge is a topset–clinoform system deposited during accelerating relative sea-level rise. It is separated from the overlying transgressive systems tract by a *maximum progradation surface*, marking a change in parasequence stacking geometry from progradational (in the lowstand wedge) to retrogradational (in the transgressive systems tract). Deposition of the lowstand prograding wedge is confined initially to the areas around the mouths of the incised rivers (Fig. 2.15). Little if any topset accommodation volume is created at this time, and the bulk of the sediment bypasses the topsets to be deposited on the clinoform slope. Slope instability and occasional fan deposition is likely to occur, and the bottom sets of the early lowstand prograding wedge may contain interbedded turbidites, which often have a characteristic 'shingled' seismic facies.

As relative sea-level begins to rise, the fluvial valleys incised into older topsets during falling relative sea-level begin to be back-filled in their lower reaches with fluvial or estuarine deposits, while topsets of the prograding wedge begin to be deposited. Accelerating relative sea-level rise results in a facies association indicative of increasing accommodation volume, such as an upwards increase in coals, overbank shales, lagoonal facies, tidal influence, etc., and a decrease in the connectivity of fluvial sandbodies. The transition to the overlying transgressive systems tract may be a gradational turnaround from progradation to retrogradation, or may, as in Fig. 2.15, be abrupt, as relative sea-level rises above the level of the offlap break of the previous sequence, resulting in a huge increase in topset accommodation volume. This boundary can be called the *maximum progradation surface*, the *transgressive surface*, or the *top lowstand surface*.

The lowstand prograding wedge is often sandier than the

preceding highstand wedges, owing to the recycling of sands from the highstand topsets. In a predominantly muddy system, sandy lowstand wedges can be sealed against underlying shales of the highstand systems tract and overlying shales of the transgressive systems tract, thus forming stratigraphic traps.

2.4.4 Transgressive systems tract

The *transgressive systems tract* is the middle systems tract of both type 1 and type 2 sequences (Figs 2.13, 2.16 and 2.18). It is deposited during that part of a relative sea-level rise cycle when topset accommodation volume is increasing faster than the rate of sediment supply. It contains mostly topsets, with few associated clinoforms, and is entirely retrogradational. The active depositional systems are topset systems; alluvial, paralic, coastal plain and shelfal. Any deltas are shelf deltas. These systems may show evidence of an undersupply of sediment, and may be rich in coals, overbank deposits and lagoonal or lacustrine deposits. Drainage systems may be flooded to form estuaries. Wide shelf areas are characteristic of transgressive systems tracts, and tidal influence may be widespread. The transgressive systems tract passes distally into a condensed section characterized by extremely low rates of deposition and the development of condensed facies such as glauconitic, organic rich and/or phosphatic shales (Chapter 11), or pelagic carbonates.

The maximum rate of rise of relative sea-level occurs some time within the transgressive systems tract, and the end of the systems tract occurs when the rate of topset accommodation volume decreases to a point where it just matches sediment supply, and progradation begins again. This point is known as the *maximum flooding surface*.

Topsets of the transgressive systems tract tend to have a lower sand percentage than those of other systems tracts, because little of the mud-grade sediment bypasses the topsets. The transgressive systems tract can therefore often host sealing horizons to topset reservoirs, and sometimes also source beds (see Chapter 11). Posamentier and Allen (1993b) proposed a new component of transgressive systems tracts, which they termed the 'healing phase' component. They showed several examples of sediment wedges banked against the foot of highstand clinoforms, which they related to sediment reworked basinwards during transgression. An alternative view of these wedges could be that they may be the lowstand components of higher order sequences within a composite systems tract, or products of retrogressive slumping of the highstand slope.

The present-day depositional systems over much of the globe form a transgressive systems tract. Wide continental shelves are common (many of which are the flooded topsets of the last lowstand). Most of the major deltas are shelf deltas and most of the major fans are inactive. Estuaries and tidal seas are common around northwest Europe, whereas the eastern USA coast is dominated by retreating barrier coastlines and lagoons, with deep-sea sedimentation generally restricted to rare turbidites sourced from retrogressive slumping of the continental slope.

2.4.5 Highstand systems tract

The *highstand systems tract* is the youngest systems tract in either a type 1 or a type 2 sequence (Figs 2.13, 2.18). It

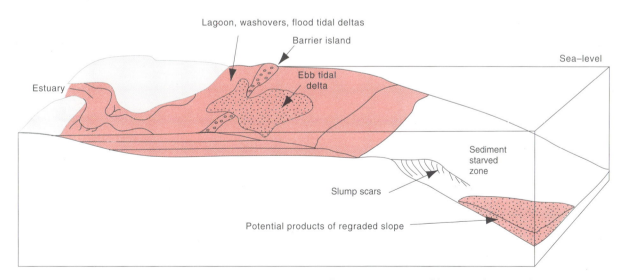

Fig. 2.16 Components of the transgressive systems tract. These are all topset systems, and here are shown to have significant tidal influence, due to the wide shelfal area of the drowned lowstand topsets. Deposition includes estuarine, lagoonal, barrier and tidal depositional systems, which pass seaward into a shelfal condensed zone

represents the progradational topset–clinoform system deposited after maximum transgression and before a sequence boundary, when the rate of creation of accommodation is less than the rate of sediment supply (Fig. 2.17). The highstand systems tract is characterized by a decelerating rate of relative sea-level rise through time, resulting in initial aggradational and later progradational architecture. Depositional systems may be similar initially to those in the transgressive systems tract, but the infill of shelf areas by progradation, and the decrease in the rate of relative sea-level rise, may lead to a decrease in tidal influence during a highstand systems tract, and a decrease in the amount of coal, and of overbank, lagoonal and lacustrine shales. Channel sandbodies will become more common and more connected.

Posamentier and Vail (1988) discuss various models which imply that the late highstand systems tract is characterized by significant fluvial deposition. They used the concept of the 'bay line', which they defined as the line to which stream profiles are graded and where fluvial processes are replaced by paralic and shelf processes. The bay line also represents the coastal onlap point during relative sea-level rise. Late in the highstand systems tract the bay line begins to migrate basinward as relative sea-level falls in the proximal part of the depositional profile, and Posamentier and Vail (1988) suggest significant alluvial accommodation will be generated. These models are an oversimplification and have been a source of considerable misunderstanding, e.g. Miall (1991), Shanley and McCabe (1994) and discussion in Chapters 7 and 8.

2.4.6 Type 2 sequence boundary and the shelf-margin systems tract

Relative sea-level may fall over the proximal area of the highstand topsets, without falling at the offlap break. A sequence boundary results, but not one characterized by fluvial incision or submarine fan deposition. The sequence boundary is recognized on seismic data by a downward shift in coastal onlap to a position landward of the offlap break, where topset reflections can be seen onlapping an older topset (Fig. 2.18). This is known as a type 2 sequence boundary, and the subsequent systems tract is known as a *shelf-margin systems tract*. It consists of prograding topsets and clinoforms, and is progradational initially but becomes aggradational upwards, passing eventually into a retrogradational transgressive systems tract. The shelf-margin systems tract may be very difficult to recognize in outcrop or on a well-log data base, and is differentiated from the underlying highstand systems tract by a subtle unconformity only, and possibly by a change in parasequence stacking pattern. It also could be recognized in a grid of wells, or a large area of outcrops, as the onlap of one parasequence on to another, and the merging of flooding surfaces (e.g. the merging of two coal seams).

The type 2 sequence boundary and shelf-margin systems tract are sometimes misused in the literature because of the difficulty in demonstrating a basinward shift in coastal onlap towards but not beyond the offlap break. Seismic resolution is often insufficient to resolve the subtle change in dip where topset onlaps topset. The change from progradation to aggradation, which is also considered characteristic of the type 2 boundary, is not definitive on its own because other factors such as decreasing sediment supply

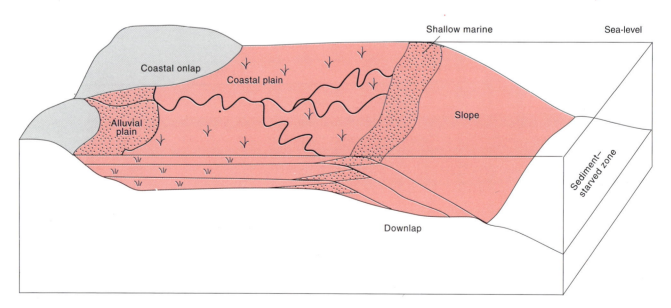

Fig. 2.17 Components of the highstand systems tract on a shelf-break margin. These include topset (alluvial, coastal), shallow marine, and slope systems

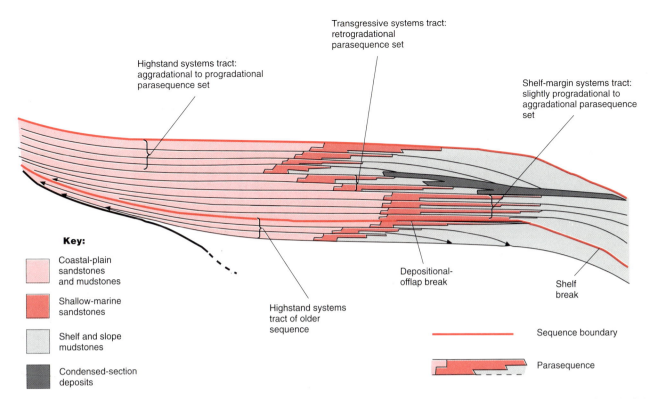

Fig. 2.18 A type 2 sequence. The type 2 sequence boundary is recognized from a downward shift in coastal onlap landward of the offlap break. This downward shift does not result in shelf bypass, fan deposition, or incision of the highstand topsets. The sequence boundary is overlain by a shelf-margin systems tract; a system tract of topsets with a predominantly aggradational stacking pattern. The rate of sea-level fall at the shoreline is equal to, or less than, the subsidence (from van Wagoner *et al.*, 1988)

rates could also affect such a pattern. In outcrop studies type 2 sequence boundaries are often used simply to distinguish minor sequence boundaries. Note also that a type 2 sequence boundary may pass laterally into a type 1 sequence boundary depending on the tectonic subsidence pattern in the basin.

2.4.7 Lowstand systems tracts on a ramp margin

The systems tracts described above are developed on a shelf-break margin, where the clinoform slope is steep enough and deep enough to allow large-scale failure and the formation of submarine fan systems. On a *ramp margin*, the lowstand systems tract was described by Van Wagoner *et al.* (1988) as consisting of a relatively thin lowstand wedge that may contain two parts (Fig. 2.19). The first part is characterized by stream incision and sediment bypass of the coastal plain. This is interpreted to occur during a relative fall in sea-level when the shoreline steps rapidly basinward until the relative fall stabilizes. The second part of the wedge is characterized by a slow relative rise in sea-level, the infilling of incised valleys, and continued shoreline progradation. This results in a lowstand wedge composed of incised valley-fill deposits up-dip and one or more progradational parasequence sets down-dip. The top of the lowstand wedge is the transgressive surface; the base of the lowstand wedge is the sequence boundary (discussed in detail in 8.3.4).

During falling relative sea-level on a ramp margin there is no bypass of sediment to the basin floor. Instead the sediment may be deposited as a set of downstepping prograding wedges, known as forced regressive wedges (Posamentier *et al.*, 1992). A number of these may be preserved between the highstand and lowstand prograding wedges. Posamentier (1993) referred to these wedges as *forced regressive wedge systems tracts*, which at the site of deposition are overlain by 'regressive subaerial surfaces of erosion', and underlain by 'regressive marine surfaces of erosion'. The latter are likely to pass landward into subaerial unconformities, and therefore both boundaries are strictly sequence boundaries. These forced regressive wedges are often sand-rich, and may form attractive stratigraphic traps where encased in shale. A number of examples of forced regressions, i.e. falls in relative sea-level on a ramp margin, are presented by Posamentier *et al.* (1992), and Posamentier and Chamberlain (1992) describe the detailed stratigraphy of a ramp-margin lowstand systems tract within the Viking Formation of Canada.

The transgressive and highstand systems tracts on a ramp margin are similar to those on a shelf-break margin,

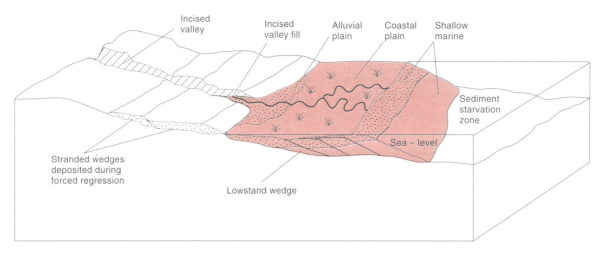

Fig. 2.19 Components of the lowstand systems tract on a ramp margin. The water depth is not great enough to allow development of a significant slope, and therefore turbidite systems are not developed during falling relative sea-level

although there is no significant clinoform component to the highstand systems tract.

2.4.8 Controls on systems tract boundaries

The actual time of initiation of a systems tract is interpreted by Van Wagoner *et al.* (1988) to be a function of the interaction between eustasy, sediment supply and tectonics. To this can be added 'topset area', which has a significant bearing on the development of the transgressive surface and the maximum flooding surface. Figure 2.15 shows the relationship between topset accommodation and systems tracts in a simple system of continuous subsidence and sinusoidally varying eustatic sea-level. The systems tract boundaries occur at the following points.
1 The type 1 sequence boundary (base of the lowstand systems tract) occurs when the rate of relative sea-level rise is zero and decreasing at the offlap break. The time at which this occurs is a function of eustasy and subsidence.
2 The boundary between the lowstand fans and the lowstand prograding wedge occurs when the rate of relative sea-level rise is zero and then increasing at the offlap break. The time at which this occurs is a function of eustasy and subsidence.
3 The boundary between the lowstand prograding wedge and the transgressive systems tract occurs when the rate of creation of topset accommodation volume equals, and is just about to exceed, the rate of sediment supply. Topset accommodation volume is the change in accommodation volume over the topset area during a rise in relative sea-level, and when this exceeds the sediment supplied, a transgression will occur. The timing of the boundary is a function of eustasy, subsidence, sediment supply, and topset area, and may occur when sea-level first floods back over the previous highstand topsets (as in Fig. 2.15).
4 The boundary between the transgressive systems tract and the highstand systems tract (the maximum flooding surface) occurs when the rate of creation of topset accommodation volume equals, and is just about to fall below, the rate of sediment supply. This is a function of eustasy, subsidence, sediment supply, and topset area.

It can be seen from the above that the timing of most of the systems tract boundaries is affected by very many factors. Those affected by fewest factors are the sequence boundary and the top of the lowstand fans.

The relative volumes of the systems tracts will be a function of their duration, and the sediment supply rate. There may be a linkage between supply rate and systems tract, for example in high latitudes where low sea-level in a glacial period may be associated with ice cover in the fluvial drainage basin. These factors, and factors such as the basin topography, can seriously distort the ideal sequence geometry shown in Fig. 2.15. It is very rare to find a seismic line that looks like this ideal model. This does not mean the model is wrong, merely that it should not be used as a template for interpretation.

2.4.9 Other possible systems tracts within a relative sea-level cycle

Van Wagoner *et al.* (1988) suggest that systems tracts should be defined objectively on the basis of the types of bounding surface, their position in a sequence (if this can be determined) and on their internal geometry. Two other theoretical systems tracts are not recognized in the original Exxon scheme. These are shown on Fig. 2.20, and described below.

The *midstand systems tract* (or *forced regressive systems tract* of Hunt and Tucker (1992); see 2.4.7) represents an entire sequence where at no time subsidence was sufficiently high to outpace sediment supply and allow transgression. This might be expected in basins with low or negative tectonic subsidence and/or high rates of sediment supply. Third-order scale midstand systems tracts on a shelf-break

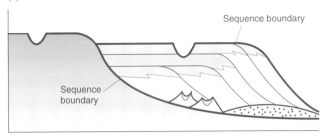

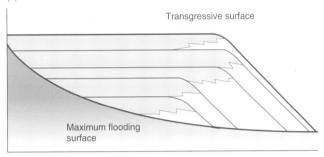

Fig. 2.20 Systems tracts not currently described in the Exxon scheme; (a) midstand systems tract; (b) regressive systems tract

margin have been described by Jones and Milton (1994) and Milton and Dyce (1995) from the Palaeogene of the North Sea at a time of basin-margin uplift. On a shelf-break margin (such as the North Sea Tertiary, or the Rhone delta) the midstand systems tract may comprise a unit of fans and a prograding wedge. On a ramp margin only prograding wedges will be developed.

The *regressive systems tract* (Fig. 2.20) is a theoretical systems tract formed between two rapid rises in relative sea-level separated by a slow rise (or by a pulse of increased sediment supply during continuous rate of rise). The systems tract is bounded below by a maximum flooding surface, and consists of a prograding wedge. The prograding wedge is bounded above by a maximum progradation surface. The internal geometry of the wedge is aggradational to progradational to aggradational again. Regressive systems tracts would be expected when eustatic cycles were superimposed on rapid background subsidence, so that not even type 2 sequence boundaries formed during the eustatic fall. Alternatively they may occur during a steady rise in relative sea-level with fluctuating sediment supply. Regressive systems tracts were predicted to occur in foreland basins by Posamentier and James (1993), although these authors termed them shelf-margin systems tracts, despite the absence of an underlying sequence boundary.

2.4.10 Composite (second and third order) sequences and systems tracts

Composite sequences were defined by Mitchum and Van Wagoner (1991) as 'successions of genetically related sequences in which the individual sequences stack into lowstand, transgressive and highstand sequence sets'. Figure 2.21 shows a composite sequence; bounded by two sequence boundaries but also containing four higher order sequence boundaries. Figure 2.22 shows the relative sea-level curve associated with Fig. 2.21, where two orders of cyclicity are apparent.

Most second- and many third-order sequences will contain higher order sequence boundaries, so it is important to state the order at which a sequence or systems tract is defined. For example, the highstand systems tract of a second-order composite sequence may, in reality, be a *highstand sequence set*, i.e. a stack of higher order sequences where the topset prograding parasequences may be dominant, but some higher order lowstand deposits may also exist. This has been demonstrated by Jones and Milton (1994), where all the systems tracts in a second-order sequence in the North Sea Tertiary contain third-order-scale lowstand fans. These fans particularly dominate the stratigraphy in the second-order lowstand systems tract. Finally, it is important to remember that systems tract boundaries in a composite sequence will be gradational, representing the interfingering of sequences and systems tracts of a higher order.

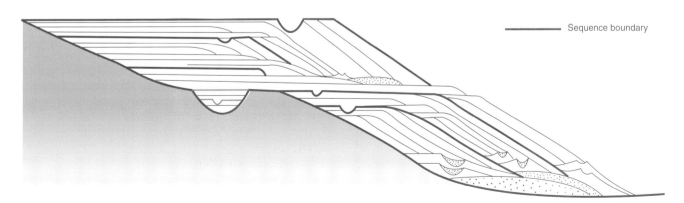

Fig. 2.21 A type 1 composite sequence, consisting of a stack of five higher order sequences. The high-frequency sequences form the building blocks of the composite sequence, and their nature is determined by their position in the composite sequence

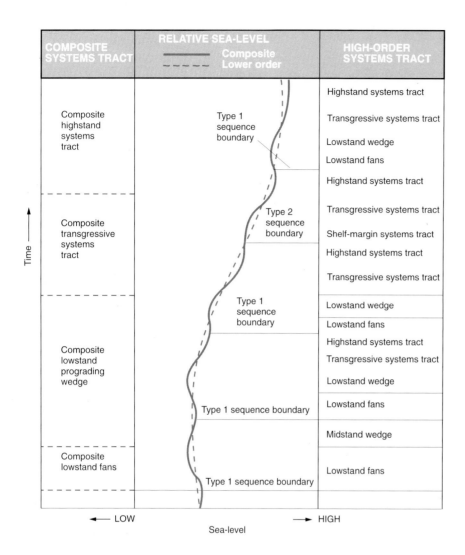

Fig. 2.22 The combination of low-order and high-order variations in relative sea-level implied in Fig. 2.21. The nature of the high-order sequences (their type, and the systems tracts within them) are determined by their position within the low-order sequence

2.4.11 Genetic stratigraphic sequences

Sequences, as discussed above, are cyclic stratigraphic units bounded by subaerial unconformities. However, because deposition is cyclic, the choice of the boundary is relatively arbitrary. Galloway (1989), after the work of Frazier (1974), suggested another means of subdividing the stratigraphy, using the maximum flooding surface as the cycle boundary. He therefore defined a *genetic stratigraphic sequence* as a package of sediments recording a significant episode of basin-margin outbuilding and basin filling, bounded by periods of widespread basin-margin flooding (Fig. 2.23).

With hindsight it is regrettable that he used the term 'sequence' here, instead of the 'depositional episode' of Frazier (1974), because this has led to a great deal of confusion. Some workers have used the Mitchum *et al.* (1977a) definition of the term, and some the Galloway (1989) definition. The sequence boundary, the maximum flooding surface and the maximum progradation surface are all valid correlation surfaces for dividing the stratigraphy. Each surface has its advantages and disadvantages as a primary stratigraphic boundary.

The sequence boundary can be recognized easily on seismic data by a downward shift in coastal onlap (as described in section 3.2.4). It indicates bypass and resedimentation into the basin, and is associated with the development of basinal reservoirs and hydrocarbon play systems. Recognition of the sequence boundary is therefore of great practical value in stratigraphic prediction for petroleum exploration. The timing of the sequence boundary is independent of sediment supply variations, so it is relatively isochronous. It is, however, difficult to see in a log and core data set, difficult to date precisely (occurring within proximal sediments potentially barren of fossils), and difficult to trace into the basin (except where associated with submarine fans).

The maximum flooding surface is also easily recognizable on seismic data, and can be identified easily on logs and in core. It is associated with topseal and often also with source-rock development. It may be represented by condensed marine facies, with a rich and easily dated fauna. It

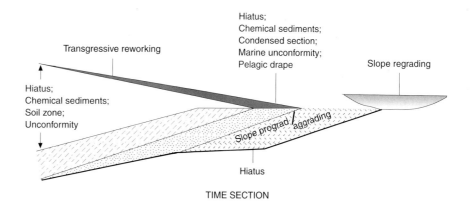

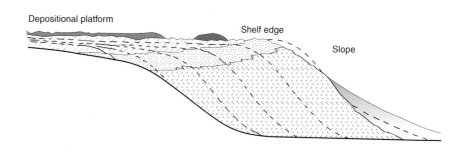

Fig. 2.23 A genetic stratigraphic sequence after Galloway (1989)

can be traced into the basin, where it correlates with a condensed interval, but may be difficult to trace into the proximal alluvial plain. Where the system is made up of several prograding lobes, it may be difficult to define precisely which lobe was most landward, and therefore where the maximum flooding surface lies. It is not associated with any particular reservoir development, and is seldom obvious in outcrop, where mudstones associated with the maximum flooding surface tend to be eroded or obscured.

The maximum progradation surface, or transgressive surface, also has been proposed by some workers as a natural surface for subdividing stratigraphy. This surface represents the furthest extent of topset reservoirs into the basin. It is recognized easily on seismic, outcrop, log and core data. It may be difficult to date precisely, and is difficult to correlate into proximal settings, and where the system is made up of several prograding lobes, it may be difficult to define precisely which lobe prograded the furthest and therefore where the maximum progradation surface lies.

The term sequence is usually now restricted to a unit bounded by subaerial unconformities. However, the most obvious surfaces in a basin are often the condensed intervals and maximum flooding surfaces (Loutit et al., 1988). These can be used for a pragmatic initial subdivision of the stratigraphy into mapping units, such as in the Jurassic North Sea studies of Partington et al. (1993a). However, this subdivision into genetic stratigraphic 'sequences' *sensu* Galloway (1989) is not an end in itself. A complete understanding of palaeogeography and facies distribution is achieved only by subdivision into systems tracts, which requires identification of the sequence boundaries and transgressive surfaces as well as the maximum flooding surfaces.

2.5 High-resolution sequence stratigraphy and parasequences

2.5.1 Introduction

The concept of stratigraphic cycles driven by rises and falls in relative sea-level was developed using seismic data. This has a relatively coarse resolution of many tens to hundreds of metres, and seismic-based stratigraphy has been termed 'low-resolution sequence stratigraphy' by Posamentier and Weimer (1993). *High-resolution sequence stratigraphy* integrates observations at a log, core or outcrop scale (Chapter 4). These detailed data sets allow gross stratal geometries to be linked with the internal facies assemblages. In addition, high-resolution shallow seismic data over modern sedimentary systems has contributed much detail to the understanding of bedding geometries within sequences and systems tracts.

High-resolution sequence stratigraphy is used increasingly as a tool in hydrocarbon reservoir description (e.g. Posamentier and Chamberlain, 1992; Reynolds, 1994). The key introductory publication covering the theory and practice of high-resolution sequence stratigraphy is that of Van Wagoner *et al.* (1990).

2.5.2 Parasequences and their continental equivalents

Shallow marine sediments are commonly arranged into regular upward-coarsening units with an upward-shoaling facies succession, separated by much thinner units representing an upwards-deepening facies succession (Fig. 2.24, see also Fig. 8.4). Often the upward-deepening component is represented only by a hardground or omission surface marking a transition from shallower to significantly deeper water facies. In sequence stratigraphic terminology, these cycles are termed *parasequences*. Van Wagoner *et al.* (1990) defined parasequences as relatively conformable successions of genetically related beds or bedsets bounded by marine flooding surfaces and their correlative surfaces. In special positions within the sequence, parasequences may be bounded either above or below by sequence boundaries.

The marine flooding surface in this definition is a surface separating younger from older strata across which there is evidence of an upward increase in water depth. This deepening is commonly accompanied by minor submarine erosion or non-deposition (but not by subaerial erosion due to stream rejuvenation or a basinward shift in facies), with a minor hiatus indicated. The marine flooding surface has a correlative surface in the coastal plain and a correlative surface on the shelf.

The recognition of parasequence boundaries, and their discrimination from sequence boundaries, was addressed by Van Wagoner *et al.* (1990), who suggest that shallow-marine parasequence boundaries are essentially flat, relatively condensed sections that represent abrupt deepening, and may be characterized by marine carbonate, phosphate or glauconite accumulations. The boundaries also mark abrupt changes in lithology and bed thickness, and are occasionally associated with lag deposits. Where lags do occur they are composed only of sediment reworked from below.

Parasequences form as a result of an oscillation in the balance between sediment supply and accommodation volume. Fluctuations in sediment supply due to autocyclic processes, such as avulsion and lobe switching, are probably the major control on parasequence formation. However, high-frequency variations in relative sea-level would produce high-order sequences that could look very similar to parasequences, especially if they were type 2 sequences.

Parasequences are defined by the marine flooding surface, and so cannot be defined in settings where changes in water depth are unrecorded. However, it is likely that parasequences have correlative equivalents in non-marine strata, such as fluvial avulsion cycles (although this has yet to be proved). Marine flooding surfaces could probably be correlated with coal beds on the coastal plain, and with widespread overbank mudstones and wet palaeosols on the alluvial plain. There are no criteria for recognizing para-

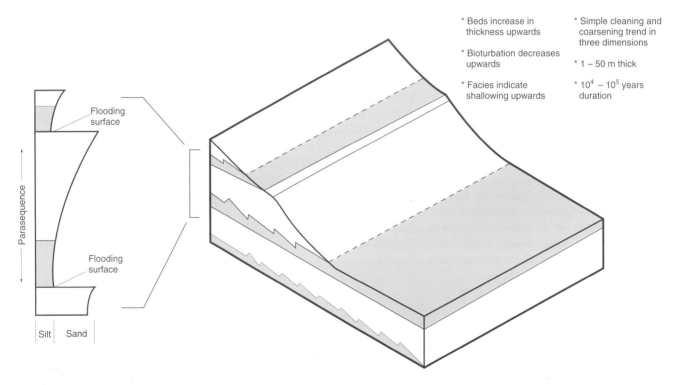

Fig. 2.24 An idealized parasequence. This represents a shallow-marine parasequence in a wave or storm-dominated setting, where upward coarsening can be related directly to upward shallowing

sequences in a deep marine setting. Parasequences will have correlative units on the clinoform slope representing the units of slope deposition fed by each delta lobe. Unless sediment is also bypassed to the basin floor, parasequences will have no equivalents in submarine fan facies. Mitchum and Van Wagoner (1991) speculated that individual fan lobes or leveed channels in the deep marine setting may reflect individual parasequences.

2.5.3 Parasequence sets

A parasequence set was defined by Van Wagoner *et al.* (1990) as a succession of genetically related parasequences forming a distinctive stacking pattern bounded by major marine-flooding surfaces and their correlative conformities (Fig. 2.25). In some cases one or both boundaries of a parasequence set will be a sequence boundary. Whilst parasequences may represent individual topset reflections within a systems tract on seismic data, parasequence sets often represent the entire topset component of that systems tract.

A stacking pattern refers here to the architecture of a vertical succession of parasequences. Progradational, retrogradational and aggradational stacking patterns can be recognized (Figs 2.25 and 2.26). In a progradational stacking pattern, the facies at the top of each parasequence become progressively more proximal higher in the succession. In a retrogradational stack the facies become more distal upwards, and in an aggradational stack the facies at the top of each parasequence is similar.

The topsets of the lowstand and highstand prograding wedges generally consist of a progradational parasequence set, whereas transgressive systems tracts consist entirely of a retrogradational parasequence set. The terms 'systems tract' and 'parasequence set' are not always synonymous (Posamentier and James, 1993), and in areas of high subsidence and sediment input, more than one parasequence set can exist in a systems tract. Parasequence sets are considered here as a class of depositional unit intermediate between parasequence and sequence, and the major marine flooding surfaces that bound parasequence sets may form subregional correlation markers.

2.5.4 Parasequence thickness trends

The thickness of a parasequence is controlled primarily by the water depth into which the shoreline progrades. This water depth represents the rise in relative sea-level since abandonment of the previous parasequence, and parasequence thickness is therefore a product of the rate of rise of relative sea-level and the periodicity of the parasequences.

If parasequence periodicity is relatively constant, then a slow rate of rise of relative sea-level results in thin parasequences, and a rapid rate of rise results in thick parasequences. Changes in the rate of relative sea-level rise should then be recognizable from trends in parasequence

PROGRADATIONAL PARASEQUENCE SET

* Net basinwards movement of the shoreline
* Characteristic of highstand systems tract and lowstand prograding wedge

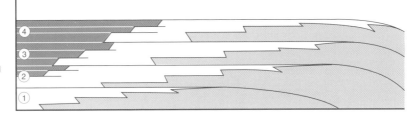

AGGRADATIONAL PARASEQUENCE SET

* No net movement of the shoreline
* Characteristic of shelf-margin systems tract

RETROGRADATIONAL PARASEQUENCE SET

* Net landwards movement of the shoreline
* Characteristic of transgressive systems tract

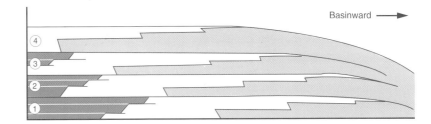

Fig. 2.25 Parasequence sets (after Van Wagoner *et al.*, 1988)

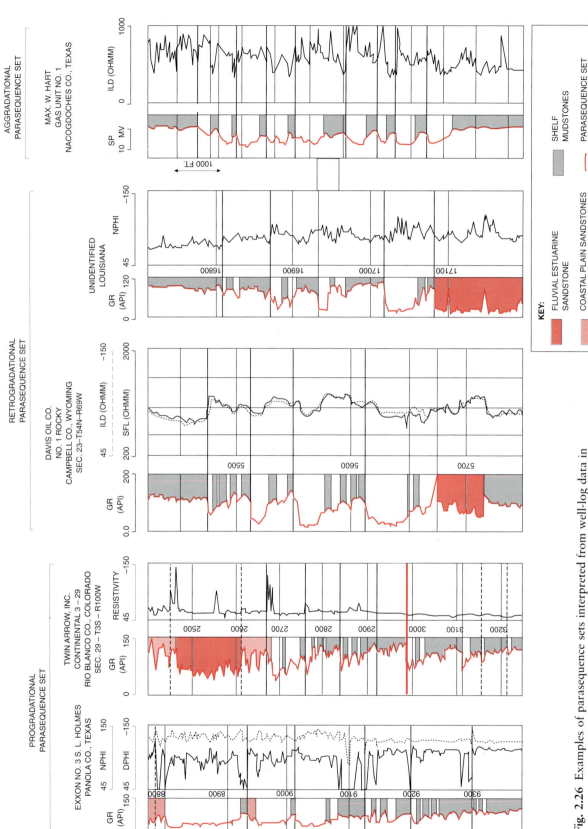

Fig. 2.26 Examples of parasequence sets interpreted from well-log data in Cretaceous-aged sections in the Western Interior Seaway, USA (from van Wagoner et al., 1990). Grain-size profiles are interpreted from an electric log suite. Individual coarsening upward cycles are interpreted as parasequences and systematic changes in the grain size of stacked parasequences allows the interpretation of their stacking pattern

thickness. These ideas were documented by Posamentier *et al.* (1988), who suggested that a lowstand prograding wedge will be characterized by upwards-thickening parasequences (reflecting accelerating relative sea-level rise) whereas a highstand prograding wedge will be characterized by upward-thinning parasequences, due to decelerating relative sea-level rise.

This thickness analysis can be applied in limited circumstances. Retrogradational parasequence sets, for example, often show a thinning upward trend, due to basinward thinning of the individual parasequences. This is not related to decreasing rates of relative sea-level rise. Thickness trend analysis also assumes a constant parasequence frequency, which may not be a valid assumption in many cases.

2.5.5 Sequence boundaries

As described earlier, sequence boundaries can be recognized on seismic data from a downward shift in coastal onlap, implying a fall in relative sea-level, with exposure and erosion of the highstand topsets. In a core, well log or outcrop data set, the downward shift in coastal onlap is rarely evident. Direct evidence for exposure, erosion and forced regression must be sought instead (Fig. 2.27; see also section 8.3.1). A *facies dislocation* is a surface where rocks of a shallower facies rest directly on rocks of a significantly deeper facies. The trend of gradual shallowing predicted by Walther's law is thus 'dislocated'. This dislocation may be obvious, such as where a coal bed overlies an outer shelf mudstone, or it may subtle, such as an upper shoreface facies overlying lower shoreface with middle shoreface absent. In shallow marine settings the facies dislocation is often associated with an abrupt grain-size increase. A facies dislocation implies a fall in relative sea-level and the development of a subaerial unconformity, although this could be up-dip from where the dislocation is observed. It thus marks a sequence boundary or its correlative conformity. Facies dislocations commonly are developed over the more distal areas of the highstand topsets, and the highstand clinoforms.

Incised valleys are described by Van Wagoner *et al.* (1990) as entrenched fluvial systems that extend their channels basinward and erode into underlying strata in response to a fall in relative sea-level. On the shelf, lowstand deposits filling the incised valleys are bounded below by a sequence boundary and above by the transgressive surface. A facies dislocation may be present at the base of the incised valley, although a grid of wells, or an outcrop data set, may be needed to prove the existence of an incised valley.

Incised valleys are differentiated from distributary channels by being deeper and wider than the scale of an individual channel or channel belt. The valley incises below the level of the distributary mouth bars, and often contains a more proximal fluvial facies deposited as part of the aggradational late lowstand prograding wedge. Alternatively they may contain estuarine to marine facies deposited as part of the transgressive systems tract.

Between the incised valleys in the more proximal areas of the highstand topsets, sequence boundaries may be very hard to recognize. Any evidence of exposure, such as major palaeosols, reddening, or weathering will be fairly superficial and may be removed by subsequent transgressive erosion. These surfaces are known as E/T surfaces, for the superposition of erosion and transgression (Walker and Eyles, 1991). The only evidence for a sequence boundary may lie in the components of the transgressive lag, which could be significantly coarser grained than the underlying succession, or may contain material that is not derived solely from the underlying succession.

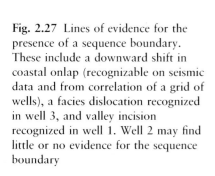
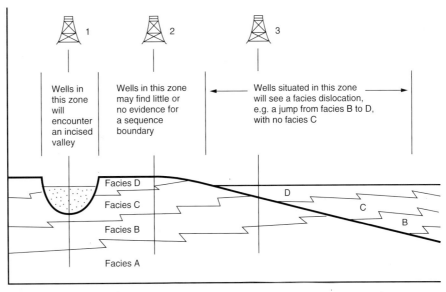

Fig. 2.27 Lines of evidence for the presence of a sequence boundary. These include a downward shift in coastal onlap (recognizable on seismic data and from correlation of a grid of wells), a facies dislocation recognized in well 3, and valley incision recognized in well 1. Well 2 may find little or no evidence for the sequence boundary

In rare cases a sequence boundary may be recognized by the truncation of underlying parasequences (e.g. fig. 24 in Van Wagoner et al., 1990). Care must be taken, however, to ensure that the limits of the parasequences are erosional, not depositional.

2.5.6 Maximum flooding surfaces

In well log, core or outcrop data sets, maximum flooding surfaces are recognized as the boundary between a transgressive unit, or retrogradational parasequence set, and an overlying regressive unit, or progradational parasequence set (Fig. 2.28). In a proximal direction the maximum flooding surface may lie within an aggradational parasequence stack, and it passes into a shelfal and basinal condensed section in a distal direction. The condensed section may be represented by a distinctive log facies or lithofacies, such as a glauconitic horizon, chert band, limestone band or high-radioactivity, low-velocity shale. The distinctive character of maximum flooding surfaces, and the widespread development in the basin of the equivalent condensed interval, makes them the easiest of the sequence stratigraphic surfaces to identify (Loutit et al., 1988). They are equivalent to the bounding hiatal surfaces that define the genetic stratigraphic units of Galloway (1989).

It should be noted that other condensed sections may form within a sequence, which are not equivalent to the maximum flooding surface; such as the boundary between the basin-floor fans and the slope fans, the boundary between the slope fans and the lowstand prograding wedge, and surfaces of major avulsion within a systems tract.

2.5.7 Ravinement surfaces

A *ravinement surface* is a surface of transgressive erosion. Swift (1968) described transgressive intervals in cratonic basins as commonly appearing to rest disconformably on underlying strata. The underlying strata sometimes may be pre-existing deposits of earlier cycles, but often are the marginal marine deposits of the contemporaneous cycle. The significance of such disconformities was first noted by Stamp (1921), who showed that the surf zone of a transgressing sea may bevel the marginal deposits of the coast being transgressed. Stamp called the resultant disconformity a ravinement.

The most widely accepted mechanism of landward beach and barrier migration is termed shoreface retreat, in which, as sea-level rises, sediment is eroded from the upper shoreface and emplaced in the lower-shoreface–offshore area as storm generated beds, or in the lagoon as a series of washover fans (Bruun, 1962; and Fig. 2.29). As the upper shoreface or breaker zone passes across the former barrier it erodes the lagoonal and washover facies deposited during the earlier stages of transgression and the lower shoreface facies therefore overlies a planar erosion surface. This erosion surface is referred to as the shoreface erosion plane, or the ravinement surface (Stamp, 1921).

The extent of erosion at the shoreface depends on the rate of rise of relative sea-level. In areas of rapid subsidence and/or relatively rapid sea-level rise, a comparatively complete transgressive unit may be preserved, whereas with slow subsidence and/or sea-level rise shoreface erosion is more pronounced and the transgressive unit is attenuated (Fischer, 1961).

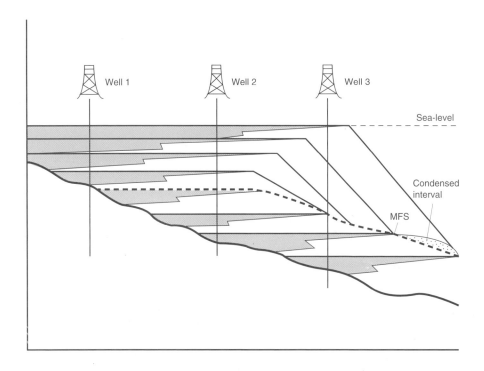

Fig. 2.28 Representation of a maximum flooding surface (MFS) in well data. In proximal wells the maximum flooding surface lies within aggradational topset facies and may be difficult to distinguish from a simple parasequence boundary. In distal wells the maximum flooding surface clearly overlies a retrogradation parasequence set

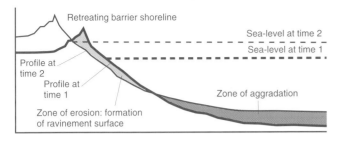

Fig. 2.29 Formation of a ravinement surface by transgressive erosion during shoreface retreat

The ravinement surface acts like a facies belt moving in parallel with other coastal facies belts during transgression. It results in a marine flooding surface; a rapid progression upwards from a supralittoral facies to a sublittoral facies with no intervening preservation of intermediate facies. This may form a striking surface in log, core or outcrop data, but is in detail diachronous and forms within a single episode of transgression. Ravinement surfaces therefore may bound every parasequence in a parasequence set. However, a more major ravinement surface may form at the transgressive surface (the boundary between the lowstand prograding wedge and the overlying transgressive systems tract). Other major ravinement surfaces may bound parasequence sets within a single systems tract.

2.5.8 Problems and pitfalls of high-resolution sequence stratigraphy

High-resolution sequence stratigraphy, performed on a subsurface data set, is not easy. The major problems can be summarized as follows:

1 Recognition of parasequences, and indeed of the depositional setting of the interval being studied, is tenuous without core control, good biostratigraphic control, or seismic indicators of basin setting (e.g. confirmation that it is a topset interval).

2 Correlation of parasequences may not be straightforward. One parasequence may look very much like another. Correlation will be easiest with closely spaced wells, or where some parasequences have a diagnostic log shape, or a marker lithology such as a prominent coal bed.

3 Recognition of sequence boundaries is not easy in the areas between the incised valleys and is not easy generally within a succession of parasequences.

4 Differentiating incised valleys from non-incised channel deposits can be very tricky. Much argument often focuses around the significance of anomalously large channel units. Van Wagoner et al. (1990) give several guidelines.

5 Systems tract boundaries can be recognized as surfaces where parasequence correlation lines terminate. Parasequences either onlap these surfaces, are truncated below them, or pinch out depositionally below them (in the case of the maximum flooding surface). It can be difficult to know which form of termination you are dealing with, and therefore the nature of the surface involved.

6 In an outcrop data set, high-resolution sequence stratigraphy is generally easier. There is abundant facies information, and surfaces can be traced laterally with relative ease. Differentiating incised from non-incised channels may still be difficult, although ideally the base of an incised valley could be traced laterally into an exposure/erosion surface. Outcrops generally are not continuous and correlating across gaps is perhaps the major problem. There will be no seismic data to allow the outcrop information to be placed in the context of stratal geometry, although sometimes large cliff sections give geometric information at a seismic scale (e.g. Bosellini, 1984, Fig. 4.5).

Sequence Stratigraphic Tools

CHAPTER THREE
Seismic Stratigraphy

3.1 **Seismic interpretation**
   3.1.1 Principles of seismic stratigraphic interpretation
   3.1.2 Resolution of seismic data
   3.1.3 Seismic processing and display for stratigraphic interpretation

3.2 **Seismic reflection termination patterns**
   3.2.1 Marking up a seismic section
   3.2.2 Categorizing reflection terminations
   3.2.3 Seismic facies and attribute analysis
   3.2.4 Recognition of stratigraphic surfaces

3.3 **Recognition of systems tracts on seismic data**
   3.3.1 Recognition of lowstand systems tracts
   3.3.2 Recognition of transgressive systems tracts
   3.3.3 Recognition of highstand systems tracts

3.4 **Pitfalls in interpretation**

3.1 Seismic interpretation

3.1.1 Principles of seismic stratigraphic interpretation

Seismic stratigraphy is a technique for interpreting stratigraphic information from seismic data. Together with its offspring sequence stratigraphy, it is acknowledged as being among the most significant developments in the earth sciences in the last 30 years. The ideas behind the technique were introduced in a number of papers in Association of American Petroleum Geologists (AAPG) Memoir 26 (Vail et al., 1977a,b,c).

The fundamental principle of seismic stratigraphy is that within the resolution of the seismic method, seismic reflections follow gross bedding and as such they approximate time lines. It is important to realize that this statement does not deny in any way the physical fact that the seismic reflections are generated at abrupt acoustic impedance contrasts, nor does it dispute the fact that variations in impedance contrast will produce reflections of varying amplitude (impedance is the product of rock density and seismic velocity). The key message is that the correlative impedance contrasts represented on seismic data come from bedding interfaces and not lateral facies changes. At the scale of seismic resolution, facies changes in time-equivalent strata are gradual and do not generate reflections (Fig. 3.1).

The axiom states that reflections can be thought of as time-lines that represent time surfaces in three dimensions, separating older rocks from younger. There are acknowledged exceptions. Some reflections, such as multiples or reflected refractions, are unfortunate artefacts of the physics of the method and need to be recognized as 'unreal' in a geological sense. Others, such as those from fluid contacts or diagenetic changes (such as bottom simulating reflectors generated from the change from Opal 'A' to Opal 'B'), are 'real' and represent genuine cross-cutting surfaces (in a chronostratigraphic sense). Finally there are 'reflections which are the outcome of lack of seismic resolution due to bed thickness' (Biddle et al., 1992), or 'tuned' lithofacies juxtaposition (Tipper, 1993).

Notwithstanding these exceptions to the rule it follows that a seismic section can be assumed to supply chronostratigraphic information, as well as lithostratigraphic information derived from reflection characteristics at impedance contrasts. This combination, along with the geo-

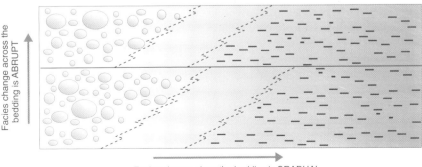

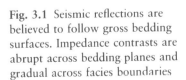

Fig. 3.1 Seismic reflections are believed to follow gross bedding surfaces. Impedance contrasts are abrupt across bedding planes and gradual across facies boundaries

metric information present on seismic sections, produces a very powerful tool for interpreting the subsurface stratigraphic record.

3.1.2 Resolution of seismic data

A key requirement for the successful application of seismic stratigraphic principles is a good understanding of the resolution of the seismic method. A geologist working on outcrop data and studying bed length and continuity is limited only by the degree and quality of the exposure. Subsurface wire-line logging tools in wells may be able to resolve beds on a centimetre to a few metres scale. Seismic reflection has, by comparison, a much coarser resolution (Fig. 3.2), which needs to be considered both vertically and laterally.

Vertical resolution

This can be defined as the minimum vertical distance between two interfaces needed to give rise to a single reflection that can be observed on a seismic section. In a single noise-free seismic trace this is governed by the wavelength of the seismic signal. At its simplest the shorter the wavelength (and hence the higher the frequency) the greater the vertical resolution. Seismic data are acquired and processed to produce as wide a range of frequencies as possible. It is the highest frequencies that constrain the resolution. If a wedge-shaped relationship is envisaged (Fig. 3.3), and the spacing between the reflective interfaces decreases below the critical wavelength of the signal, interference begins to cause the formation of a composite wavelet with anomalous amplitude. Reflectors that are spaced more closely than one-quarter of the wavelength have responses which begin to add constructively to produce a reflection with a high amplitude, known as the thin bed effect or 'tuning'.

In addition to the bed thickness constraints there are three other factors that limit the final resolution of the seismic data. Firstly, the Earth acts like a giant filter that progressively attenuates the high-frequency components of the seismic pulse. Secondly, there is a general trend towards increasing acoustic velocity with depth due to compaction and increased cementation. This effectively increases the wavelength of the signal, with detrimental effect on the

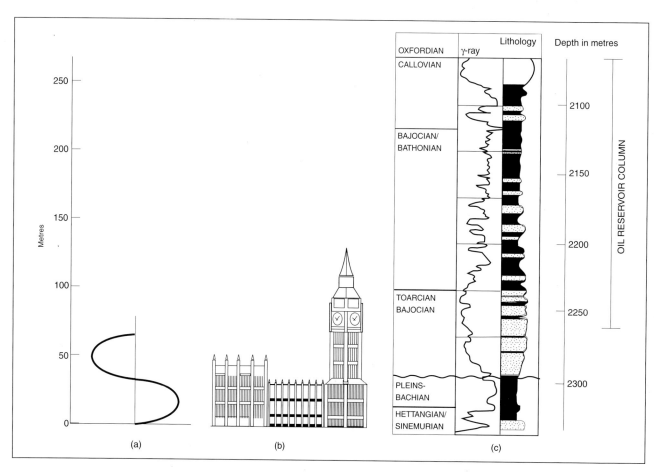

Fig. 3.2 A comparison of resolution of interpretation tools for the Beatrice Field, North Sea. (a) A single cycle sine wave of 30 Hz in medium of velocity $2000\,\text{ms}^{-1}$ (or 60 Hz; $4000\,\text{ms}^{-1}$); (b) Big Ben, London, c. 380 ft; (c) A γ-ray log through the Beatrice Oil Field

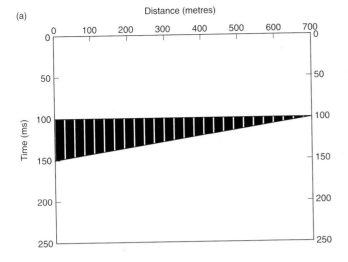
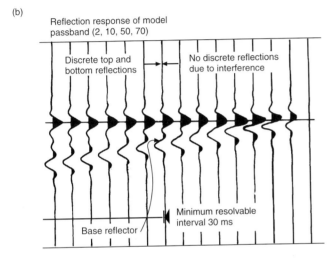
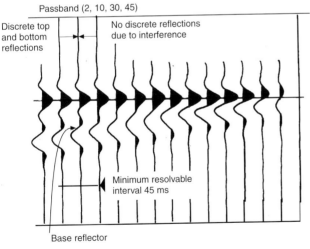

Fig. 3.3 The effects of bed thickness and frequency on vertical seismic resolution. (a) Single layer wedge: acoustic impedance model with laterally varying bed (or time) thickness, i.e., 0–50 ms. (b) Reflection response of model

resolution. Finally, if there is high ambient noise on the raw data, the processing stream may include a high-cut filter, which has the effect of removing the high frequencies necessary for finer resolution.

Lateral resolution

Seismic energy travels through the subsurface and comes into contact with the reflecting surfaces over discrete areas much in the same way that a spot-light travels through the darkness and illuminates a particular area. The energy travels as wave fronts and the region on the reflector where the seismic energy is reflected constructively is known as the *Fresnel Zone* (Sherrif, 1977). Lateral resolution is determined by the radius of the Fresnel Zone, which itself depends on the wavelength of the acoustic pulse and the depth of the reflector (Fig. 3.4). Thus in non-migrated seismic data, lateral resolution is dependent on the seismic bandwidth, on the interval velocity and on the travel time to the reflector (Fig. 3.5). The procedure of migrating seismic data considerably enhances resolution. For two-dimensional migration there is still the problem of line orientation relative to actual dip, but this is resolved on three-dimensional data. Thus for migrated data, lateral resolution depends on trace spacing, the length of the migration operator, time/depth of the reflector and the bandwidth of the data.

3.1.3 Seismic processing and display for stratigraphic interpretation

There is no single processing stream that can be recommended as optimal for seismic stratigraphic interpretation. Different acquisition parameters, different sources and above all variable geology mean that each case should be treated carefully and as an individual project. Seismic interpretation is about the recognition of familiar patterns in the data. Processing can either enhance or obscure the seismic representation of the geology. Processors and interpreters need to work together to produce the best finished article. It is important that the processor knows the type of geological problems that the interpreter is trying to resolve and it is equally important that the interpreter be aware of what has been done to the data prior to the 'finished section'. Even after careful demultiple, migration and amplitude control, there is still the matter of choosing the best display parameters. Much can be done to improve the interpretability of the data by choosing the optimum display format. Moreover the redisplay of an older data set is also a comparatively inexpensive way of bringing new life to an old data set.

There are four approaches to displaying the data. The first involves altering the *trace shape* (i.e. the shape of the 'wiggle') to emphasize or enhance aspects of the reflectivity of the stratigraphy. The second method relates to the form of *trace equalization* used, to compensate for the inevitable

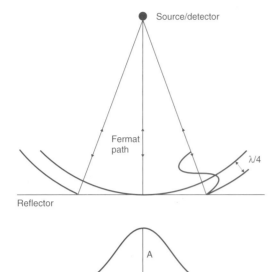

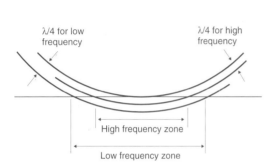

Fig. 3.4 The Fresnel Zone and constraints on lateral seismic resolution

loss of reflective energy with depth. The third technique relies on displaying the other aspects of the data, the so-called *complex attributes*. Finally there are the simple, inexpensive, but nevertheless effective, techniques that rely on *visual enhancement* of the final processed data.

Trace shape

Seismic energy returning to the geophones is stored as a series that records time and amplitude, normally sampled at 4 or 2 ms intervals. Post-processing, these data are displayed as a continuous series with interpolation between data points.

The choice of parameters governing the display of the interpolation trace is critical for the final appearance of the section. The simplest method is to show the data as a '*wiggle*' trace, where the displacement from the centre line of the trace represents reflection amplitude, and polarity is expressed by deflection to the right or left of this baseline (Fig. 3.6). This format allows precise observation of the change of wavelet shape from trace to trace and it emphasizes high-amplitude anomalies where traces overlap. It is thus very useful for stratigraphic interpretation, particularly at a reservoir scale, where information on bed thickness, lithology and fluid content is required. Unfortunately the wiggle-only trace format is sensitive to dip and tends to emphasize steeply dipping events, especially diffractions.

An alternative to the simple wiggle is the '*variable area*' format, in which there is no continuous trace but the magnitude of associated values is displayed and shaded (usually black) (Fig. 3.6). These displays bring out reflector continuity, but care is required to avoid loss of information on the shape of the wavelet and this problem, plus the stark black and white appearance of the section, means that variable area plots are rarely used on their own. A similar approach is the '*variable intensity (density)*' display, in which variations in reflector strength are indicated by varied shading or different colours. This is the standard colour display in most workstations and it permits much greater resolution of real events against high background noise levels than the variable area format.

Conventional seismic commonly uses the combination '*variable area and wiggle trace*' display, in which wiggle traces are superimposed on to variable area or sometimes variable density displays (Fig. 3.6). This provides information on the waveform and emphasizes continuity, but the section appearance is very sensitive to parameters such as display gain. There is also the danger of the loss of

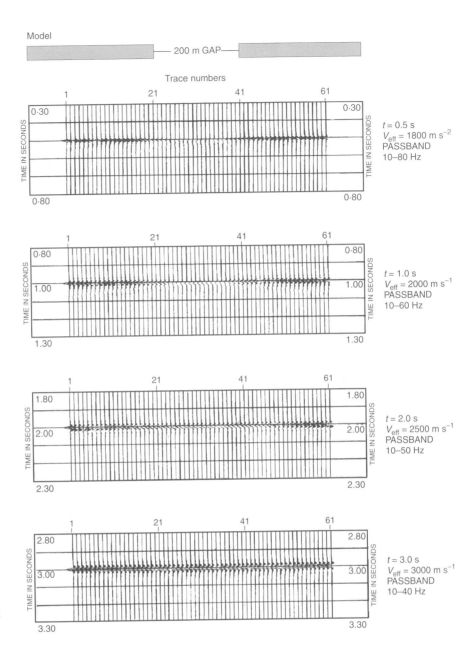

Fig. 3.5 An example of variations in lateral seismic resolution with increasing depth, seismic velocity and decreasing bandwidth

information on amplitudes as traces cross or overlap at peaks. This can be controlled by constraining the maximum deflection of the trace within set-limits (clipping), but this distorts the amplitude information and also can lead to large areas of white on the section where high-amplitude troughs are not displayed.

Once the trace type is chosen there are a number of parameters that can be altered to govern the trace shape, and these can alter the appearance of the section dramatically.

Swing controls the amount of deflection of a peak or trough as a percentage of the trace spacing. If it is optimized it can emphasize lateral continuity of weak reflectors.

Bias controls the position of the zero baseline between positive and negative reflections. By adjusting the bias it is possible to choose to emphasize peaks at the expense of troughs or vice versa (a positive bias moves the baseline to the left and enhances the peaks). A high bias, either positive or negative, tends to emphasize continuity but very high negative bias may lead to loss of correlatability of the troughs.

Clip is used to control the maximum deflection of troughs and peaks from the baseline. It is normally set as a multiple of trace spacing. Low values can be used to smooth out or homogenize amplitudes.

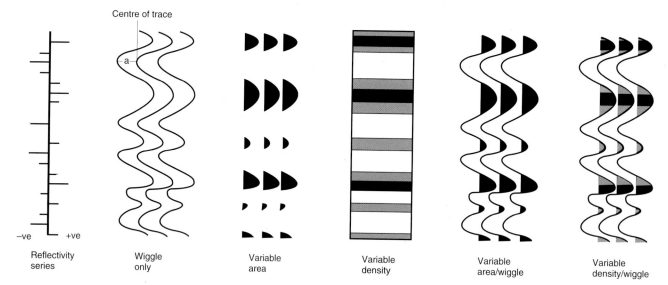

Fig. 3.6 Examples of various types of seismic trace display. a, amplitude

Gain is set at the plotting stage of the seismic line and it is a very important parameter for emphasizing amplitude variations. It affects the 'darkness' of the section and so low values of gain may be used to help bring out amplitude anomalies from the background signal.

Trace equalization

The pronounced loss of amplitude with depth on seismic data can be overcome by application of a technique called equalization, the aim of which is to try to produce a more balanced and interpretable section. The process adjusts the amplitude of the section to make the average amplitude constant over a predetermined interval or window on the trace. This window may be the whole section, as in single gate equalization, or multiple, as in the procedure termed *automatic gain control* (AGC).

Fast AGC is a special application of equalization in which the gate is set at a particularly short interval, perhaps one to ten times the sample rate (4–40 ms). It is applied to equalize all amplitudes as a way of enhancing reflector continuity and terminations. It should be used with some caution because fast AGC distorts the wavelet and emphasizes noise. Nevertheless the technique does have value, especially when applied to bring out detail adjacent to high amplitudes.

Complex attributes

Conventional seismic information is displayed as frequency and amplitude plots. However, these data, which have been recorded at the geophone or hydrophone, can be manipulated mathematically to produce an additional series of attributes. Although no new information is produced in this process, the display and interpretation of these complex seismic trace attributes can sometimes give new insights into the geology that would not have been obtained from conventional data. Their potential use in seismic stratigraphic interpretation is shown in Table 3.1, and described in more detail by Tanner and Sherrif (1977).

Table 3.1 Seismic stratigraphic application of seismic attributes (after Sonneland *et al.*, 1989).

Attribute	Application
Instantaneous frequency	Bed thickness
	Lithological contrasts
	Fluid content
Instantaneous phase	Bedding continuity
Cosine of instantaneous phase	Bedding continuity
	Identifying sequence boundaries
Reflection strength	Lithological contrast
	Bedding continuity
	Bed spacing

Visual enhancement techniques

This term covers some simple, cheap but often very effective redisplay options, which can be readily applied even when tapes are not available.

Colour. There is no doubt that the use of colour improves the interpretability of sections. Historically this was an expensive option for paper, but with the advent of inexpensive plotters this is no longer true. Moreover, workstations are eminently suited to colour display and allow easy experimentation to find the best scales to suit the user's preferences. Colour is probably best used as a background intensity display on which wiggle traces are superimposed.

The increase in interpretability associated with colour is clear in the presence of noise. It is better not to use too many colours and always to be aware that a significant proportion of the male population is colour blind.

Squash plots. Seismic sections were displayed originally for structural interpretation and the aim was to have as near a true scale, with no vertical distortion, as the time display would allow. Although this may be an ideal scale for trap definition it is often less than optimum for seismic stratigraphy, where sometimes the interpreter is interested in extremely subtle geometric relationships. Sedimentary dips are generally low. Submarine fans may exhibit dips of $1-3°$ on the levees and depositional lobes. Clastic shallow-marine prograding complexes rarely have dips greater than a few degrees on the slope and the coastal plain is very close to horizontal. In order to see marine onlap, downlap and coastal onlap relationships in these environments, especially when there is little syn-sedimentary structuring, it is necessary to enhance the apparent dips. Indeed this is what the interpreter has been doing optically by squinting obliquely along the section. The redisplay can be done very simply and cheaply by foreshortening the horizontal scale, while keeping the vertical scale fixed, commonly 5 or $10 \, \text{cm s}^{-1}$. The advantage of keeping the vertical scale fixed is that non-squeezed lines still tie. However, with modern data the close trace spacing means that the reduction in horizontal scale must be accompanied by dropping traces or trace summing. An alternative approach to resolve this problem is to reduce the horizontal scale until the minimum horizontal trace spacing is reached. The horizontal scale can be kept fixed and further vertical exaggeration is achieved by increasing the vertical scale. In the past the interpreter could request sections to be displayed at compressed horizontal scales (e.g. 1 : 100 000 or 1 : 200 000) or the data could be optically compressed in a squeeze camera in which the film and the section pass the lens at different speeds. Nowadays, workstations permit easy scale changes and all interpreters should be encouraged to experiment to find the vertical exaggeration that best brings out the stratigraphic relationships in their data set.

3.2 Seismic reflection termination patterns

3.2.1 Marking up a seismic section

The first step in the stratigraphic interpretation of a seismic line is to determine the vertical and horizontal scale of the section. An appreciation of both is essential to constrain the geological models that will be constructed later. It is also worthwhile to find out from the header or the seismic data itself if the section has been migrated, and whether it is marine or land data. Both are rich in multiples, even after optimum processing, but it is generally easier to identify multiples on marine data, where the acoustic contrast at the sea-floor generates both simple water-bottom multiples and peg-leg multiples. If a reflection's origin as a multiple is suspected then it can be marked in some distinguishing colour, usually light blue by convention. In Fig. 3.7, from the Tertiary succession of the Outer Moray Firth, North Sea, a reflection is seen at the right-hand (eastern end) side of the data, intersecting the edge of the section at 0.3 s. This reflection is probably a water-bottom multiple, caused by the sound waves bouncing twice between the sea-surface and sea-bed, and being recorded at a two-way-time (TWT) twice that of the sea-bed. This reflection does not have any geological significance, and can be ignored in the rest of the interpretation.

The next step is to divide the seismic data into the discrete natural stratigraphic packages that make up the section. To do this, first identify and mark reflection terminations. It will be obvious that all reflections do not go on for ever; most stop, often against another reflection. Mark these terminations with an arrowhead (by convention use a red pencil). This is not always a straightforward procedure; sometimes two reflections can appear to merge, and it may be unclear which of the reflections terminates against which. In noisy, chaotic, multiple-prone or low-amplitude data it may be unclear whether reflections terminate, fade out, and/or reappear somewhere else. Generally it is a good idea to start by ignoring zones of broken or chaotic reflections and to concentrate on the better data areas. The chaotic zones and poor data zones can be interpreted later with the aid of the model derived from the good data. On Fig. 3.7, the terminations of the continuous, high-amplitude flat reflections above 0.7 s are hard to see. These reflections look quite continuous, except where they are truncated by channels. However, on closer inspection, often helped by looking along the seismic line at a low angle, subtle terminations can be recognized within this interval. In contrast, the zone between 0.7 and 1 s contains numerous dipping reflections, which consistently terminate up-dip to the left and down-dip to the right.

Where reflections terminate in a consistent manner they define a line on the section (and a surface in three dimensions). This is known as a *seismic surface*. The next step is to use the red arrowheads to identify and highlight the seismic surfaces on the line. The number of surfaces will vary depending on the stratigraphic complexity, and there generally will be several major seismic surfaces of consistent reflection termination, and other more minor ones. These surfaces are by convention coloured yellow in the initial stages of a seismic stratigraphic exercise prior to them being assigned an identification colour based on their type or age. In Fig. 3.7 these surfaces of consistent reflection termination have been marked. The most prominent are one or two surfaces of high relief in the zone 0.2–0.5 s. However, a good seismic surface is also seen at around 0.5 s. Others are present in the interval of flat-lying reflections, and a complex set of surfaces are present within the interval of dipping reflections from 0.7 to 1 s. A strong reflection at around 0.7 s is also classified as a seismic

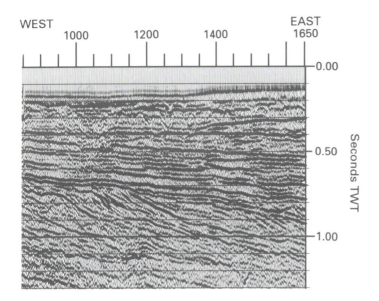

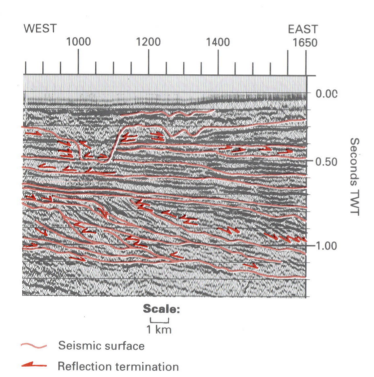

Fig. 3.7 Seismic data from the Outer Moray Firth, central North Sea, showing the seismic stratigraphy of the post-Palaeocene section; (top) uninterpreted; (bottom) interpreted. The surface at around 0.7 s is dated as close to the top Eocene, the surface at 0.5 s may mark the top of the Miocene, and the high-relief surfaces in the shallower section are interpreted as glacial lowstand surfaces in the Plio-Pleistocene

surface, owing to the onlap of overlying reflections around shotpoint (SP) 9000.

Once all the seismic surfaces have been picked it is necessary to perform a similar exercise on the other lines in the seismic data set, and to tie the interpretations (i.e. ensure that the interpretation is consistent where lines cross) to generate a three-dimensional grid of surfaces. Some of the seismic surfaces so defined will be important regional surfaces, others will be more local and may be less significant.

By picking the surfaces of reflection termination, the interpreter has divided the stratigraphy on the seismic data set into a number of depositional packages containing relatively conformable reflections of a similar or gently changing character and geometry, which are bounded by surfaces that mark the reorganization of reflection geometry.

3.2.2 Categorizing reflection terminations

Reflection terminations are characterized on a two-dimensional seismic section by the geometric relationship between the reflection and the seismic surface against which it terminates. Mitchum *et al.* (1977a) introduced the terms 'lapout, truncation, baselap, toplap, onlap and downlap' to describe reflection termination styles (Fig. 3.8). These terms are described below. Most are based purely on geometry, but others involve some interpretation of whether the reflection termination is an original depositional limit. The terms are applied to describe the reflection's current configuration.

Lapout is the lateral termination of a reflector (generally a bedding plane) at its depositional limit, whereas *truncation* implies that the reflector originally extended further but has either been eroded (erosional truncation) or truncated by a fault plane, a slump surface, a contact with mobile salt or shale, or an igneous intrusion (Mitchum *et al.*, 1977a,b).

Baselap is the lapout of reflections against an underlying seismic surface (which marks the base of the seismic package). Baselap can consist of *downlap*, where the dip of the surface is less than the dip of the overlying strata, or *onlap*, where the dip of the surface is greater (Fig. 3.8).

Downlap commonly is seen at the base of prograding clinoforms, and usually represents the progradation of a basin-margin slope system into deep water (either the sea or a lake). Downlap therefore represents a change from marine (or lacustrine) slope deposition to marine (or lacustrine) condensation or non-deposition. The surface of downlap represents a marine condensed unit. It is extremely difficult to generate downlap in a subaerial environment. Note, however, that it may be easy to confuse true depositional downlap and original onlap rotated by later tectonism.

The reflection terminations interpreted as downlap may in many cases be apparent terminations, where the strata thin distally below seismic resolution.

Onlap is recognized on seismic data by the termination of low-angle reflections against a steeper seismic surface. Two types of onlap are recognized; *marine* and *coastal*. Marine onlap is onlap of marine strata, representing a change from marine deposition to marine non-deposition or condensation, and results from the partial infill of space by marine sediments. Patterns of marine onlap cannot be used to determine changes in relative sea-level, because the level of marine onlap is not related directly to relative sea-level.

Marine onlap reflects a submarine facies change from significant rates of deposition to much lower energy pelagic drape. Wells drilled beyond the marine onlap limit of a rock unit will encounter a time-equivalent condensed unit or hiatus (time gap). The seismic surface of marine onlap represents a marine hiatus or condensed interval.

Coastal onlap is onlap of non-marine, paralic, or marginal marine strata and represents a change from a zone of deposition to basin-margin (subaerial or shelf) erosion and non-deposition. Coastal onlap generally is inferred from seismic data as the landward onlap of topset reflections (section 2.4), where these are assumed or are demonstrated to represent littoral, paralic or non-marine deposits. These topset deposits are assumed to have accumulated close to sea-level, and patterns of coastal onlap with respect to the onlapped surface indicate changes in relative sea-level. A landward progression in coastal onlap results from rising relative sea-level, whereas a downward or basinward shift in coastal onlap results from a fall in relative sea-level (described in more detail in Chapter 2).

Coastal onlap does not necessarily occur at the coastline, and progressive landward coastal onlap may accompany either regression or transgression depending on sediment supply. Wells drilled landward of the coastal onlap limit of a rock unit do not encounter a time-equivalent succession, but may encounter an unconformity, a palaeosol, or a karst horizon.

Toplap is the termination of inclined reflections (clinoforms) against an overlying lower angle surface, where this is believed to represent the proximal depositional limit. In marginal marine strata, it represents a change from slope deposition to non-marine or shallow marine bypass or erosion, and the toplap surface is an unconformity. An apparent toplap surface can occur, where the clinoforms pass upwards into topsets that are too thin to resolve seismically. In a deep marine setting, an apparent toplap surface is much more likely to be a marine erosion surface, such as seen in contourites. In this case the surface is localized and rarely flat over a large area.

Erosional truncation is the termination of strata against an overlying erosional surface. Toplap may develop into

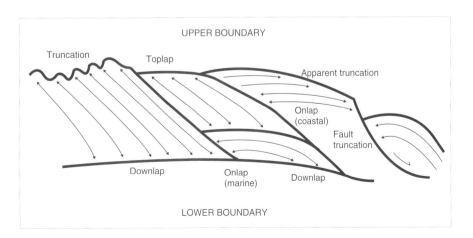

Fig. 3.8 Types of reflection termination

erosional truncation, but truncation is more extreme than toplap, and implies either the development of erosional relief or the development of an angular unconformity. The erosion surface may be marine, such as at the base of a canyon, channel or major scour surface, or a non-marine erosion surface developed at a sequence boundary.

Apparent truncation is the termination of relatively low-angle seismic reflections beneath a dipping seismic surface, where that surface represents marine condensation. The terminations represent a distal depositional limit (or thinning below seismic resolution), generally within topset strata, but sometimes also within submarine fans. Many reflection terminations in marine strata fall under the heading 'apparent', because it is likely that thin condensed units extend beyond the limit of seismic resolution (Fig. 3.9).

Fault truncation represents the termination of reflections against a syn- or post-depositional fault, slump, glide or intrusion plane. Termination against a relict fault scarp is onlap.

Frequently it appears that reflections overlying a seismic surface terminate against it, whereas those below appear to be conformable (or vice versa). This *conformity* is often apparent only, and is due to a very low angle between the surface and the apparently conformable reflections or to condensation along the surface.

Several types of reflection termination can be seen on Fig. 3.7. There is obvious truncation beneath a high-relief channelled surface between 0.3 and 0.5 s. Short reflections within the channels onlap the channel margins. The dipping reflections in the Eocene package between 0.7 and 1 s downlap to the right and either toplap or are truncated to the left. The overlying seismic surface is relatively flat between shotpoints 950 and 1100, and is a toplap surface. A unit of reflections below 0.8 s and east of shotpoint 1200 can be seen to onlap to the west and downlap to the east.

3.2.3 Seismic facies and attribute analysis

Once the seismic data has been divided into its component depositional packages, using the procedure described above, further geological interpretation of the stratigraphy may be attempted. This usually is done through seismic facies mapping, described by Sangree and Widmier (1977) as the interpretation of depositional facies from seismic reflection data. It involves the delineation and interpretation of reflection geometry, continuity, amplitude, frequency, and interval velocity, as well as the external form and three-dimensional associations of groups of reflections. Each of these seismic reflection parameters contains information of stratigraphic significance.

One of the easiest attributes to map and define is the geometry of the reflections. Prograding basin-margin units are commonly seen on seismic data to consist of topsets and clinoforms (Fig. 3.10; and as described in Chapter 2, section 2.1.2). Examples are shown on Fig. 3.7, with the late Eocene package east of SP 1200 and below 0.7 s displaying well-developed topsets and clinoforms. The dipping reflections to the west of this package are clinoforms, with minor or absent topsets. A well-developed topset–clinoform package can be interpreted as representing a systems tract of paralic to shelfal units (the topsets) and slope sediments (the clinoforms). The break in slope between the two is the offlap break (Fig. 2.3).

Other sedimentary units, such as submarine fan lobes, may sometimes have a similar shape, with relatively flat reflections becoming steeper in a basinward direction. The key to recognizing a true topset–clinoform package is to recognize a clear offlap break, and for the topset reflections to be concordant and parallel. Both these criteria are met by the late Eocene package in Fig. 3.7. Occasionally, clinoform reflections can be seen to decrease in slope downward, and pass basinward into flatter lying reflections known as bottomsets. In other cases the more distal low-angle reflections are not in depositional continuity with the clinoforms, but form separate depositional packages onlapping the clinoform front.

Ramasayer (1979) presented a rigorous methodology for two-dimensional seismic facies mapping, known as the 'A,B,C technique'. Three characteristics of each seismic package are recorded, given code letters (Table 3.2) and mapped. These are the nature of the reflection terminations against the upper boundary, the nature of the reflection

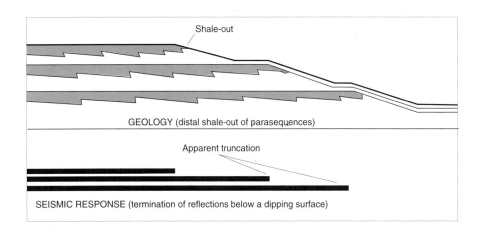

Fig. 3.9 Apparent truncation; the termination of reflections against an overlying surface, which looks like erosional truncation, but represents the original depositional limit of the strata

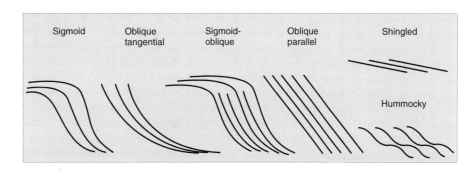

Fig. 3.10 Types of clinoform profile (after Mitchum *et al.*, 1977a)

Table 3.2 Seismic facies classification (from Ramasayer, 1979)

Code system A-B/C	
Upper boundary (A)	*Internal configuration (C)*
Te, erosional truncation	P, parallel
Top, toplap	D, divergent
C, concordant	C, chaotic
	W, wavy
Lower boundary (B)	DM, divergent mounded
On, onlap	M, mounded
Dwn, downlap	Ob, oblique progradational
C, concordant	Sig, sigmoid progradational
	Rf, reflection free
	Sh, shingled

terminations against the lower boundary, and the internal configuration of the reflections. Thus the late Eocene package in Fig. 3.7 mentioned above would be characterized in its proximal portion as: C-On/P; and in its distal portion as: C-Dwn/Ob.

These codes can be marked on a map, and distributions of the various seismic facies can be constructed using the entire seismic grid. With calibration from well data, it is often possible to make a reasonable geological facies map from seismic facies. The late Eocene seismic package described above has not been drilled, but the seismic stratigraphic and seismic facies analysis presented here predicts that it contains a basin-margin–slope assemblage. However, the nature of the facies within the topsets is still uncertain, and could be either alluvial plain, coastal plain, paralic or shelf facies. An example of a seismic facies map is shown as Fig. 3.11, reproduced from Mitchum and Vail (1977).

There is no unequivocal link between seismic facies and depositional systems, with the probable exception of the link between clinoforms and slope systems. Sangree and Widmier (1977) list seismic facies with their geological interpretation, but without well control this link is tenuous. Continuous flat-lying reflections may, for example, reflect deep-marine shales, coastal-plain topsets, alluvial plain, or lacustrine facies. However, a seismic facies map may be used to construct one or more tentative geological models, which should then be tested against and calibrated by well penetrations through the mapped interval. Without well control, a seismic facies map generally remains open to several geological interpretations.

Using modern geophysical workstation technology a range of parameters can be quantified and mapped for any seismic package. The *amplitude* of the seismic reflections at the top or base of the package can be mapped (Enachescu, 1993). Workstations can 'snap' a pick to the seismic data, ensuring that the horizon interpreted always lies on a seismic trough or peak, and can map the amplitude of that horizon through the data set. An amplitude map can be read directly as a geological facies map where the amplitude of the reflection is related to the geology. For example, underlying sands may result in a low-amplitude reflection, whereas underlying shales result in a high amplitude. The amplitude map would thus show the sand–shale distribution. However, in a seismic stratigraphic interpretation, where the top of a seismic package is not a reflection but a surface of reflection terminations, it may not be possible to 'snap' the pick to any one reflection, and amplitude maps of seismic surfaces may not be possible.

The *average amplitude* of the whole package is often a useful attribute. This generally is measured as a root-mean-square of the amplitude within the package (RMS amplitude), or as a mean square amplitude ('average energy'). This attribute can be quantified, mapped and contoured using a workstation, and used to differentiate zones of different seismic amplitude. Seismic amplitude is a function of density and/or velocity contrasts in the bedding, and often is related closely to the depositional facies. In submarine fans, for example, the channels may be recognized as linear zones of high-amplitude reflections and the depositional lobes as zones of low amplitude (or often vice versa). For example, Den Hartog Jager *et al.* (1993) show a 'gated amplitude extraction' (RMS amplitude) of an interval which includes the channel within the reservoir of the Forties Field. The outline of the channel is clearly seen as a linear zone of anomalous amplitude.

3.2.4 Recognition of stratigraphic surfaces

The key surfaces that divide stratigraphy into component systems tracts are sequence boundaries, transgressive surfaces, maximum flooding surfaces and marine onlap/downlap surfaces between the lowstand fans and the lowstand

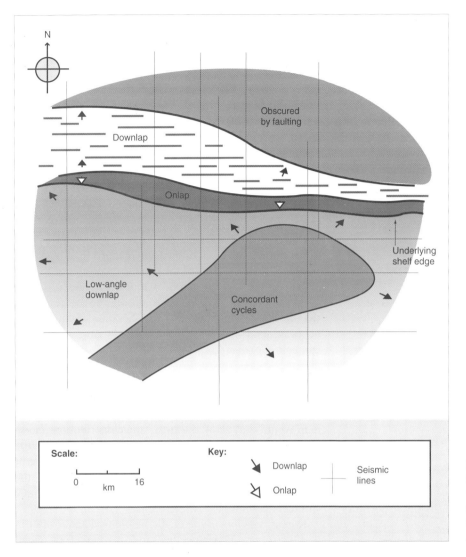

Fig. 3.11 An example of a seismic facies map (from Mitchum and Vail, 1977). This map shows reflection terminations at the base of a Lower Cretaceous package, offshore western Africa

wedge (Chapter 2). Most of these can be recognized as seismic surfaces (Fig. 3.12).

A *sequence boundary* (Chapter 2) can be recognized on seismic data in two ways; from the development of a high-relief truncation surface, particularly one which erodes the topsets of older units, and by a downward shift in coastal onlap across the boundary.

High-relief erosion surfaces can be seen on Fig. 3.7 at around 0.2–0.3 s. These are sequence boundaries, associated with glacial lowstands and fluvial erosion, possibly at the base of the ice cover.

Coastal onlap is the proximal onlap of topset reflections. As these are believed to have formed at or close to sea-level and certainly within the reach of shallow marine processes, a downward (and thus basinward) shift in coastal onlap implies a fall in relative sea-level, which can be assumed to have been accompanied by subaerial exposure and erosion over the topset area. Where coastal onlap falls below the previous offlap break, the topset reflections onlap an older clinoform, and the sequence boundary is a type 1 sequence boundary. Where coastal onlap does not fall below the previous offlap break, the topset reflections onlap an older topset, and the sequence boundary is a type 2 sequence boundary. The difference between type 1 and type 2 sequence boundaries is discussed in more detail in Chapter 2.

The late Eocene package on Fig. 3.7, lying below 0.7 s and to the east of shotpoint (SP) 1200, overlies a sequence boundary. This is shown by the onlap of three topset reflections against an older clinoform, around SP 1200. The lowest topset onlaps at 0.8 s, whereas the offlap break of the underlying clinoform lies at 0.7 s. This sequence boundary is type 1, representing a fall in relative sea-level of around 100 m (equivalent to 0.1 s TWT).

A *transgressive surface* marks the end of lowstand progradation, and the onset of transgression. It need not be associated with any reflection terminations, but will mark the boundary between a topset–clinoform interval, and an interval of only topsets (see Fig. 3.7).

A *maximum flooding surface* is recognized on seismic data as a surface where clinoforms downlap on to under-

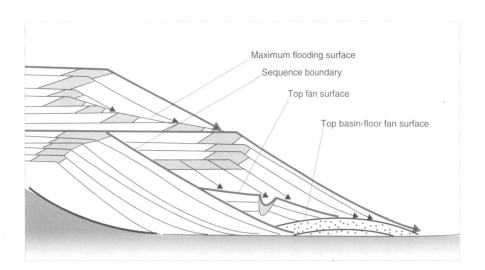

Fig. 3.12 Seismic surfaces within a sequence

lying topsets, which may display backstepping and apparent truncation. Note that not every downlap surface is a maximum flooding surface. An important downlap surface usually can be mapped at the base of the clinoforms of the lowstand prograding wedge. This is the *top lowstand fan surface* (as the lowstand wedge often downlaps on to the fan). The difference is that the facies below this downlap surface are basinal deposits, not topsets. An additional complication can occur in a generally transgressive setting, where both the highstand and lowstand wedge downlap on to topsets of an older sequence. If it is possible to correlate the downlap surface landward, it should pass laterally into either a sequence boundary (in which case it is the top lowstand fan surface) or into a topset assemblage (in which case it is the maximum flooding surface).

In a basinal setting, packages of reflections are bounded by marine onlap surfaces. Ideally these surfaces should be correlated landward, into a basin margin setting, and identified as one of the four surfaces above. This will not always be possible, particularly in a generally retrogradational setting, where older slopes form zones of bypass. In a distal setting in a basin, where the only deposition consists of lowstand fans, the marine onlap surfaces between the fans will represent condensed intervals time-equivalent to the lowstand wedge, highstand, and transgressive systems tracts, and will contain the correlative conformities to all four of the surfaces mentioned above.

3.3 Recognition of systems tracts on seismic data

It was suggested in Chapter 2 that systems tracts should be identified by the nature of their boundaries, and by the stacking pattern of their internal stratigraphy. This principle applies to their recognition from seismic data. If the nature of the boundaries has been interpreted, the systems tract has been identified. Examples of systems tracts on seismic data are shown on Figs 3.14–3.17, from the Palaeogene succession of the central North Sea. This succession is ideal for performing seismic stratigraphy, being shallow, well resolved by seismic data, and complex. Details of the stratigraphy, derived from the same data set, are presented by Jones and Milton (1994), who discuss how the sequences and systems tracts are distorted by the effects of high rates of uplift of the basin margins.

3.3.1 Recognition of lowstand systems tracts

A lowstand systems tract is bounded below by a sequence boundary, and above by a transgressive surface. These surfaces are recognized by the criteria mentioned above. Figure 3.13 shows a lowstand systems tract, already discussed as part of Fig. 3.7. The lower boundary is identified as a sequence boundary by the coastal onlap of three topset reflections against an older clinoform (a downward shift in coastal onlap of around 100 m). The nature of the top surface cannot be determined from Fig. 3.13, because we cannot see the eastern limit of the systems tract. However, other data in the area confirm this to be a lowstand prograding wedge. This seismic package is the T98 unit of Jones and Milton (1994). It is latest Eocene (or possibly early Oligocene), but is too shallow to be prospective due to biodegradation of oil at shallow depths of burial in this area. Submarine fans also are developed in this lowstand systems tract, but are not intersected by this particular seismic line.

Figure 3.14 shows a lowstand systems tract in the same area, but deeper in the succession. It is recognized as a lowstand systems tract because it is underlain by a sequence boundary, recognized by the downward shift in the toplap level of the clinoforms within the systems tract. It is also overlain by a transgressive surface (a transition to a retrograding topset unit, shown in more detail in Fig. 3.15), and it contains a basinal fan unit, recognized as a mounded unit of a couple of reflections lower, and more distal, than the clinoforms. The details of the relationship between the clinoform reflections and the basinal reflections is not clear, but a possible interpretation has the clinoforms down-

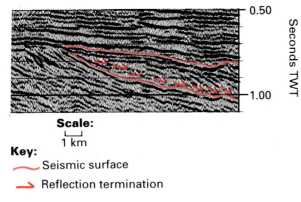

Fig. 3.13 A lowstand systems tract on seismic data. On this part of this line, only the lowstand prograding wedge is seen. The underlying sequence boundary is recognized by a downward shift in coastal onlap. Late Eocene, Outer Moray Firth, central North Sea

lapping on to a top fan surface (Fig. 3.14). This lowstand systems tract therefore may be divided into two parts, a lowstand fan and a younger lowstand wedge.

This systems tract is dated as late Palaeocene, and forms part of the T45 unit of Jones and Milton (1994). The clinoforms would represent the Dornoch Formation, whereas the basinal shales would be the Sele Formation. Larger Sele Formation fans form attractive reservoir targets in the North Sea. This lowstand represents the time of maximum uplift of the Scottish Mainland and the adjoining North Sea during the Palaeocene uplift episode (Jones and Milton, 1994).

3.3.2 Recognition of transgressive systems tracts

Transgressive systems tracts are bounded below by a transgressive surface and above by a maximum flooding surface and consist of retrograding topset parasequences. Transgressive systems tracts are often very thin, and may consist of no more than one reflection. Figure 3.15 shows a transgressive systems tract from the early Eocene of the central North Sea. It is recognized as a transgressive systems tract, as its base marks the transition from an underlying interval of mainly clinoforms, to an interval of mostly or entirely topsets. It also clearly shows internal retrogradational geometries. Two high-amplitude reflections are seen within the systems tract, the higher (younger) of which is displaced landward with respect to the lower. These reflections come from two retrograding coal intervals; one or both of which is equivalent to the Top Dornoch Formation coal of Deegan and Scull (1977). This figure shows the difficulty of using time-transgressive facies such as coals to subdivide stratigraphy.

The transgressive systems tract is overlain by a maximum flooding surface, recognized by the downlap of overlying clinoforms. A further coal overlies these clinoforms (the Top Beauly Formation coal), marking the next transgression. No clear apparent truncation can be seen below the maximum flooding surface, except perhaps at the extreme right-hand end of the systems tract (Fig. 3.15).

3.3.3 Recognition of highstand systems tracts

Highstand systems tracts are bounded below by a maximum flooding surface and above by a sequence boundary and exhibit progradational geometries. Figure 3.16 illustrates a highstand systems tract from the early Eocene of the central

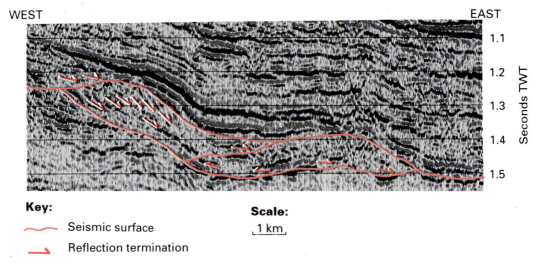

Fig. 3.14 A lowstand systems tract on seismic data. Here the lowstand wedge can be seen as a clinoform unit, with a toplap level significantly below the previous highstand offlap break. A lowstand fan unit is seen as a mounded unit with bi-directional downlap. Earliest Eocene, Outer Moray Firth, central North Sea

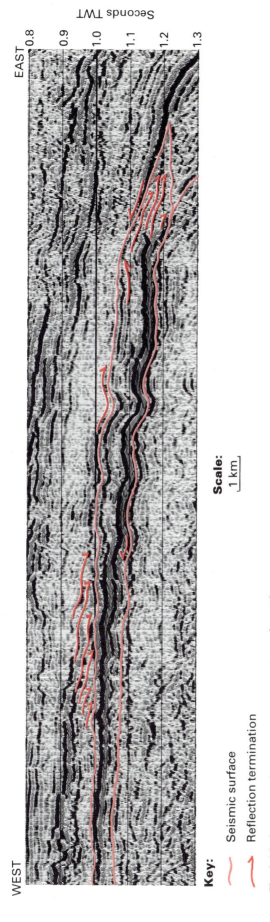

Fig. 3.15 A transgressive systems tract on seismic data. This systems tract contains two prominent retrograding reflections with high amplitudes, caused by two coal beds. The systems tract is overlain by a downlap surface to an overlying clinoform unit. Early Eocene, Outer Moray Firth, central North Sea

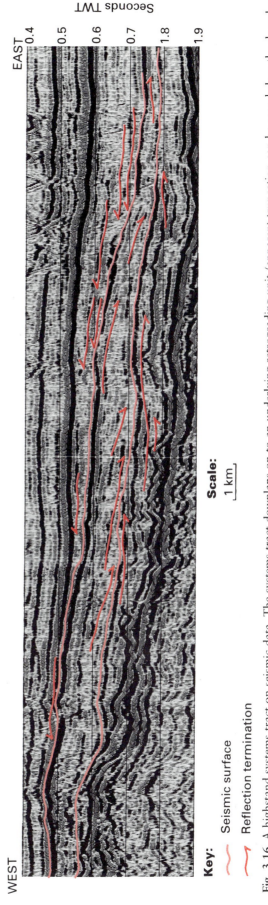

Fig. 3.16 A highstand systems tract on seismic data. The systems tract downlaps on to an underlying retrograding unit (apparent truncation can be seen below the downlap surface). It is overlain by a sequence boundary. Early eocene, Outer Moray Firth, central North Sea

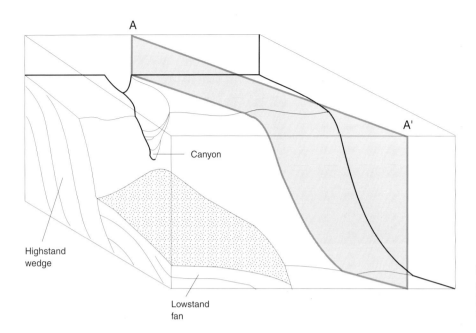

Fig. 3.17 Not every seismic line encounters every systems tract. Seismic line A–A' here misses the incised valley and lowstand fan

North Sea. It consists of prograding topsets and clinoforms, representing progradation, and it overlies a maximum flooding surface. Clinoforms within the systems tract downlap topsets of the underlying systems tract. Apparent truncation can be seen beneath this surface, and the underlying systems tract infills erosional relief on an older sequence boundary. The systems tract also underlies a sequence boundary. Flat-lying topsets of the subsequent systems tract onlap the last highstand clinoform (representing a downward shift of about 100 m). The nature of the two boundaries and the progradational architecture confirm this as a highstand systems tract. In detail, this highstand systems tract could be interpreted to consist of two or perhaps three down-stepping progradational wedges (each one displaced basinward and below the previous wedge). These are probably higher order (fourth order) sequences, and the systems tract is a composite systems tract, as discussed in Chapter 2.

3.4 Pitfalls in interpretation

There are many pitfalls and ambiguities inherent in seismic stratigraphic interpretation. The more important ones are outlined below.

1 Seismic data have a relatively coarse resolution, and stratal relationships in thin successions may be impossible to resolve.

2 Not every systems tract will be present on every line. Any one line may, for example, completely miss the lowstand fan system. This is illustrated in Fig. 3.17, where a lowstand fan is shown developed at the mouth of a canyon incised into a highstand slope. A seismic section along line A–A' would encounter neither the incised valley nor the lowstand fan.

3 A common mistake is to assume that all seismic surfaces which have been identified by reflection terminations must be sequence boundaries *sensu* Van Wagoner *et al.* (1988).

4 The key to successful seismic stratigraphy is the appreciation of the significance of coastal onlap and being able to recognize it on seismic data. However, it is easy to confuse marine and coastal onlap. Although coastal onlap is confined to topset reflections, it is not always easy to determine which reflections are truly topsets. Topsets can be identified confidently only where they are parallel and lie landward of an offlap break.

5 Fluvial incision and marine canyons may be easily confused. Fluvial incision is an indicator of a sequence boundary; canyon cutting is not necessarily so.

6 In a clinoform succession with extensive bottom sets, it is easy to misidentify the downlap surface. Many of the clinoforms will terminate against older bottomsets, and the true downlap surface is at the bottomset terminations.

CHAPTER FOUR

Outcrop and Well Data

4.1 Introduction and historical perspective

4.2 Resolution of well data

4.3 Sequence stratigraphy of outcrops and cores
    4.3.1 Parasequences in outcrops and cores
    4.3.2 Parasequence stacking patterns and systems tracts
    4.3.3 Key stratigraphic surfaces in outcrops and cores

4.4 Sequence stratigraphy of wireline logs
    4.4.1 Log suites used in sequence stratigraphy
    4.4.2 Log trends
    4.4.3 The log response of clinoforms
    4.4.4 The log response of parasequences
    4.4.5 Log responses from basinal environments
    4.4.6 Estimation of depositional controls and sequence stratigraphy from log response

    4.4.7 Key surfaces
    4.4.8 Identification of systems tracts from log response
    4.4.9 Pitfalls and ambiguities in sequence analysis of log data
    4.4.10 Check-list for sequence stratigraphic interpretation of a well-log data base

4.1 Introduction and historical perspective

The sequence stratigraphy of outcrop, core and wireline log data was developed in the early 1980s with the testing of seismic stratigraphic techniques against rock data. These techniques have evolved considerably over the last decade, especially in the last few years, with the publication of two major volumes. Firstly, Society of Economic Paleontologists and Mineralogists (SEPM) Special Publication 42 (Wilgus *et al.*, 1988) contains a series of papers in which Exxon workers laid out a series of key sequence stratigraphic definitions and conceptual models. The definitions and models together provide a framework linking basin-scale seismic stratigraphic models and sedimentary facies models. In the second significant publication, Van Wagoner *et al.* (1990) dealt specifically with high-resolution sequence stratigraphy from outcrop, core and wireline log data. With the publication of these volumes, outcrop, core and wireline log sequence stratigraphy has reached a wider audience and much progress has been made in the last few years. However, at the same time a range of views, sometimes contradictory, has emerged. Miall (1991), for example, has questioned many aspects of the models presented in SEPM Special Publication 42, and Walker (1990) has documented many challenging stratigraphic geometries from well-constrained subsurface data sets in western Canada that do not fit conventional models easily.

This chapter is organized into an outcrop and drill-core section, material with which the reader is likely to have some familiarity, followed by a longer section on wireline logs and combinations of logs and drill-core. The outcrop and core section is deliberately brief as it is covered in greater depth in Chapters 7–10.

4.2 Resolution of well data

Not all the techniques described in this chapter yield the same scale or resolution of stratigraphic information. Figure 3.2 shows a gamma-ray log and simplified lithological log through the reservoir section of the North Sea Beatrice Oilfield compared with a seismic waveform. The logs can resolve bedding detail that the seismic cannot, and hence permit a more detailed stratigraphic analysis. However, the gamma-ray log and other electric logging tools described in more detail in section 4.4 do not have perfect resolution, and cannot 'see' beyond the vicinity of the well bore. Furthermore, converting log properties to lithology does not always provide a unique geological answer, and wherever possible logs should be calibrated by core data. Core data from wells is good material for sequence stratigraphic analysis, but is rarely acquired in significant quantity. Usually the core is cut in porous units over hydrocarbon-bearing intervals, which usually means several tens of metres of core at most. In addition, cores, like wireline logs, see only a one-dimensional view of the geology at the well bore. As well as log and core data, wells also yield biostratigraphic information, discussed more fully in Chapter 6.

4.3 Sequence stratigraphy of outcrops and cores

4.3.1 Parasequences in outcrops and cores

Parasequences are defined in section 2.5 as relatively conformable successions of beds and bed sets bounded by marine flooding surfaces and their correlative surfaces. At present, parasequences can be recognized with confidence

61

only in shallow marine and shoreline successions, and are much more difficult to identify in deeper shelf and basinal settings, and in non-marine environments.

The precise nature of the parasequence geology depends on the facies associations, and several end-members are described in section 2.5 and by Van Wagoner *et al.* (1990). The most common is the coarsening-up signature (Fig. 4.1), widely recognized in parasequences from marine settings. Here, the shale content decreases upwards, but sand content and bed thickness may increase upwards. The marine flooding surfaces can be recognized by abrupt deepening, for example marine shales lying erosively on sandstones with rootlets. In addition, a number of non-diagnostic features characterize the outcrop and core expression of parasequence boundaries:
1 marine carbonate, phosphate or glauconite may be present, suggesting low siliciclastic sedimentation rates;
2 lags, which may record the transgression of the shoreface, are common, but are generally thin (< 10 cm thick) and contain only sediment reworked from below;
3 where flooding surfaces pass into amalgamated marine sandstones they sometimes can be traced as zones of preferential marine sedimentation;
4 if outcrop exposure is good enough, or cores are spaced sufficiently closely, parasequence boundaries may be recognized as essentially flat, with only a few centimetres (but rarely up to a metre or two) of sediment eroded at the boundary.
Dirtying-up parasequences may rarely also be recognized from tidal estuary facies, and parasequences that shallow-up to a non-marine facies may have a fine-grained unit of paralic shale at the top, and possibly a coal bed.

Parasequences should have correlative units in contemporaneous alluvial systems. Cyclicity in alluvial/fluvial systems generally results in stacked channel facies, which individually may show fining upward trends in core and outcrop. The link between these channel cycles and the parasequences further down the systems tracts is discussed in more detail in Chapter 8.

4.3.2 Parasequence stacking patterns and systems tracts

The stacking patterns (or 'architecture') of parasequences are discussed in section 2.5.3. The three patterns observed are progradational, in which the facies at the top of each parasequence becomes progressively more proximal, aggradational, in which the facies at the top of each parasequence is essentially the same, and retrogradational, where the facies become progressively more distal. These stacking patterns can be recognized both in outcrop and in core, and the position of the parasequences with respect to a major stratigraphic surface can help to constrain the systems tract represented by the parasequence architecture.

The example illustrated in Fig. 4.2 is from the Viking Formation, Alberta, Canada. A core cut in well 6-11-48-21–W4 shows two complete parasequences, considered to have been deposited in a tidal sand-sheet environment (Chapter 8, and Reynolds, 1994a). At its base the core penetrates the top of a strongly bioturbated, fine-grained sandstone with rare ripple-scale cross-lamination. An abrupt decrease in the sandstone content marks the first flooding surface, which is also bioturbated and contains scattered medium-grade sand grains. Upward, there is a steady increase in sandstone content and change from horizontal to vertical burrows. These changes, along with an increase in the preservation of unburrowed cross-beds, are interpreted to indicate shallowing and progradation within a parasequence. The top of this parasequence is marked by a decrease in sandstone content and by scattered medium-grade sand grains, followed again by burrowed systems

Fig. 4.1 Coarsening upward parasequence in outcrop, Panther Tongue Formation Cretaceous, Midwest USA. Telegraph pole for scale

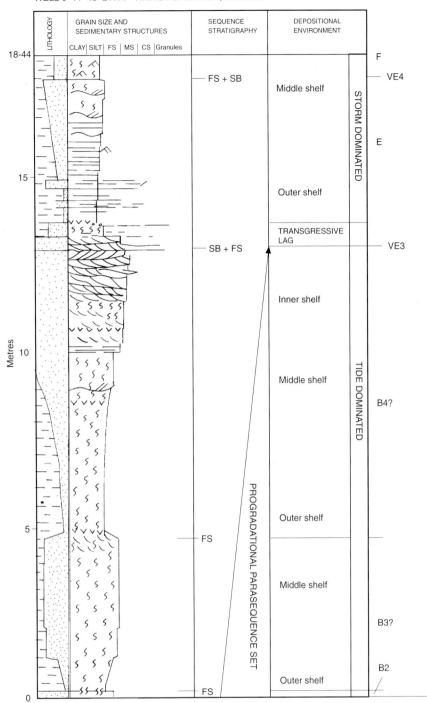

Fig. 4.2 Parasequences in core, Viking Formation, Alberta, Canada. See text for details. FS, flooding sequence; SB, sequence boundary

similar to the first parasequence, but towards the top of this second parasequence, dune-scale cross-bedding (*sensu* Ashley, 1990) predominates. This cross-bedding is intepreted to indicate daily current activity, the development of a more proximal facies and hence a progradational parasequence set. Above the second parasequence is a third flooding surface, marked by a thick lag deposit, then a dramatic change in grain size and sedimentary structures. The change in sedimentary processes suggest that this surface may be an interfluve sequence boundary (Chapter 8), which from regional mapping evidence correlates laterally with the development of an incised valley. The most likely interpretation of the progradational parasequence set in this example is as a highstand systems tract, overlain by a sequence boundary.

Retrogradational parasequence architectures are well-demonstrated in an outcrop example from the Jurassic Scarborough Formation in Yorkshire (Fig. 4.3, and Gowland and Riding, 1991). Below the Scarborough Formation, the Gristhorpe Member is characterized by rooted crevassesplay sandstones, which record a lower delta-plain environment, capped by a coal, which is in turn overlain by a dark mudstone. The coal is interpreted to record the cutting off of clastic sediment supply during initial relative sea-level rise, with the base of the mudstone forming the flooding surface followed by mudstone deposition in a transgressive lagoon. The mudstone is overlain by two irregular coarsening upward parasequences, with the base of the first mudstone exhibiting bioturbation and a sharp contact with the underlying mudstone. The sharp contact is interpreted to represent the transgression of a low-energy shoreline, and the parasequences themselves are considered to have been deposited in a wave-influenced brackish embayment in a lower delta-plain setting. Three further parasequences can be distinguished, each of which has an upward-coarsening signature, but in contrast with the first two parasequences, they show bioturbation and the development of carbonates on the flooding surfaces. The upward increase of bioturbation and carbonate content, along with an increase in marine fauna, indicate that the parasequences comprise a retrogradational set. Regional geological evidence also shows that the fifth parasequence is the most marine portion of this particular section, following which there is some evidence for parasequence progradation. This retrogradational parasequence set is hence interpreted as a transgressive systems tract.

4.3.3 Key stratigraphic surfaces in outcrops and cores

As described above, parasequence architecture can help to indicate the status of a stratigraphic surface. However, it may be difficult to distinguish a flooding surface separating parasequences from a more significant flooding surface in the absence of continuous core coverage in the subsurface, or from limited exposure. In the absence of interpretable wireline-log evidence or a regional geological context, such surfaces should not be overinterpreted beyond naming them flooding surfaces. Such caveats also apply to sequence boundaries, where lack of exposure or core coverage can give rise to the overinterpretation of boundaries. Sharp-based channel sandstones cutting into flood-plain deposits may simply represent a river meandering across its flood-plain, a normal sedimentary process, rather than abrupt fluvial incision caused by sea-level fall. In cases where a sequence boundary is suspected but not proven, it can be referred to as a candidate sequence boundary.

The recognition of a sequence boundary from outcrop or core requires the recognition of a facies dislocation; the superposition of a relatively proximal on a significantly more distal facies without the preservation of the intermediate facies (Fig. 4.4). This is not likely to be obvious in all locations; in the core example described above, the sequence boundary was represented by a lag deposit that could be correlated regionally into an incised-valley fill. However, if an incised-valley fill is cored or exposed, the rapid jump from fully marine to fluvial or estuarine valley-fill deposits may be dramatic. This is particularly well illustrated by the Scarborough Formation outcrop described in the preceding section. Above the retrogradational parasequence set a thin progradational unit of fully marine highstand parasequences is evident. Above this unit is a thick, coarse sandstone, the Moor Grit, which cuts down into the underlying Scarborough Formation (Fig. 4.3). The Moor Grit is an incised-valley fill of tidally influenced sandstones, with minor mudstone drapes, showing a facies dislocation with the underlying fully marine mudstones and sandstones.

In areas of spectacular mountainside exposure it may be possible to recognize the major stratigraphic surfaces from large-scale geometries. This has been possible in the Italian Dolomites, where progradational Triassic carbonate platforms show slope sediments of the highstand systems tract downlapping deeper water mudstones and carbonates of the transgressive systems tract. The line drawing of Fig. 4.5, from Bosellini (1984), shows the outcrop expression of a maximum flooding surface at the base of prograding clinoforms of the Catanaccio.

4.4 Sequence stratigraphy of wireline logs

Sequence stratigraphic analysis of wireline logs is an important component of analysing a subsurface data set. Log data allow lithology and depositional environment to be placed on the seismic section, thus linking seismic facies, rock properties, and sedimentological facies. Tying seismic to well data is not a trivial exercise, however, and the reader is referred to McQuillin et al. (1984) for further details.

Sequence stratigraphic analysis of wireline log data is neither easy nor unambiguous. Some systems tract boundaries may have a subtle expression on logs, and may even be hard to recognize in core (section 4.3). Correlation between individual wells is often ambiguous. Where wells are closely spaced and core control is good, for example in a data set of production wells from an oil or gas field, the data coverage may be sufficient to resolve the sequence stratigraphy. However, a more sparse well-data-set may not allow a single stratigraphic model to be derived.

4.4.1 Log suites used in sequence stratigraphy

Sequence analysis is concerned with the inference of depositional controls on sedimentary successions. Sequence analysis of a log suite must therefore concentrate on logging tools that measure depositional parameters. It is important to be aware of the potential pitfalls involved in attempting to obtain stratigraphic information from log responses. A more comprehensive discussion of the use of logs is provided by Rider (1986).

A suite of wireline logs over an interbedded succession of siliciclastics is shown in Fig. 4.6, and the individual log curves are discussed below.

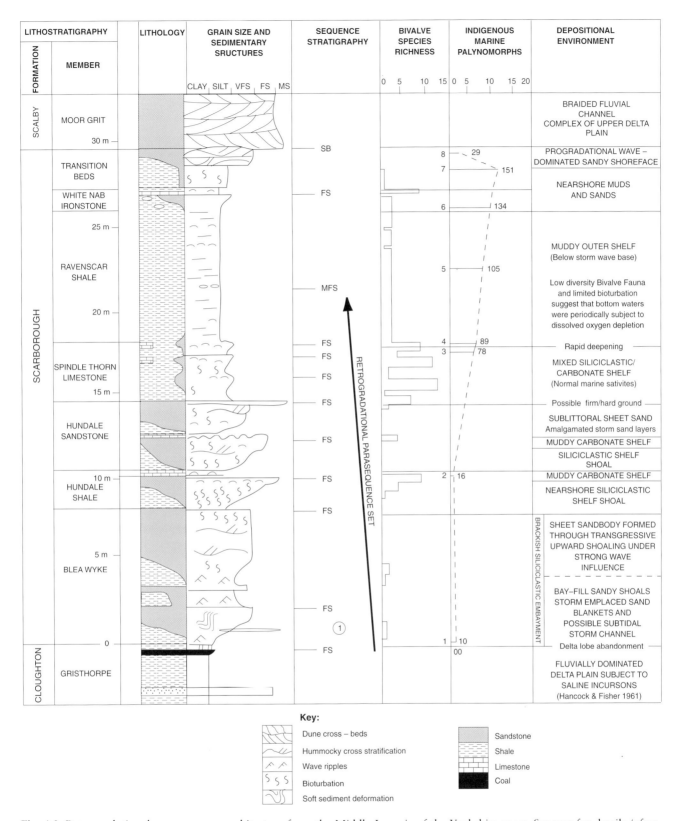

Fig. 4.3 Retrogradational parasequence architecture from the Middle Jurassic of the Yorkshire coast. See text for details (after Gowland and Riding, 1991). FS, flooding surface; MFS, maximum flooding surface; SB, sequence boundary

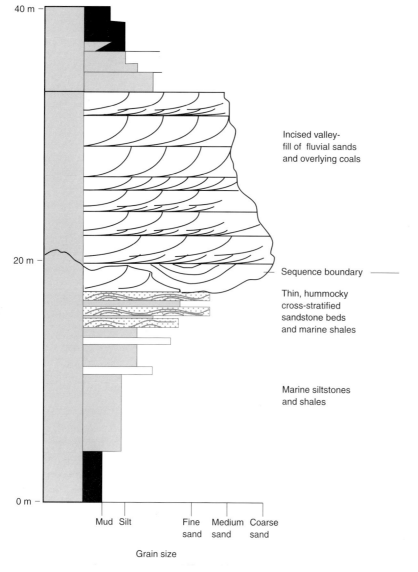

Fig. 4.4 Schematic log of a sequence boundary showing a major facies dislocation between marine and fluvial sediments

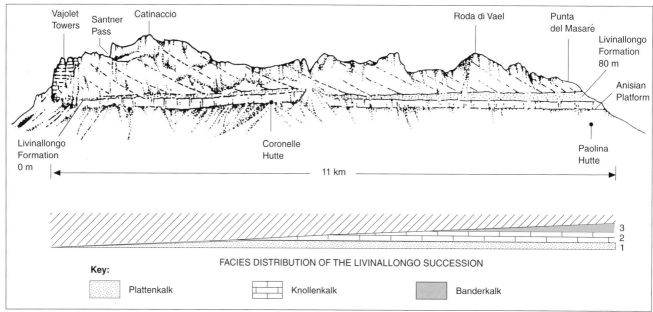

Fig. 4.5 Line drawing of prograding Triassic carbonates, Catanaccio, Italian Dolomites. The downlap surface of the progrades is close to the maximum flooding surface, which lies within the Livinallongo Formation. The Livinallongo Formation lies above an Anisian carbonate platform

Gamma-ray logs

The gamma log is one of the most useful logs for sequence stratigraphic analysis, and is run in most wells. The radioactivity of the rock, measured by the gamma tool, is generally a direct function of the clay-mineral content, and thus grain size and depositional energy. Gamma-ray logs are often used to infer changes in depositional energy, with (for example) increasing radioactivity reflecting increasing clay content with decreasing depositional energy. Although this usually is the case, there are exceptions where this rule breaks down.

Uranium in organic-rich anoxic shales, or precipitated post-depositionally in sandstone aquifers, may give anomalously high gamma readings. The radioactivity from feldspar in arkosic sandstones may give a high gamma reading, as may concentrations of heavy minerals in lags, particularly monazite and thorite. Some of these effects can be distinguished by using a spectral gamma log.

Most of the variations on the gamma log shown on Fig. 4.6 are related to depositional parameters, and to the sand : shale ratio. Exceptions are the cemented zones and coals, which have low gamma readings without being necessarily more sand rich.

Sonic

Sonic logs measure the sonic transit time through the formation. Transit time is related to porosity and lithology. Shales will have a higher transit time (lower velocity) than sandstones of a similar porosity, which sometimes allows the sonic log to be used as a grain size indicator. High concentrations of organic matter in coals and black shales will result in very long travel times, and these 'troughs' on the sonic log are often indicators of organic-rich condensed sections. The log is also affected by post-depositional cementation and compaction, and by the presence of fractures. The sonic log in Fig. 4.6 does not differentiate very well between sandstones and mudstones, but clearly indicates the cemented zones and coals.

SP log

SP (spontaneous potential) logs measure the difference in electrical potential between the formation and the surface. They are sensitive to changes in permeability, and are good at distinguishing trends between permeable sands and impermeable shales. The SP log works best where there is a good resistivity contrast between the mud filtrate and the formation water. Opposite impermeable shales the SP curve usually shows a more-or-less straight line on the log, known as the shale base line, and any differentiation within the shales is best done on the gamma or resistivity logs. Spontaneous potential is affected by hydrocarbons, cementation and changes in formation water salinity. The SP log in Fig. 4.6 differentiates between the sandstone–mudstone interbeds in the lower part of the section, but is of little use for determining trends in the upper mudstone.

Density–neutron suite

The density–neutron suite (the Schlumberger FDC–CNL suite, and other similar curves) is the best indicator of lithology and thus can be used to link lithology and depositional trends. It is one of the best log suites for sequence stratigraphic analysis, but not as commonly run as a gamma tool. The density (FDC) log measures the electron density of the formation via the backscatter of gamma rays, which is related to the true bulk density. The neutron log (CNL) attempts a measurement of formation porosity by using the interaction between neutrons emitted from the tool and hydrogen within the formation.

The logs are scaled to approximately overlie in clean carbonate lithologies. In clean sandstones there will be a small separation (larger if the sand is feldspathic). An increase in shale content will result in an increasing neutron reading (from hydrogen in bound water within the clays) with no apparent change in density. The resulting crossover and separation between the curves can be a sensitive and useful grain-size indicator. In addition, coals are easily identified on the density–neutron suite. The density log is affected by caved hole (oversized borehole due to erosion or collapse of the walls), and by heavy minerals such as pyrite and siderite. The presence of gas increases the neutron response, owing to the high proportion of hydrogen atoms within methane.

The density–neutron suite in Fig. 4.6 is as good as the gamma log for determining depositional trends, with the added bonus that the coals and cemented zones are differentiated clearly. The density log in particular is a good indicator in this well of small-scale upward-cleaning cycles (e.g. in the water-bearing sandstone).

Resistivity suite

Resistivity logs measure the bulk resistivity of the rock, which is a function of porosity and pore fluid. A highly porous rock with a conductive (saline) pore fluid will have a low resistivity, whereas a non-porous rock, or a hydrocarbon-bearing formation, will have a high resistivity. Resistivity trends may be excellent indicators of lithology trends, *provided the fluid content is constant* (i.e. in the oil leg or in the water leg). Resistivity logs often are excellent for correlating within shale successions, or within clean sandstones with uniform gamma response. Different resistivity logs give different scales of bed resolution, and the raw resistivity traces from dip-meter logs, measured every 2.5 or 5 mm, provide geological information on a bed-scale. In Fig. 4.6, the effect of the oil leg in the upward-cleaning sandstone masks any depositional trends.

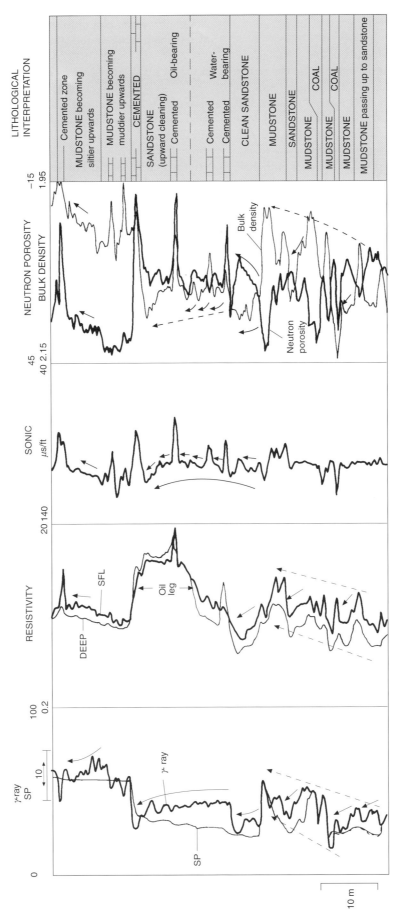

Fig. 4.6 A wireline log suite suitable for lithological and stratigraphical interpretation. This suite is based on real data from the Brent Group, North Sea, Middle Jurassic. It is a siliciclastic assemblage with minor coals and limestones, and here contains an oil column

4.4.2 Log trends

Log response may be used to estimate lithology, subject to the caveats above. Trends in log response (at any scale) therefore may equate with trends in depositional energy, and thus with patterns of sedimentary infill. In shallow marine successions, for example, increasing depositional energy is related directly to decreasing water depth. The literature is full of studies relating log trends to deposition; for example linking point-bar deposits to trends of upward-increasing gamma, or mouth-bar deposits to trends of upward-decreasing gamma (e.g. Pirson, 1977; Coleman and Prior, 1980; Galloway and Hobday, 1983; Cant, 1984; Rider, 1986).

A number of distinctive trends are recognized frequently on wireline logs, and generally are described from their gamma-log expression. Log trends may be observed as a change in the average log reading, or from shifts in the sand or shale base line (where the sand base line on a gamma log is the line marking the minimum gamma locus of the curve over an interval, and the shale base line is the line marking the maximum gamma locus). Typical log trends are illustrated on Fig. 4.7, and their significance is discussed below.

The Cleaning-up trend shows a progressive upward decrease in the gamma reading (generally seen in both the sand base line and the shale base line; Fig. 4.7) representing a gradual upward change in clay-mineral content. This is either a progressive change in lithology (seen in many intervals in Fig. 4.6), or a gradual change in the proportions of thinly interbedded units below the resolution of the logging tool.

The trend of the other log curves depends on the log response of the cleaner lithology. In a water-bearing clay–silt–porous sand assemblage, the sonic transit time often decreases upward initially from shale to silt, then increases upward as sand porosity increases, while the resistivity decreases upward and the FDC–CNL curves cross over each other. The pattern may be significantly different where the sands are progressively cemented or are hydrocarbon-bearing.

In shallow marine settings the cleaning-up motif is usually related to an upward transition from shale-rich to shale-free lithologies, owing to an upward increase in depositional energy, upward shallowing and upward coarsening. This interpretation should be cross-checked with any available palaeobathymetric data (core, biostratigraphic, presence

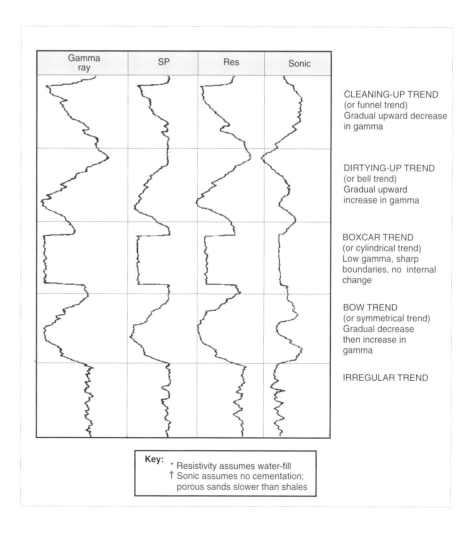

Fig. 4.7 Idealized log trends, assuming saltwater-filled porosity

of marker lithologies such as coal, etc.), and seismic data should be used to place the well in depositional context. Figure 4.8 shows a cleaning-up gamma trend, from the Middle Jurassic Tarbert Formation, of a well in the northern North Sea (reproduced from Mitchener *et al.*, 1992). Core control demonstrates that this corresponds with an upward coarsening and an upward shallowing of the facies.

There is a general coincidence between seismic clinoforms and large-scale cleaning-up motifs, discussed in more detail in section 4.4.4.

In deep marine settings the cleaning-up motif is seen generally as part of a more symmetrical bow trend (discussed below), related to an increase in the sand percentage of thinly bedded turbidites.

Occasionally, cleaning-up units also may be a result of a gradual change from clastic to carbonate deposition, or a gradual decrease in anoxicity, neither of which need be related necessarily to upward-shallowing or to progradation of a depositional system.

The 'dirtying-up' trend shows a progressive upward increase in the gamma reading (generally seen in both the sand base line and the shale base line; Fig. 4.7), related to a gradual upward change in the clay-mineral component. This could be a lithology change, for example from sand to shale, or an upward thinning of sand beds in a thinly interbedded sand–shale unit. Both of these imply a decrease in depositional energy. As in the cleaning-up trend discussed above, the trend on the other curves depends on the log response of the cleanest sand.

Upward-fining predominates within meandering or tidal

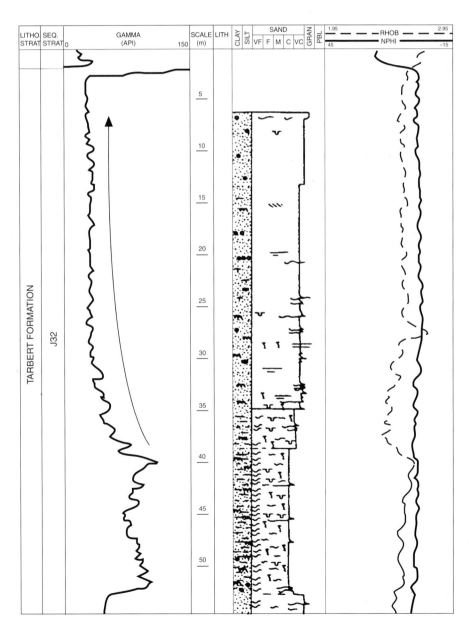

Fig. 4.8 A cleaning-up log unit from the Tarbert Formation, Middle Jurassic Brent Group, northern North Sea. This unit represents an upward-coarsening depositional unit, produced by upwards shallowing during progradation in a shallow marine setting (after Mitchener *et al.*, 1992)

channel deposits, where it represents an upward decrease in fluid velocity, and thus energy, within the channel. Larger fining-up units are commonly recorded in coarse fluvial successions and in estuarine fill (Chapter 8). Channel deposits often have a basal lag, which may affect the gamma response if the lag contains shale clasts or heavy minerals.

In shallow marine settings the dirtying-up trend often reflects the retreat or abandonment of a shoreline–shelf system, resulting in upward deepening and a decrease in depositional energy. In the case of Fig. 4.9 (reproduced from Mitchener et al., 1992), the dirtying-up log trend corresponds with a transgressive shoreline–shelf unit, assigned to the Middle Jurassic Tarbert Formation, yet clearly in a different systems tract to the Tarbert Formation shown in Fig. 4.8. Larger scale shallow-marine fining-up units may occur through the stacking of smaller coarsening-up parasequences.

In deep marine settings the dirtying-up motif may be a result of a decrease in the sand percentage of thinly bedded turbidites, and so may record the waning/abandonment period of submarine fan deposition (Fig. 4.11).

Dirtying-up successions also may be a result of a gradual increase in anoxicity, or a gradual change, perhaps climatically controlled, from carbonate to clastic deposition.

Boxcar log trends (also known as the cylindrical motif) are sharp-based low-gamma units with an internally relatively constant gamma reading, set within a higher gamma background unit. The boundaries with the overlying and underlying shales are abrupt. The sonic reading from the sands may be either a higher or a lower transit time than from the shales, depending on cementation and compaction.

The interbedding of the two contrasting log units implies the existence of two contrasting depositional energies, and an abrupt switching from one to the other. Boxcar log trends are typical of some types of fluvial channel sands, turbidites, and aeolian sands. Several of the sandstone units in Fig. 4.10 can be described as boxcar sands, and the J64 unit has a boxcar motif. This log is from a turbidite sandstone reservoir of the North Sea Miller Field, and is reproduced from Garland (1993).

Turbidite boxcar units generally show a much greater range of thickness than boxcar fluvial channel units. Trends of upward-thickening or upward-thinning of boxcar sands within a depositional unit often can be recognized in turbidites (e.g. upward-thickening sandstones at the base of J64 in Fig. 4.10), and these trends show no systematic shift in the shale base line or sand base line provided the units are thick enough to be resolved properly.

Note that shallow-marine sand bodies may have truncated bases due to faulting, or sharp bases due to falls in relative sea-level or other factors. These therefore may have a boxcar appearance, although systematic shifts in the shale line may be observable. In addition, evaporites often have a boxcar response on the gamma log.

The Bow trend (also known as barrel trends, symmetrical trends) consists of a cleaning-up trend, overlain by a dirtying-up trend of similar thickness and with no sharp break between the two. A bow trend is generally the result of a waxing and waning of clastic sedimentation rate in a basinal setting, where the sediments are unconstrained by base level, as for example during the progradation and retrogradation of a mud-rich fan system. Figure 4.11 shows gamma bows in Jurassic submarine fans in the area of Ettrick Field, central North Sea. In some of the Ettrick wells, turbidite sandstone beds are developed in the centre of the 'bow'.

The bow trend is seldom developed in a shallow marine setting, where base-level constraints usually lead to thicker progradational and thinner transgressive units. However, shallow-marine bow motifs are not unknown, especially where rift topography or growth faulting may have constrained transgression and allowed a thick transgressive body to develop.

Irregular trends have no systematic change in either base line, and lack the clean character of the boxcar trend. They represent aggradation of a shaley or silty lithology, and may be typical of shelfal or deep water settings, a lacustrine succession, or muddy alluvial overbank facies. Often there may in fact be a subtle and systematic shift in the base lines in what appears to be an irregular trend. Redisplay of the logs by increasing the horizontal scaling

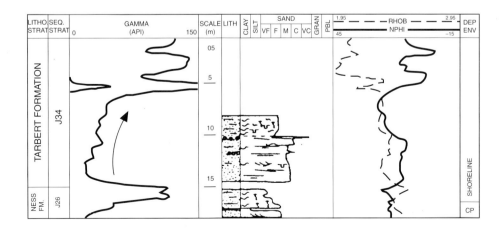

Fig. 4.9 A 'dirtying-up' log unit from the Tarbert Formation, Middle Jurassic Brent Group, northern North Sea. This unit represents an upward-fining depositional unit, produced by upwards deepening during retrogradation in a shallow marine setting (after Mitchener et al., 1992)

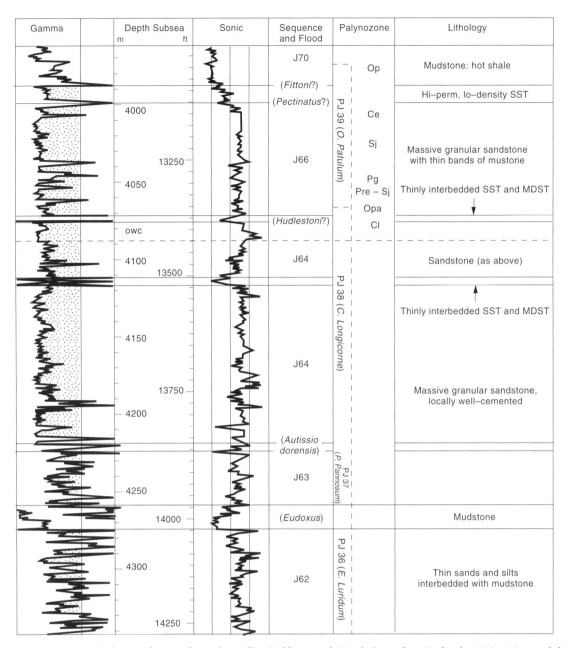

Fig. 4.10 Late Jurassic turbidite sandstones from the Miller Field, central North Sea (after Garland, 1993). Many of the depositional packages can be described as 'boxcar'; having abrupt bases and tops, and relatively consistent internal gamma response. OWC, oil-water contact; SST, sandstone; MDST, mudstone

and/or decreasing the vertical scaling may make these subtle trends visible. Irregular log trends are unlikely in any shelfal or paralic facies, where cyclic changes in water depth are likely to be recognized as cyclic log trends, and identified as parasequences.

4.4.3 The log response of clinoforms

The log response of a systems tract varies significantly, depending on whether the well passes through clinoforms, topsets, or basinal deposits. The procedures for sequence analysis of the log curve will be different in each case, and analysis of logs from a large number of wells may be needed before a coherent picture of the depositional architecture can be built up.

On well logs, a clinoform unit is inferred from a cleaning-upwards pattern that is suspected to reflect upward-shallowing. The base of the cleaning-up pattern will be equivalent to a downlap surface. Confirmation that this log response represents a prograding clinoform pattern could come from core or biostratigraphic data supporting upward-shallowing, because upward shallowing in clastic systems can occur only through progradation, except in special circumstances. Additional confirmation is best sought from seismic data, with subsidiary data from dip-meter studies (although the dips involved may be small), and the inferred

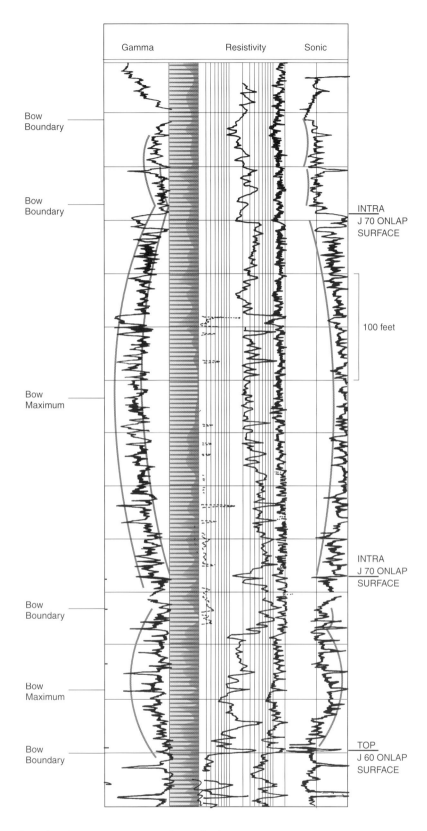

Fig. 4.11 Late Jurassic submarine depositional units from the Ettrick Field, central North Sea. These units have a 'bow' log motif, with a gradual decrease then increase in gamma response, reflecting a waxing and waning of sedimentation on a mixed sand–mud fan. The best sand development tends to be in the middle of the bow

setting within the basin and within the stratigraphy (e.g. between basinal facies below and basin-margin facies above).

The base of a clinoform unit is a downlap horizon. This may be recognized as a distinct base to the cleaning-up unit, often with a log facies diagnostic of marine condensation, such as a high-gamma shale (Fig. 4.8), or a cemented horizon. In other cases the downlap surface is more difficult to identify. The top of a cleaning-up clinoform trend may be marked either by an abrupt increase in shale content (gamma reading), resulting from abrupt deepening across the transgressive surface, or it may be overlain by topsets (see below).

In units thick enough to be resolvable on seismic data, breaks in the cleaning-up log profile generally tie with breaks within the clinoforms. An abrupt increase in shale content within the clinoform trend implies an abrupt upward jump to a deeper facies, resulting from lobe switching or transgression during relative sea-level rise. Similarly an abrupt decrease in the gamma response may imply an abrupt jump to a shallower facies, and thus a sequence boundary, a normal fault or a slump. A sequence boundary may be distinguishable from a normal fault in these circumstances by using seismic data, and possibly dip-meter data, or by mapping it out. Further discussion on the recognition of sequence boundaries on logs is found in section 4.4.7.

Two clinoform units commonly are present within a sequence, the highstand prograding wedge and the lowstand prograding wedge (Chapter 2). It is not always possible from log response to tell which systems tract the clinoform unit represents, although some indicators are mentioned in section 4.4.7. The thickness of the clinoform interval on the log gives an approximate measure of the clinoform height, and thus the basin water depth (modified by compaction, the effects of syn-depositional subsidence and other factors).

4.4.4 The log response of parasequences

Topset parasequences are formed by repeated cycles of filling of accommodation space between the offlap break and the coastal onlap point, and are seen as small-scale cycles on logs.

The precise nature of the log response of parasequences depends on the facies, as described in section 4.4.1. The most common is the cleaning-up motif (Fig. 4.8), widely recognized in parasequences from marine settings. In the cleaning-up motif the shale content decreases upwards, whereas primary porosity and bed thickness (as determined from microresistivity traces) may increase upwards. The marine flooding surfaces are recognized as abrupt upwards increases in shale content (and hence gamma and SP readings, and other diagnostic and indicative log responses). Figure 4.12 shows a series of shallow marine to paralic parasequences developed in the Middle Jurassic Ness Formation in the northern North Sea (reproduced from Mitchener *et al.*, 1992). The parasequences are recognized in core as upward-shoaling cycles on a 1–5 m scale, often topped with coal beds. On the gamma log the parasequences are small-scale upward-cleaning log units, and the marine flooding surfaces are abrupt upward increases in gamma reading. Similar parasequences can be seen on Fig. 4.6.

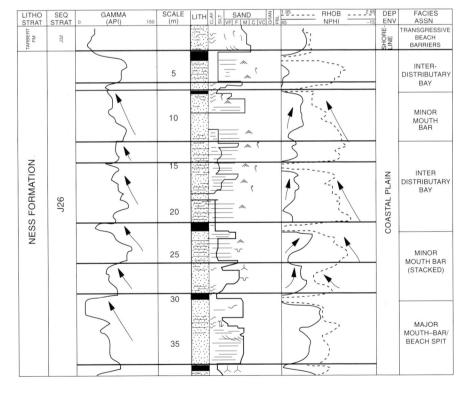

Fig. 4.12 Stacked topset parasequences from the Ness Formation, Brent Group, northern North Sea. These are recognized as small cleaning-up log units, often overlain by coals. The marine flooding surfaces generally immediately overlie the coal beds, and show abrupt upward increases in gamma (after Mitchener *et al.*, 1992)

Key:
— Marine flooding surface

Parasequence stacking patterns

Progradation, aggradation and retrogradation of the basin-margin system may be recognizable from the way parasequences are stacked into sets. Examples of parasequence stacking patterns are shown in Fig. 2.26, reproduced from Van Wagoner *et al.* (1990). Parasequence stacking, and parasequence sets, are discussed in section 2.5.3.

4.4.5 Log responses from basinal environments

The log response of basinal units tends to be more symmetrical than the log response of clinoforms or topsets. The exact nature of the log response depends on the nature of the sediment; mud-rich basinal units tend to show a symmetrical bow response (Fig. 4.11), whereas sand-prone systems tend to show a box-car or cylindrical log trend (Fig. 4.10). Often wells through different parts of the same depositional system show different log trends, a key feature of submarine fans.

These log trends are separated by log markers that represent background pelagic sedimentation, uninterrupted by sediment gravity flows from the basin margin. These markers are usually thin shales with little or no silt and sand, which may have anomalously high gamma response, low density, low resistivity and low sonic velocity. These shales represent marine condensed intervals (Fig. 4.10). In some environments the condensed intervals are chert-rich or carbonate-rich, and appear as sonic spikes. The marine condensed intervals shown on Fig. 4.10 are well-developed gamma 'spikes' representing pauses in submarine fan sedimentation. These are used to subdivide the stratigraphy of the Miller Field of the North Sea Jurassic into component depositional units (Garland, 1993).

Marine condensed intervals represent a cut-off in sediment supply at the well location, and may be a result of autocyclic sediment switching on the submarine fan, a change from 'basin-floor fan' to 'slope fan' *sensu* Posamentier *et al.* (1988) (see Chapter 9), or cessation of basinal deposition during a phase of rising relative sea-level.

The most prominent log markers are likely to be those related to a complete cessation of basinal deposition, and these are also likely to be significant surfaces of seismic onlap or downlap. They may contain a significant peak abundance of planktonic fossils, which provide convenient chronostratigraphically correlatable events (Chapter 6). They are time-equivalent to the lowstand prograding wedge, transgressive systems tract and highstand systems tract. The other condensed intervals described above are likely to be relatively local and represent limited time intervals.

It is not possible to perform a sequence stratigraphic analysis from basinal sediments alone, as the information needed to define systems tracts and sequence boundaries is confined to the basin margins.

4.4.6 Estimation of depositional controls and sequence stratigraphy from log response

A sequence analysis of a well-log suite is concerned with the identification of periods of basin-margin progradation and retrogradation, and the recognition of variations of relative sea-level. The logging suite shown on Fig. 4.6 is interpreted in Fig. 4.13, using the sequence stratigraphic methodology described in this chapter.

Progradation can be recognized from either a clinoform log response (large-scale cleaning- and shallowing-up unit), or from the progradational stacking of topset parasequences (as in Fig. 2.25). Evidence of basin-margin progradation will be found only within basin-margin units (topsets, clinoforms and toesets). Basinal logs may record only a time-equivalent condensed section. Two prograding units are interpreted on Fig. 4.13. A prograding shale unit is recognized as a log unit with upward-decreasing gamma and density, and upward-increasing resistivity and velocity, representing an upward increase in the silt fraction. A lower prograding sandstone unit consists of a stack of parasequences seen most easily on the neutron-density logs.

Retrogradation of the basin margin is recognized from retrogradational stacking of topset parasequences, or from interpretation of a log unit implying significant upward-deepening (Fig. 4.9). Two retrograding units are recognized on Fig. 4.13. The upper consists of a stack of very thin parasequences with prominent hardground flooding surfaces (recognized as sonic spikes). The lower is a stack of paralic parasequences, some of which are topped with coals.

Rising relative sea level is demonstrated by the recognition of stacked parasequences in basin-margin wells. For example, the stacked parasequences of Fig. 4.13 imply cyclic filling of accommodation volume during continued relative sea-level rise. Accelerating relative sea-level rise is suggested (but not proven, see arguments above) by a thickening-up parasequence stack, especially if this thickening upwards is regionally recognizable. However, this reasoning can be applied only to parasequences that aggrade to base level. The response of a prograding unit, with constant sediment supply, to accelerating relative sea-level rise is a gradual upward change from progradation to aggradation, leading eventually to transgression. This may result in a log motif that changes from cleaning-up to aggradational, typical of many lowstand prograding wedges. Decelerating relative sea-level rise is suggested by a thinning-up stack of parasequences that aggrade to base level, especially if this is regionally recognizable.

4.4.7 Key surfaces

A number of key stratigraphic surfaces need to be identified on the wireline log data base before the stratigraphy can be subdivided into its component systems tracts.

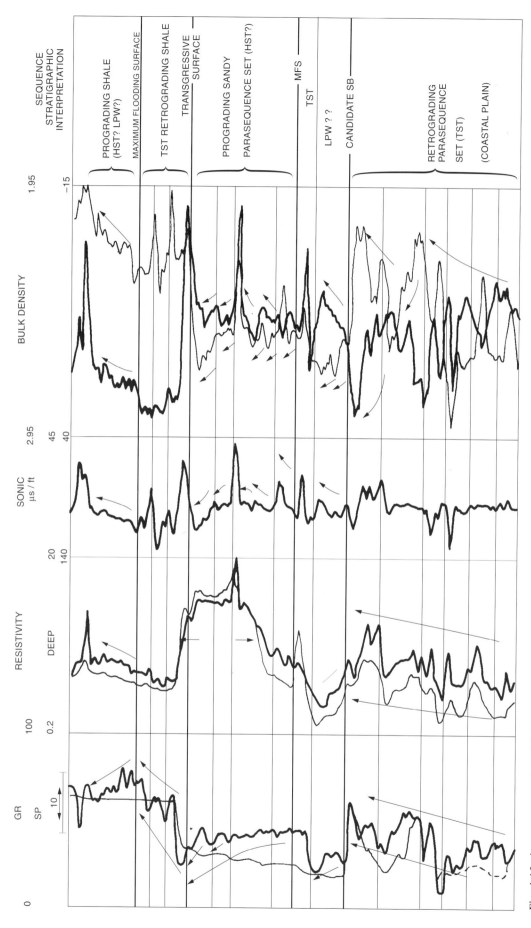

Fig. 4.13 A sequence stratigraphic interpretation of the log suite shown as Fig. 4.6. Although this log is a composite section, chosen to illustrate a variety of lithologies and fluid contents, it is derived from real data, and the sequence stratigraphic interpretation shown here is a valid exercise. HST, highstand systems tract; LPW, lowstand prograding wedge; TST, transgressive systems tract; MFS, maximum flooding surface; FS, flooding surface; SB, sequence boundary

A maximum flooding surface can be recognized in proximal locations as the surface between a retrograding unit and an overlying prograding unit. Where these are dirtying-up and cleaning-up units respectively, the maximum flooding surface will be a gamma-maximum. Maximum flooding surfaces pass laterally into shelfal condensed intervals. These may have a distinctive log response, such as a gamma peak, a resistivity trough, or a density maximum or minimum. Shelfal condensed sections often contain abundant fossils, thus providing a key biostratigraphic correlation surface. It should not be assumed that every gamma peak is a maximum flooding surface. The key point is that the surface lies above a retrograding interval and below a prograding interval.

Figure 4.14 shows a well log with well-developed cycles of progradation and retrogradation, some of which consist of stacked parasequences. A clear maximum flooding surface is shown, between an overlying progradational parasequence stack and an underlying retrogradational parasequence stack.

A maximum progradation surface can be recognized in proximal locations as the surface between a prograding unit and an overlying retrograding unit. Where these are cleaning-up and dirtying-up units respectively, the maximum progradation surface may be a gamma-minimum. The maximum progradation surface is the top of the parasequence that progrades furthest into the basin, and marks the time of turn-around between progradation and retreat. If the turn-around is gradual, the maximum progradation surface may be very difficult or impossible to define; the surface also may be vulnerable to erosion during transgression. The maximum progradation surface in a type 1 sequence lies at the top of the lowstand prograding wedge.

A clear maximum progradation surface is marked on Fig. 4.14, marking the boundary between a retrogradational parasequence stack and an underlying progradational parasequence stack.

A marine condensed interval in the basin is recognized as the shale-break between basinal log motifs (Figs 4.10 and 4.11). As explained above, hierarchies of marine condensed intervals should be expected, and it is not always easy to identify which are the most significant. Plots of faunal abundance may help (Chapter 6).

A downlap surface is recognized as the base of a clinoform (large-scale cleaning-up) log motif. The downlap surface below the highstand prograding wedge is correlatable with the maximum flooding surface, and the downlap surface below the lowstand prograding wedge is correlatable with the top fan surface or the sequence boundary.

A sequence boundary, resulting from a fall in relative sea-level, may be difficult to recognize from well data alone. It requires the recognition of a facies dislocation; the superposition of a proximal on a significantly more distal facies without the preservation of the intermediate facies. This is likely to be obvious only in two locations, on the front of the highstand clinoforms, and where shelfal erosion has been significant (i.e. in the incised valleys). In other localities there may be no significant facies dislocation, and the sequence boundary will coincide with a later flooding surface. In cases where a sequence boundary is suspected but not proven it can be referred to as a candidate sequence boundary, for the purposes of constructing predictive models. On the clinoform slope, a type 1 sequence boundary would result in a jump to a significantly cleaner log response within the cleaning-up motif. As discussed above, this easily could be confused with normal faulting. Only those

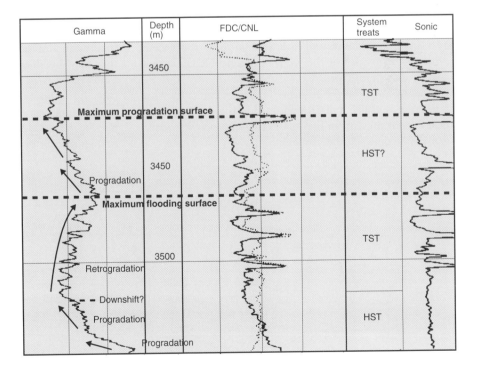

Fig. 4.14 A well log through the Late Jurassic reservoir of the Ula Field, offshore Norway. The reservoir consists of a series of shallow marine parasequences arranged in progradational and retrogradational stacks. The maximum flooding surfaces and maximum progradation surfaces are used to guide the reservoir stratigraphy. HST, highstand systems tract; TST, transgressive systems tract

wells that pass through the final clinoform of the highstand systems tract will see this abrupt jump.

In topsets a type 1 sequence boundary also could be manifested as a jump to a far cleaner facies, with, for example, fluvial deposits overlying distal shelf parasequences, or clean topset strata overlying offshore mud. The sharp base to a sand is shown on Fig. 4.13 as a candidate sequence boundary, with clean sand abruptly overlying shale. An alternative interpretation would have this surface as a ravinement surface, with an offshore sand overlying a lagoonal shale. Two sequence boundaries within the Middle Jurassic Brent Group of the northern North Sea are shown on Fig. 4.15 (reproduced from Mitchener et al., 1992). The lower of the two is recognized as an abrupt upwards cleaning on the logs, and in core as a very coarse to pebbly transgressive beach facies overlying shelf mudstone. Note the core-log shift in this well. The second sequence boundary is more subtle, but again marks an abrupt upwards cleaning and the transition from bioturbated lower shoreface sandstone to much cleaner (and slightly coarser) beach-barrier sandstone.

Wells through incised valleys may encounter a sharp-based clean sand (with a fining-up or blocky trend), representing an abrupt upwards shallowing from shelfal facies to sandy fluvial or estuarine valley-fill. Alternatively the valley-fill may be shaley, perhaps part of the transgressive systems tract. In this case it would be difficult or impossible to recognize the incision, and thus infer the sequence boundary from wireline logs in a single well. A grid of wells may allow recognition of the erosional feature, or core may demonstrate the estuarine nature of the valley-fill muds.

Sequence boundaries often mark an abrupt upward change from a progradational (cleaning-up) log motif to an aggradational or retrogradational log motif. A candidate sequence boundary is marked as 'downshift?' on Fig. 4.14, identified from the abrupt change in stacking pattern accompanied by a subtle shift in sand and shale base lines. Corroborative evidence is provided by the presence of a pebble lag at this horizon in several cores.

In many locations the sequence boundary coincides with a flooding surface, and a downward (basinward) shift in facies is not present. In these cases the sequence boundary is invisible on a log data set alone.

A type 2 sequence boundary (after van Wagoner et al., 1988, described in Chapter 2, section 2.4.6) is difficult or impossible to demonstrate from well logs alone. The definition criterion is a downward shift in coastal onlap to a position landward of the offlap break. This would be recognized in a well log data set, from correlation of a close grid of wells, as the coastal onlap of one parasequence on to the top of another. As type 2 sequence boundaries are not generally associated with hydrocarbon play geometries, their subtle nature is not a great worry to the explorer. Van Wagoner et al. (1988) suggest that a type 2 sequence boundary could be inferred from the recognition of a time of minimum rate of rise of relative sea-level at the boundary between a thinning-up parasequence trend and an overlying thickening-up trend, although it is debatable whether such a minimum rate of rise need be accompanied by any sequence boundary.

4.4.8 Identification of systems tracts from log response

Recognition of the surfaces above allows subdivision of the stratigraphy into systems tracts. Naming of the systems tracts allows them to be placed in the context of cycles of basin fill, providing the defining criteria exist. Identification of a systems tract by the nature of its boundaries is discussed in Chapter 2.

A lowstand fan is recognized as a fan unit bounded by marine condensed intervals, where the fan package could be demonstrated to correlate with a basin-margin sequence boundary. Where this correlation cannot be demonstrated (e.g. in a single well), it is impossible to decide whether or not the fan is related to a lowstand.

A lowstand prograding wedge is recognized as a prograding basin-margin unit succeeding a sequence boundary, and bounded above by a maximum progradation surface. Recognition of the sequence boundary is necessary before the lowstand nature of the unit can be demonstrated, and this may be difficult from well-log data alone. Topset parasequences in a lowstand wedge ideally would be expected to show a thickening upward trend, indicating accelerating relative sea-level, and this acceleration also may be indicated by an upward change from progradation to aggradation in the wedge.

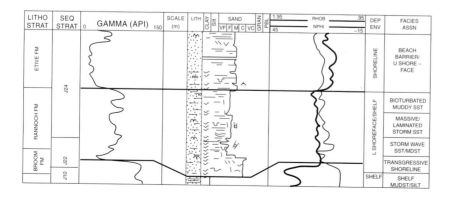

Fig. 4.15 Two sequence boundaries within the Middle Jurassic Brent Group, northern North Sea. Both are obvious as facies dislocations (an upward jump to a significantly more proximal facies), and abrupt upward decreases in gamma reading. Both are regionally recognizable (after Mitchener et al., 1992)

A *transgressive systems tract* is recognized as a retrogradational parasequence set (Fig. 4.14). It is bounded below by a maximum progradation surface (often coincident with the sequence boundary) and above by a maximum flooding surface or its correlative condensed interval.

A *highstand systems tract* is recognized as a prograding basin-margin unit bounded below by a maximum flooding surface and above by a sequence boundary. Recognition of the sequence boundary is necessary before the highstand nature of the unit can be demonstrated, and this may be difficult from well-log data alone. Topset parasequences in a highstand wedge ideally would be expected to show a thinning upward trend, indicating decelerating relative sea-level rise.

A *shelf-margin systems tract* would be recognized as a prograding basin-margin unit succeeding a type 2 sequence boundary, and bounded above by a maximum progradation surface. A type 2 sequence boundary, and therefore a shelf-margin systems tract, is difficult or impossible to recognize from well data alone.

4.4.9 Pitfalls and ambiguities in sequence analysis of log data

Several pitfalls in the link between log response and depositional parameters have been described in section 4.4.1, and in the interpretation of log trends in section 4.4.2. Additional, more general, pitfalls and guidelines are described here.

1 Use core control wherever possible. Log response, even when calibrated with core, is not an infallible guide to depositional environment or to systems tract.
2 Do not expect to find every systems tract in every well. Systems tracts are local and their areas of deposition often mutually exclusive.
3 Do not expect to recognize the sequence boundary in every well. It should be obvious where the lowstand wedge onlaps the highstand front, and at the base of any incised valley. Elsewhere it may coincide with the transgressive surface, or lie within an equivalent marine condensed interval, and so be effectively invisible on logs.
4 Other sharp breaks in log motif may resemble sequence boundaries (such as faults, slump scars, and the bases of channels).
5 The best first-pass horizons for correlating between wells are maximum flooding surfaces and their correlative marine condensed intervals. These are fairly unambiguous on logs, and are usually sufficiently rich in fossils to be dated reliably.
6 Correlating within the units bounded by maximum flooding surfaces may be extremely difficult. A sequence boundary may be recognized in two widely spaced wells, but it need not be the same sequence boundary if more than one higher order sequences are present.
7 Systems tracts cannot be named unless the defining criteria exist. For example, a progradational systems tract can be recognized on log data, but if no sequence boundary is recognized it cannot be assigned to any particular systems tract. The defining criteria are often not present in a single well, but may exist on a grid of wells, or on seismic data if resolution permits.
8 The choice of a hanging-datum (that horizon in the wells which ends up as a horizontal line on the final plot) is very important when creating a well-log correlation panel, as this often influences the way the correlation lines are drawn. Examples of potential miscorrelation due to the choice of the wrong datum are shown by Van Wagoner *et al.* (1990) in their Fig. 18. The ideal hanging datum is a horizon that was relatively flat at the time of deposition. A widespread coal, or a major flooding surface, would make a satisfactory hanging datum landward of the offlap break.

4.4.10 Check-list for sequence stratigraphic interpretation of a well-log data base

1 Display the log suite at a consistent scale, chosen to enhance the log trend. The standard log scale is often far from ideal for trend interpretation.
2 Mark the major trends on to the logs. Interpret the gamma log first, and confirm your interpretation on the other logs. Use core control to correlate facies with log response. Watch out for cemented horizons on the sonic log, hydrocarbon legs on the resistivity log, changes to non-clastic facies and casing shoes (which often look like major breaks in the log trends).
3 Interpret the gross depositional setting (prograding clinoforms, topset parasequences, basinal, etc.) using log trends and marker lithologies (e.g. coals).
4 Use any additional environmental information (seismic data, core, biostratigraphy).
5 Interpret the major condensed sections from log trend boundaries and/or log character. Use faunal abundance to corroborate this.
6 Determine intervals of progradation and retreat from stacking patterns in parasequences and major clinoform trends. Identify maximum flooding surfaces and maximum progradation surfaces. Corroborate with seismic data.
7 Interpret candidate sequence boundaries from facies discontinuities, evidence of incision of topsets, etc. Corroborate with seismic data and core. Watch out for normal faults, casing shoes, etc.
8 Interpret trends of parasequence thickening and thinning to suggest variations in the rate of relative sea-level rise.
9 Interpret systems tracts, if the defining criteria exist, using parasequence stacking patterns and the nature of the systems tract boundaries. Sedimentology from core may help, because certain facies may be suggestive of certain systems tracts (e.g. coals and tidal deposits in the transgressive systems tract). Cross-check with the seismic data.
10 Continue the interpretation around the entire well-grid. Tie carefully to seismic data using synthetic seismograms, and correlate with biostratigraphic data. Correlate sequences, systems tracts, and parasequences (if possible).

CHAPTER FIVE

Chronostratigraphic Charts

5.1 The purpose of chronostratigraphic charts

5.2 Construction of chronostratigraphic charts from seismic data
   5.2.1 Picking reflection terminations
   5.2.2 Identification of seismic surfaces
   5.2.3 Numbering seismic packages and their component reflections
   5.2.4 Transferring reflections to a time-scale
   5.2.5 Filling in the chronostratigraphic chart

   5.2.6 Adding wells on to the chart
   5.2.7 Scaling the chart to absolute time

5.3 Interpreting a chronostratigraphic chart
   5.3.1 Chronostratigraphic expression of sequence boundaries
   5.3.2 Chronostratigraphic expression of condensed sections
   5.3.3 Chronostratigraphic expression of depositional styles
   5.3.4 Chronostratigraphic expression of a seismic example

5.4 Coastal onlap curves and relative sea-level curves
   5.4.1 Coastal onlap curves
   5.4.2 Relative sea-level curves
   5.4.3 Examples from the Tertiary of the North Sea

5.5 Constructing chronostratigraphic charts from other data
   5.5.1 Producing chronostratigraphic charts from well data

5.1 The purpose of chronostratigraphic charts

Sequence stratigraphy involves the interpretation of the relationships of depositional systems in time as well as in space. Chronostratigraphic charts are a means of showing the time relationships of these systems, and their relationships to surfaces of non-deposition, condensation and erosion. These surfaces have little or no thickness in the rock record, and their true significance can be appreciated better by considering them in the time dimension.

The construction of chronostratigraphic charts provides a check for the interpreter, ensuring that the interpretation makes sense in time as well as in space. It also ensures construction of a time-framework for measuring variables such as sediment flux, subsidence, etc. In addition, chronostratigraphic plots of the proximal limit of topset deposition (known as coastal onlap charts) provide direct measurements of the frequency (but not the magnitude) of changes in relative sea-level.

The basic procedures for producing chronostratigraphic charts and coastal onlap plots from seismic data were described initially by Wheeler (1958; chronostratigraphic charts are also known as Wheeler diagrams), Mitchum *et al.* (1977a,b) and Vail *et al.* (1977b). A terminology for the description of rock units in time was proposed by Schultz (1982), who introduced the term *chronosome* for a rock unit bounded by time planes.

A chronostratigraphic chart has time as the vertical axis, with a spatial horizontal axis. On the chart are plotted the distribution of systems tracts, bounded by surfaces of onlap, toplap, downlap, etc. Within the systems tracts, the limits in time and space of the facies units may be shown.

The remainder of the chart represents the position and duration of the areas of non-deposition, hiatus, bypass, erosion and/or condensation.

Chronostratigraphic charts are most easily and accurately constructed from seismic data, where the relative positions of the depositional units in time and space can be determined more clearly. The principles of chronostratigraphic construction, like many of the principles of seismic stratigraphy, are based on the assumption that seismic reflectors follow bedding, that bedding planes are isochronous, and hence seismic reflections approximate to time-lines (section 3.1.1; Vail *et al.*, 1977a). The corollary of these assumptions is that reflection-bounded packages are chronosomes.

5.2 Construction of chronostratigraphic charts from seismic data

Once the interpreter has chosen a representative seismic section through the stratigraphy, displayed it optimally, and marked the non-stratigraphic reflections, the conversion into a chronostratigraphic chart can begin. This process will be illustrated by reference to Fig. 5.1, where a schematic seismic section (Fig. 5.1a) is translated into a chronostratigraphic section (Fig. 5.1e). Note that any one seismic line does not necessarily sample all of the stratigraphy, and the chronostratigraphic section constructed from it will not be a complete representation of deposition within the basin. A full representation of the changing patterns of deposition through time will be achieved only from a three-dimensional chronostratigraphic chart, or a grid of two-dimensional charts.

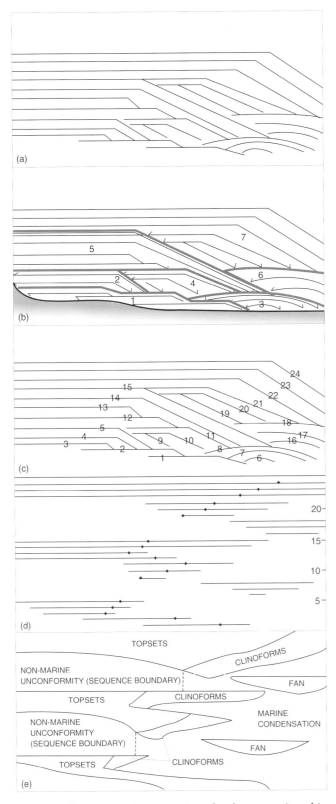

Fig. 5.1 The process of construction of a chronostratigraphic chart from seismic data; (a) a sketch seismic line; (b) a seismic stratigraphic breakdown of the line; (c) numbering of the reflections; (d) transfer of the reflections to a time axis, in numeric order; (e) a chronostratigraphic interpretation of the seismic data

5.2.1 Picking reflection terminations

Seismic reflections are not of infinite extent. When reflections appear to terminate consistently against a specific surface, it is termed a seismic surface (section 3.2.1). The main types of reflection termination against a seismic surface have been described in chapter 3, and include downlap, onlap (coastal and marine), toplap, truncation (real and apparent) and termination against a fault plane. These terms refer to present-day geometries on the seismic section, which are the product of original depositional geometry modified by compaction and tectonic activity. The seismic surfaces have been marked on Fig. 5.1b, defined by reflection terminations. Many of the seismic surfaces pass laterally into apparent conformity.

5.2.2 Identification of seismic surfaces

Termination of a reflection on seismic data is due to the termination of the bedding plane, or thinning of the bedding to below seismic resolution. A seismic surface therefore represents a facies change from appreciable sedimentation rates to very low, zero or negative sedimentation rates (erosion). There are three main types of seismic surface, with distinct chronostratigraphic expression (Fig. 5.1b and 5.1e). It is important to differentiate between these surfaces and termination relationships on a chronostratigraphic diagram.

Non-marine seismic surfaces represent the product of non-marine erosion, bypass and/or non-deposition. They are characterized above by coastal onlap (which may be removed later by erosional shoreface retreat), and below by toplap or erosional truncation. The space representing the surface on a chronostratigraphic diagram will include an area representing strata deposited, but since eroded, and a space representing non-deposition.

Marine seismic surfaces represent times of marine non-deposition, condensation and/or erosion. They will be characterized above by marine onlap, and below by apparent truncation, marine erosional truncation, or possibly by apparent conformity. The space occupied on the chronostratigraphic diagram will include an area of marine hiatus and condensation, and may include an area representing strata deposited but since eroded.

Fault-plane surfaces represent the dislocation of strata by extension, diapirism or compression. The space occupied on the chronostratigraphic diagram represents an area of rock missing due to extension or duplicated due to shortening.

The seismic surfaces bound the depositional packages (or systems tracts), which are the building blocks of the stratigraphy we wish to plot on the chronostratigraphic chart. On Fig. 5.1b non-marine seismic surfaces are present on the left of the plot, passing to the right (basinwards) into apparent conformity. The boundary between units 2 and 4 is a non-marine seismic surface, bounded above by coastal onlap and below by toplap. Seismic surfaces on the right of the figure tend to be marine seismic surfaces, either onlapping out landward (the boundary between 2 and 4) or passing landward into apparent conformity (the boundary between 1 and 2).

5.2.3 Numbering seismic packages and their component reflections

The chronostratigraphic chart has a vertical time axis. The seismic reflections need to be plotted on the chart in order of age, and so need to be numbered from the oldest to the youngest. The first step in this process is to number the reflection packages (systems tracts), bounded by the seismic surfaces, in stratigraphic order. Generally this is straightforward, especially if well data on the ages of the packages exists. However, in many cases it may be impossible to determine precisely the relative ages of two systems tracts that do not overlap in space. For example, the age of systems tract 2 relative to systems tract 3 on Fig. 5.1b is unknown. Some degree of interpretation of the time relationships is required, or an arbitrary choice may have to be made in the absence of conclusive data. Once the packages are numbered in stratigraphic order, the reflections within them also can be numbered (Fig. 5.1c). Again, this may not always be possible to do objectively.

5.2.4 Transferring reflections to a time-scale

Once the reflection packages are identified and placed in stratigraphic order, the remainder of the procedure for constructing a chronostratigraphic chart is simple. The horizontal axis of the chart corresponds to the horizontal axis of the seismic section, and the seismic reflections are plotted, in order of age, as horizontal lines on the chronostratigraphic chart in the same lateral position as they appear on the seismic line. For example, reflection 1 from Fig. 5.1c, chosen as the oldest reflection, is plotted at the bottom of Fig. 5.1d, in the same horizontal position as it appears in Fig. 5.1c.

It is possible to mark on the chart any properties of the seismic reflections you wish to record, such as their gross geometry (topset, bottomset, clinoform), or attributes such as seismic facies. The position of the offlap break is marked on the topset/clinoform reflections in Fig. 5.1d as dots on the horizontal time-lines. Similarly the thickness variations in the strata can be represented on the chart by, for example, colour coding or line thickness. Do not transfer the seismic surfaces to the chart as horizontal lines. Only the reflections should be transferred.

5.2.5 Filling in the chronostratigraphic chart

The area of the chart covered by horizontal lines represents the pattern of deposition through time and space. Discrete depositional packages of seismic reflections (systems tracts) appear as discrete areas of horizontal lines. The boundaries of these areas should be drawn in on the chart, and each boundary may be labelled, coloured or annotated according to the type of reflection termination (onlap, downlap, truncation, fault, etc.). The spaces on the chart not covered by horizontal lines represent times and locations of non-deposition, erosion, condensation below seismic resolution, or fault separation. These spaces correspond to the seismic surfaces described above. It is obviously important to differentiate these areas. When reading a chart you need to know whether any particular non-depositional zone represents condensed marine deposition, or subaerial exposure and erosion, both in order to understand the basin history and to place the systems tracts within the context of relative sea-level cycles. A large degree of interpretation may be required at this stage. Figure 5.1e represents a fully interpreted chronostratigraphic chart of the sketch seismic line of Fig. 5.1a. The chart is fully annotated and comprises a record of deposition and non-deposition through time along this seismic profile.

5.2.6 Adding wells on to the chart

Any wells that tie the seismic line should be drawn on the chart, and used in the following ways:
1 intervals of progradation and retrogradation below seismic resolution (identified using well-log and core sequence analysis as described in Chapter 4) can be added to the chart;
2 surfaces of hiatus (maximum flooding surfaces, condensed intervals) and erosion (sequence boundaries, submarine channels) identified in the wells can be added to the chart;
3 reflection characteristics, such as slope angle, seismic facies, or seismic attributes can be translated into facies information;
4 age data from the well can be used to scale the chart in absolute time.

5.2.7 Scaling the chart to absolute time

The chart constructed above has a vertical time axis, but with a non-linear scale. The seismic reflections have been plotted in order of age, with an equal time increment given to each reflection. If data exist on the absolute ages of any parts of the stratigraphy, the chart can be scaled to absolute time. However, this is not a trivial exercise. Absolute ages probably have been measured for a few key boundaries only, and the scaling of the chart between these boundaries still has to be decided. The reflections can be given equal time increments between the boundaries, or alternatively the reflections can be given some weighting, so that (for example) the more laterally extensive reflections are considered to represent more time. There almost certainly will be depositional packages not represented on the chosen seismic line, as they were deposited out of the plane of the section. The time of deposition of this package will be represented on the seismic section as a time of hiatus, condensation, bypass or erosion. A two-dimensional chronostratigraphic construction on a single line will not assign enough time to this hiatus, but will distribute the time through the deposition represented on the line. Proper weight will not be given to the time gaps unless chronostratigraphic construction is carried out in three dimensions; a problem too complex to be attempted without using computer methods such as those of Nordlund and Griffiths 1993b.

5.3 Interpreting a chronostratigraphic chart

5.3.1 Chronostratigraphic expression of sequence boundaries

A sequence boundary, comprising a subaerial unconformity and its correlative conformity, will be represented on a chronostratigraphic chart as an area, with both extent and duration. The boundaries of an unconformity are coastal onlap above and toplap or truncation below. The duration of the unconformity will be greatest towards the basin margin and least towards the basin centre, so the area on the chart will be tapered towards the basin centre. Figure 5.2 shows the chronostratigraphic expressions of a type 1 and a type 2 sequence in the Exxon model. The type 2 sequence boundary is represented by a wedge of time, and the correlative conformity to the sequence boundary can be considered time-equivalent to the apex of the wedge; the time at which the unconformity was at its maximum extent. The type 1 sequence boundary also is represented by a tapering area, but includes the time during which sediment bypassed the basin margin to be deposited in the basin. The area on the chronostratigraphic diagram has no apex, and it is difficult to choose a correlative conformity. It is accepted practice to choose the correlative conformity to a type 1 sequence boundary at the time of onset of shelf bypass (although the time of shelf bypass will be affected by the local subsidence rate as well as the timing of eustatic variation).

5.3.2 Chronostratigraphic expression of condensed sections

Marine condensation is also a process, often with a significant duration. The duration is greatest towards the basin and least towards the land. In the basinal succession the boundaries of the marine condensed interval are marine onlap boundaries. On the basin margin the condensed interval is bounded above by downlap and below by apparent truncation. The condensed interval has a tapering wedge shape on a chronostratigraphic chart, and the apex of the wedge corresponds to the time of maximum flooding (shown as MFS on Fig. 5.2). The shape will be more complex in the basinal succession, where the locus of deposition through time is affected by the shape of the basin, and the growth patterns and autocyclic switching of the submarine fans.

5.3.3 Chronostratigraphic expression of depositional styles

The most distinctive depositional expression on a chronostratigraphic chart is that of prograding clinoforms. The continuous progressive lateral migration of deposition is easily seen on the chronostratigraphic chart as a diagonal zone of deposition. Toplapping clinoforms are bounded below by a diagonal downlap line and above by a parallel diagonal toplap line (e.g. the early part of the lowstand

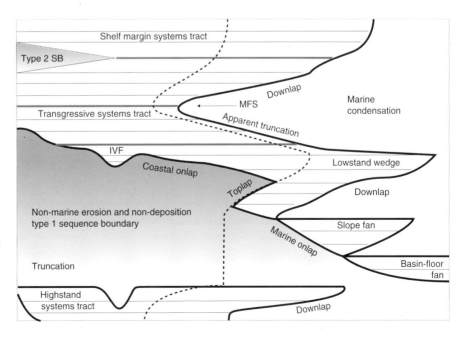

Fig. 5.2 Chronostratigraphic representation of a type 1 and a type 2 sequence boundary (SB). The dashed line represents the movement of the offlap break through time. MFS, maximum flooding surface; IVF, incised valley fill

wedge on Fig. 5.2). Topset reflectors show an expanding zone of deposition, with downlap in one direction and coastal onlap in the other (e.g. the later part of the lowstand wedge, and the shelf-margin systems tract, on Fig. 5.2).

5.3.4 Chronostratigraphic expression of a seismic example

Figure 5.3 is a chronostratigraphic expression of a regional seismic line through the Palaeogene section in the Outer Moray Firth area of the UK North Sea. Several examples of seismic data from this interval are shown in Chapter 3, and the following features are evident from the chronostratigraphic chart of this stratigraphy.

Firstly, it is striking how much of the chart is covered by periods of non-deposition. At any one point, relatively brief periods of deposition are separated by long periods of hiatus, bypass or erosion. The zone of clinoform deposition is relatively narrow, and in the lower part of the diagram moves rapidly out into the basin, overstepping the submarine fans below. These clinoforms are intermittently associated with minor topsets. Later (towards the top of the diagram) topsets are well developed, with thick coal beds, and clinoforms represent a small component of the stratigraphy.

The zone of non-marine hiatus moves 150 km basinward during the time represented. This can be due only to a major fall in relative sea-level, and is not simple progradation. There is also a decrease in the width of the clinoform belt during time, representing a decrease in water depth in the basin, due partly to infill and partly to relative sea-level fall.

Transgressions in the T45 and T50 units are rapid, representing flooding across older topsets. The T45 floods back as far as the older T30 clinoform front. This implies some relict topography on this clinoform front.

Several other subtleties can be seen from this plot. It is a representation of the data with almost as much information as the original, in a format that emphasizes the time succession.

5.4 Coastal onlap curves and relative sea-level curves

The construction of chronostratigraphic diagrams is obviously not an end in itself. A chronostratigraphic examination of seismic data can allow a reconstruction of variations in relative sea-level; one of the key controls on stratigraphic development. Curves of relative sea-level variations through time can be used to date local tectonic movements, differentiate regional depositional events, and predict stratigraphy (using sequence stratigraphic models) in non-drilled parts of the basin. Relative sea-level curves can be produced directly from the patterns of vertical movement of the offlap break, if seismic resolution allows.

More usually, relative sea-level curves are derived from coastal onlap curves.

5.4.1 Coastal onlap curves

Clinoform strata in a basin-margin setting commonly have relatively flat topsets, sloping clinoforms and relatively flat toesets. Topset facies are believed to be controlled by shallow marine, paralic and non-marine processes, whereas clinoform facies are believed to be controlled by slope processes. Topsets are separated from clinoforms by the offlap break. Onlap of strata landward of the offlap break is known as coastal onlap. As these strata are controlled by shallow marine processes, variations in relative sea-level will influence patterns of coastal onlap and patterns of movement of the offlap break.

A plot of coastal onlap through time is produced as a part of a chronostratigraphic diagram, or can be produced directly from the seismic data without constructing the rest of the plot. The variation of coastal onlap through time in a type 1 sequence (similar to that shown in Fig. 5.2) is shown as Fig. 5.4. This represents the lateral movement through time of the proximal depositional edge of the topset deposition. The lateral movement of the offlap break (the break in slope between the topsets and clinoforms; see section 2.1) is shown on Fig. 5.4 as a dashed line, and differentiates progradation (seaward movement of the offlap break) from retrogradation (landward movement of the offlap break) and aggradation (no net landward or basinward movement of the offlap break).

Variations in relative sea-level can be read from the coastal onlap curve. Landward movement of coastal onlap implies rising relative sea-level, whereas seaward movement implies static or falling relative sea-level. This can be seen on Fig. 5.4, where landward shifts in coastal onlap within the lowstand wedge, highstand and transgressive systems tracts are associated with rising relative sea-level, whereas the fall in relative sea-level accompanying deposition of the fans is associated with a basinward movement in coastal onlap. The magnitude of the lateral movement in coastal onlap depends partly on the magnitude of the relative sea-level change, and partly on the basin topography, and is equal to the relative sea-level change divided by the tangent of the topographic slope. It is difficult therefore to deduce the magnitude of relative sea-level changes from a coastal onlap curve, although the frequency of the changes is recorded. Note, however, that these changes may not all be recorded on any one seismic line. A particular line may not intersect a lowstand wedge, for example, and the relative sea-level history associated with that lowstand would not be apparent on the chronostratigraphic chart constructed from that line. A three-dimensional chronostratigraphic analysis is needed to understand fully the relative sea-level history of an area. Also a long-duration period of erosion may remove evidence for previous relative sea-level variations.

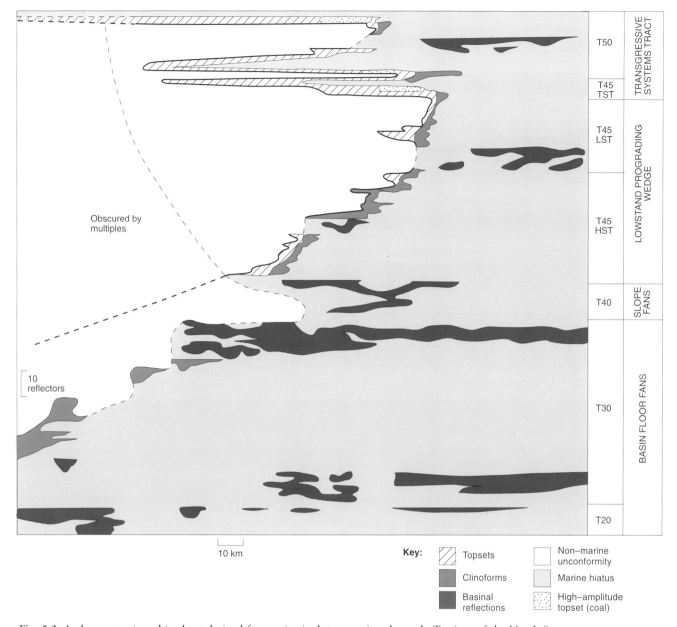

Fig. 5.3 A chronostratigraphic chart derived from seismic data covering the early Tertiary of the North Sea

Coastal onlap curves were produced for a number of basins by workers in Exxon in the 1970s, and used to construct the composite global coastal onlap curve published by Haq *et al.* (1988). The 'saw-tooth' nature of this curve is an artefact of the plotting process, and the abrupt basinward shifts in coastal onlap represent the boundaries between onlapping topsets of the highstand systems tract and onlapping topsets of the lowstand systems tract (although in some cases the onlap patterns in the lowstand fans were also included in the coastal onlap plot). This asymmetry in the variations in coastal onlap should not be construed as an asymmetry in relative sea-level variations, with rapid falls and slow rises. Asymmetric coastal onlap cycles will result from sine-wave variations in relative sea-level, as a function of the sedimentary response.

5.4.2 Relative sea-level curves

A relative sea-level curve may be constructed by plotting the vertical movement of the topsets through time. This is based on the assumption that topsets were deposited as relatively flat units close to sea-level. Obviously this

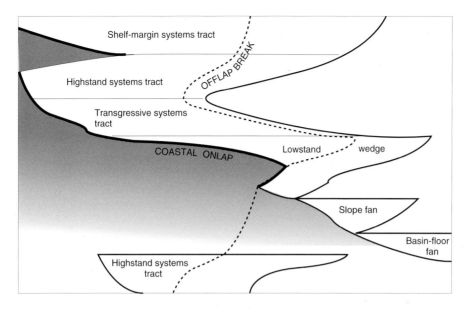

Fig. 5.4 Coastal onlap and offlap break curves for a type 1 sequence. The offlap break curve is shown as a dashed line

assumption is not valid in detail, because even paralic topsets will have had some depositional slope. It is probably a good enough assumption in practice for measuring large-scale variations in relative sea-level. For more accuracy, the vertical movement of the offlap break could be measured, based on the assumption that the offlap break occurred at a constant depth during relative sea-level changes. Again this probably is not valid in detail, because the depth of the offlap break is a function of the balance between fluvial supply and marine reworking, a balance that will change during a relative sea-level cycle.

Rising relative sea level causes the topsets and the offlap break to rise with time, with respect to the underlying surfaces, whereas falling relative sea-level causes the offlap break to fall, and the topsets to be displaced downwards. A plot of relative sea-level therefore can be constructed from the vertical movement of the topsets and/or the offlap break. Details of the relative sea-level history will be lost if any of the topsets are removed by erosion.

Variations in the rate of change of relative sea-level can give characteristic patterns in the topset–clinoform system. Static relative sea-level results in no vertical movement of the offlap break or the coastal onlap point. The coastal onlap point may, however, move basinward, with the development of a toplap surface. The resulting offlapping clinoform configuration is known as oblique offlap and is shown on Fig. 3.10.

An accelerating rate of rise of relative sea-level causes offlap to build outward and upward, and become more aggradational upward. This is known as aggradational offlap. Eventually this may lead to transgression.

A decelerating rate of rise of relative sea-level causes offlap to build upward and outward, and become more progradational upward. This is known as sigmoidal offlap (Fig. 3.10).

5.4.3 Examples from the Tertiary of the North Sea

Figure 5.4 has been presented already as a chronostratigraphic representation of the Palaeocene succession on a central North Sea seismic line. A coastal onlap curve of the Palaeocene and Eocene from a nearby line is shown in Fig. 5.5, reproduced from Jones and Milton (1994). This plot has the following features.

1 Movements in coastal onlap are evident with a variety of frequencies and magnitudes. Two large-scale basinwards movements are seen; one from T30 to T45, and the second from T60/70 to T98. Coastal onlap moved in the order of 100 km during these cycles. The first equates to the basinward movement in coastal onlap seen on Fig. 5.3.
2 Higher frequency movements in coastal onlap generally have a lower magnitude, of the order of 10–40 km.
3 Unlike conventional saw-tooth curves of coastal onlap, with rapid basinward shifts and slower landward shifts, this plot shows some more gradual basinward movements in coastal onlap, which here represent toplapping clinoform units.

Figure 5.6 is a relative sea-level plot for the same seismic line, reproduced from Jones and Milton (1994). It is scaled in milliseconds, and has not been depth converted, nor have the effects of compaction been allowed for. However, a rough depth conversion of 1 m per millisecond gives an idea of the magnitude and frequency of relative sea-level variations in this area during the Palaeogene. Relative sea-level changes of a variety of magnitude and frequency can be seen, which are responsible for the movements in coastal onlap shown in Fig. 5.5. The magnitudes of the lower frequency cycles are several hundred metres.

The Palaeocene of the North Sea was a time of rapid influx of sediment into a deep-water basin that was previously relatively starved of clastic input, and a site of chalk accumulation. Historically the influx has been attributed to

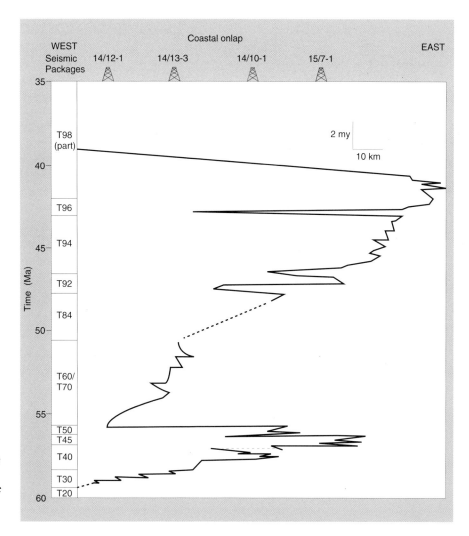

Fig. 5.5 A coastal onlap curve for the Palaeocene and Eocene of the central North Sea, measured from seismic line BP87-303 (from Jones and Milton, 1994)

uplift of the Scottish mainland, associated with development of a mantle hotspot, resulting in a massive increase in sediment supply. Milton *et al.* (1991) argued from basin reconstructions that the uplift affected not only the hinterland but the basin as well. The chronostratigraphic analysis of the seismic data, presented here and in Jones and Milton (1994), allows the uplift to be demonstrated, timed and measured. The increase in clastic sediment supply can be shown, therefore, to be coincident with a massive, long-duration, tectonically induced fall in relative sea-level. It is hardly surprising therefore that the sedimentary record associated with the Palaeocene is dominated initially by submarine fans. Superimposed third-order relative sea-level cycles also can be recognized, and these control the individual stratigraphic packages within the Palaeogene succession.

5.5 Constructing chronostratigraphic charts from other data

Chronostratigraphic charts are constructed most easily from a data source where the time relationships of deposition are clear. On a basin scale this often means seismic data, but chronostratigraphic representations of outcrop or aerial photograph data are quite possible, although rarely published. Charts produced from well data, or from a few isolated logged sections, are very hard to produce objectively. Without the cross-sectional view of the age relationships that a seismic line or photomontage provides, the interpreter can only really draw a chronostratigraphic chart of his or her depositional model. These charts are necessarily simplistic, and may become quite stylized. The production of these charts from well data alone is quite common, and some guidelines for their production are shown below.

5.5.1 Producing chronostratigraphic charts from well data

Chronostratigraphic diagrams can be constructed from well data alone. This involves the sequence analysis of the well data, and the identification of intervals of progradation and retrogradation, and surfaces of hiatus and erosion, both marine and non-marine (Chapter 4). The intervals and surfaces are then correlated using biostratigraphic data (Chapter 6), and a chronostratigraphic diagram is constructed using biostratigraphic age as the vertical axis. The

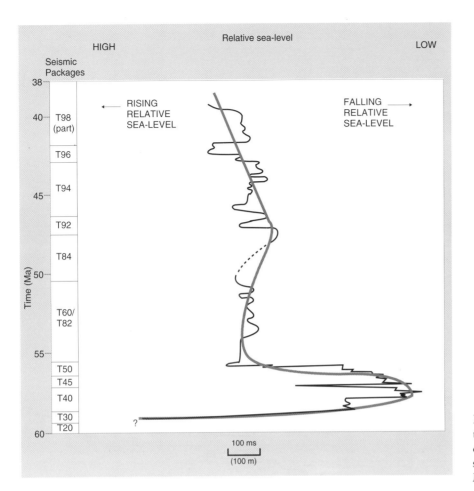

Fig. 5.6 A relative sea-level curve for the Palaeocene and Eocene of the central North Sea, measured from seismic line BP87-303 (from Jones and Milton, 1994)

surfaces and events appear on the chronostratigraphic diagram as they would if constructed from seismic data, as follows:

1 surfaces representing isochronous processes (e.g. sequence boundaries, condensed intervals) are wedges of time;
2 surfaces or beds representing diachronous processes (e.g. ravinement erosion, sandstone deposition, clinoform deposition) are roughly parallelograms;
3 surfaces representing events (such as the resumption of deposition) are non-horizontal lines.

Perhaps the safest way to attempt a chronostratigraphic representation of a basin cross-section from well data alone is to first draw the cross-section as it would appear on seismic data, and then translate this into a chronostratigraphic diagram.

CHAPTER SIX

Biostratigraphy

6.1 Introduction

6.2 Fossil groups and zonal schemes
    6.2.1 Fossil groups
    6.2.2 Fossil zonation schemes and biochronostratigraphic resolution

6.3 Palaeoenvironmental analysis
    6.3.1 Benthos and palynofacies
    6.3.2 Plankton
    6.3.3 Biofacies

6.4 Biostratigraphy and sequence stratigraphy
    6.4.1 Sequence boundaries and their correlative conformities
    6.4.2 Lowstand systems tract
    6.4.3 Transgressive surfaces
    6.4.4 Transgressive systems tracts
    6.4.5 Maximum flooding surfaces
    6.4.6 Highstand systems tracts

6.5 Conclusions

6.1 Introduction

Biostratigraphy is a well-established branch of stratigraphy based on the palaeontology of rocks. It uses the chronostratigraphic range of fossil species to correlate stratigraphic sections, and their palaeoenvironmental preference to provide information on depositional setting.

Prior to the advent of seismic data, biostratigraphy was the only means by which geologists could correlate time-equivalent depositional sections accurately. However, many of the fossil groups favoured by early palaeontologists, certainly prior to the middle of this century, tended to be forms that inhabited the sea-bed (benthos) rather than the water column (plankton). As a consequence, these benthic correlations often had a greater affinity to palaeoenvironmental conditions and depositional facies rather than chronostratigraphy (Loutit et al., 1988). Indeed the close relationship between the presence of these benthic markers and lithofacies was probably responsible for prolonging the common practice of treating lithostratigraphic correlations as chronostratigraphic events.

Seismic stratigraphy has now largely superceded biostratigraphy as the primary correlative tool in subsurface basin analysis. However, biostratigraphy, together with other dating methods, such as isotope stratigraphy (Emery and Robinson, 1993) and magnetostratigraphy, play an important role in providing the chronostratigraphic control to seismic correlations (Armentrout, 1987, Loutit et al., 1988, McNeil et al., 1990). Also, without the aid of biostratigraphy, seismic stratigraphy is limited in areas with complex structure.

This chapter will show how biostratigraphic data can be integrated with other techniques to enhance and constrain sequence stratigraphic interpretations, beginning with an introduction to the main fossil groups and biostratigraphic techniques.

6.2 Fossil groups and zonal schemes

6.2.1 Fossil groups

All fossils have potential application to sequence stratigraphy, although to date sequence boundaries and maximum flooding surfaces accurately a high frequency of chronostratigraphically significant fossil events is required. This is best obtained through the integration of marker taxa from several different fossil groups. The most useful fossils are those which, as they evolved, exhibited distinct and rapid morphological change, so as to be unequivocally identifiable. It is also important that they are distributed widely, and therefore correlateable within and between basins, and occur in sufficient abundance that their presence is a statistically viable event. Several macrofossil groups, such as the ammonites and goniatites, together with the larger foraminifera, are of particular value, but the limitation of sample size in the petroleum industry has meant that small fossils, usually less than a few millimetres in diameter and as small as 4 µm, are the most widely used in biostratigraphy. The three most useful groups include: *microfossils* (e.g. foraminifera, ostracods, diatoms, calpionellids, Radiolaria, calcareous algae and conodonts), *nannofossils* (e.g. coccoliths and discoasters) and *palynomorphs* (e.g. dinoflagellates, chitinozoa, acritarchs, tasmanitids, pollen and spores). The advantage of using small fossils is that, in favourable palaeoenvironments, they usually occur in abundance. Figure 6.1 shows the stratigraphic range for some of the more important fossil groups used in the oil industry.

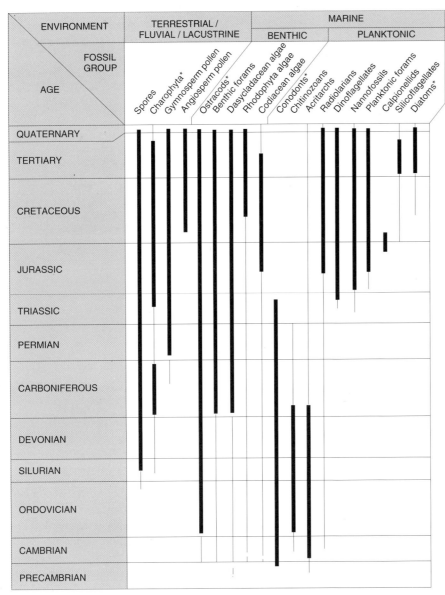

Fig. 6.1 Chronostratigraphic range of stratigraphically useful microfossil groups for both terrestrial and marine environments

* Freshwater ostracods, brackish water charophytes, benthic and freshwater diatoms and benthic conodonts also exist

The availability of plants and animals for fossilization is a function of evolution, geography and environmental conditions. The actual fossil preservation of plants and animals is dependent upon the mineral and organic construction of the organism, the environment it was deposited in and diagenetic history after burial. The absence of expected marker fossils, either through biofacies restriction or non-preservation, is a limitation common to biostratigraphic studies and a major hindrance to interpretation.

6.2.2 Fossil zonation schemes and biochronostratigraphic resolution

Fossil organisms evolved, diversified and became extinct in response to new environmental opportunities and pressures. The first appearance datum (inception or 'base') and the last appearance datum (extinction or 'top') of a fossil species in the rock record are useful markers for biostratigraphic correlation. Other events, such as maximum abundance, are also often used, although these should be treated with caution because local factors such as sedimentation rates can affect their occurrence.

Biostratigraphic time is measured in biochronozones which are based on the global inceptions and extinctions of fossils. Bolli et al. (1985) provided a comprehensive synthesis of the marine planktonic fossil groups used to define these biochronozones. However, the full global range of a fossil species may not be represented in every basin, owing to local environmental or geographical constraints. In such circumstances biozones based on these inception or extinction events only have local correlative value. This means that global correlation of the type required to construct sequence stratigraphic charts like that of Haq et al. (1987), or any comparisons with such charts, can be made only with carefully scrutinized data.

The chronostratigraphic resolution obtainable from fossil markers depends on the geological period, the number of fossil groups used and the type of depositional environment. The resolution of a fossil group is calculated by dividing the geological period by the number of biozones in a particular global scheme. Average chronostratigraphic resolutions for different fossil groups are shown on Table 6.1.

Published biozonation schemes use both fossil inception and extinction events to define biozones. In contrast, biozone tops in the oil industry are defined preferentially on fossil extinction events, whereas inceptions are relegated to defining subzones. This is because the commonest samples available to the industrial biostratigrapher are ditch cuttings, and as these are pumped to the surface in a mud slurry they are subject to lag effects and open up-hole contamination. Detailed reservoir studies, on the other hand, use inceptions for high-resolution biozonation schemes, because core and side-wall core samples usually are available. These are used to produce more detailed correlations in order to better understand lateral reservoir variations and connectivity.

Local schemes usually are more refined and have greater chronostratigraphic resolution. For example, the nannofossil biozonation of the Late Miocene through to Pleistocene of the Gulf of Mexico has an average resolution of 0.375 million years. The combined resolution from several fossil groups gives even greater resolution. For example the combined nannofossil and foraminiferal resolution for the Late Miocene through to Pleistocene of the Gulf of Mexico is approximately 0.2 million years.

6.3 Palaeoenvironmental analysis

6.3.1 Benthos and palynofacies

Organisms living on or within the sea-bed are called benthos. In the oil industry, the foraminiferal benthos is used most commonly to define marine palaeoenvironments (Van Gorsel, 1988), although other organisms, such as

Table 6.1 Examples of the resolution of fossil groups by age and by geography

Fossil group	Age range	Geography	Average resolution (million years)	References
Planktonic Foraminifera	Neogene	Tropical	1.2	1
Planktonic Foraminifera	Neogene	Subtropical	1.4	1
Planktonic Foraminifera	Palaeogene	Tropical	1.7	2, 3, 4
Planktonic Foraminifera	Palaeogene	Southern temperate	3.0	5
Nannofossils	Neogene	Undifferentiated	1.0–1.3	6, 7
Nannofossils	Palaeogene	Undifferentiated	1.3–1.6	6, 7
Radiolaria	Neogene and Palaeogene	Undifferentiated	1.9–2.0	8
Diatoms	Neogene and Palaeogene	Undifferentiated	1.4–2.4	9, 10
Dinoflagellates	Neogene and Palaeogene	Undifferentiated	5.7	11
Dinoflagellates	Neogene	North Sea	3.3	
Dinoflagellates	Palaeogene	North Sea	1.1	
Planktonic Foraminifera	Cretaceous	Tropical	2.5	12
Planktonic Foraminifera	Cretaceous	Temperate	4.0	13
Nannofossils	Cretaceous	Undifferentiated	3.0	14
Radiolaria	Cretaceous	Undifferentiated	10.0	15
Palynomorphs	Cretaceous	Undifferentiated	6.5	11
Palynomorphs	Late Jurassic	North Sea	1.0	
Palynomorphs	Early–Middle Jurassic	North Sea	2.0–2.5	

1, Kennett and Srinivasan (1983); 2, Bolli and Saunders (1985); 3, Banner and Blow (1965); 4, Berggren and Van Couvering (1974); 5, Jenkins (1985); 6, Bukry (1973); 7, Martini (1971); 8, Sanfilippo et al. (1981); 9, Fenner (1985); 10, Barron (1985); 11, Williams (1977); 12, Caron (1985); 13, Bolli et al. (1985); 14, Sissingh (1977); 15, Sanfilippo and Riedel (1985).

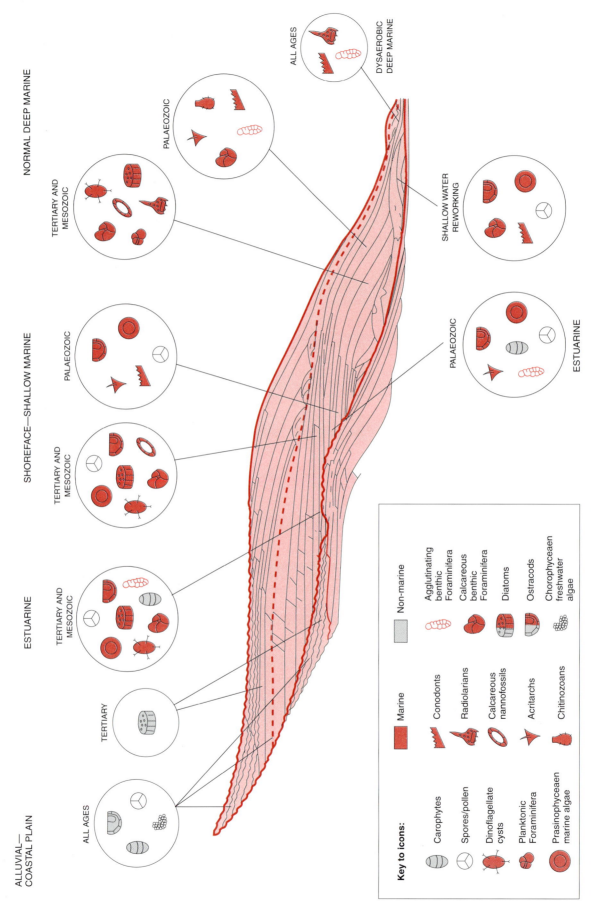

Fig. 6.2 Palaeoenvironmental distribution for some of the main microfossil groups

calcareous benthic algae, conodonts and ostracods also are important (Fig. 6.2). Benthic foraminifera have evolved to exist in a range of environments, from marginal marine settings to the deep ocean (Murray, 1973, 1992). This benthos also can withstand a variation of environmental conditions, including temperature, oxygen, salinity, substrate and light penetration (Fig. 6.3). In deep water bathyal and abyssal environments the physical properties of the stratified oceanic water masses, such as nutrients, oxygen, salinity and temperature, control benthic distribution, whereas in shelf seas, current energy, substrate type, salinity, temperature and photic intensity are important. There is, therefore, a general relationship between benthic organisms and water depth (Fig. 6.4).

Another method of determining depositional setting is palynofacies analysis (Fig. 6.5). This has proved to be particularly useful in fluvio-deltaic systems (e.g. Brent Province, North Sea; Denison and Fowler, 1980; Hancock and Fisher, 1981; Parry et al., 1981; Nagy et al., 1984).

6.3.2 Plankton

Organisms that live suspended in the water column are called plankton. The distribution of marine plankton is also controlled by environmental parameters, such as salinity, oxygen supply, temperature and nutrient availability. Phytoplankton are also controlled by photic intensity, which decreases with water depth and water clarity. They are inhibited therefore by turbid water in the vicinity of muddy delta systems. Environmental parameters vary with respect to water mass origin, climate, geography and water depth. The different groups of plankton also have varying degrees of tolerance to these environmental parameters. Some are more sensitive than others. For example, Radiolaria and planktonic foraminifera are generally uncommon in shelfal environments, whereas dinoflagellates and acritarchs may live in marginal marine to open marine environments (Fig. 6.6). Consequently the distribution of some planktonic fossils can be related broadly to water mass, water depth, and proximity to the land. The planktonic:benthic microfossil ratio (Murray, 1976) and the ratio of 'deep' marine dinocysts to 'shallow' marine dinocysts provides some measure of 'oceanicity' or upwelling.

6.3.3 Biofacies

An association of organisms representing a particular depositional environment is referred to as a *biofacies*. The

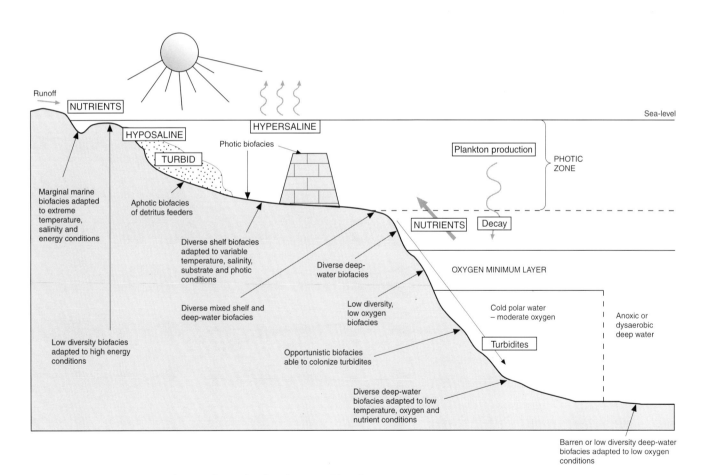

Fig. 6.3 Main environmental controls on the distribution of benthic organisms

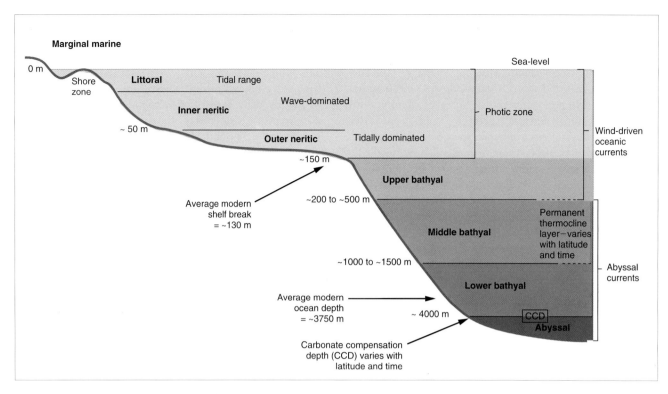

Fig. 6.4 Modern marine bathymetric classification and environmental controls on water column stratification. These controls vary seasonally and with geographic location. They will also have varied in the geological past, particularly with respect to base level change and the presence and size of polar ice-caps

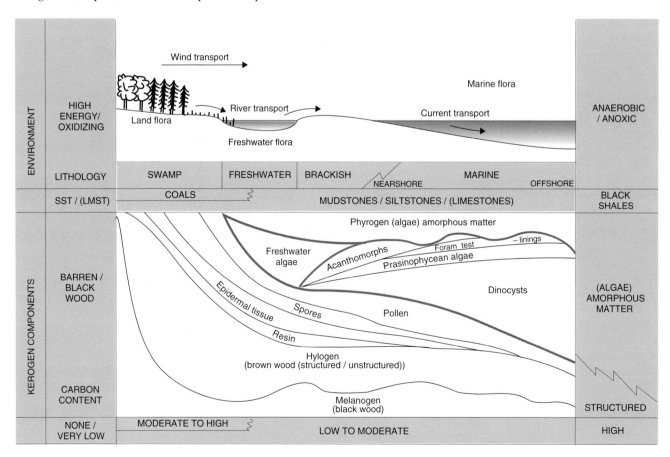

Fig. 6.5 Palynofacies: the distribution of palynological particle types in specific environments

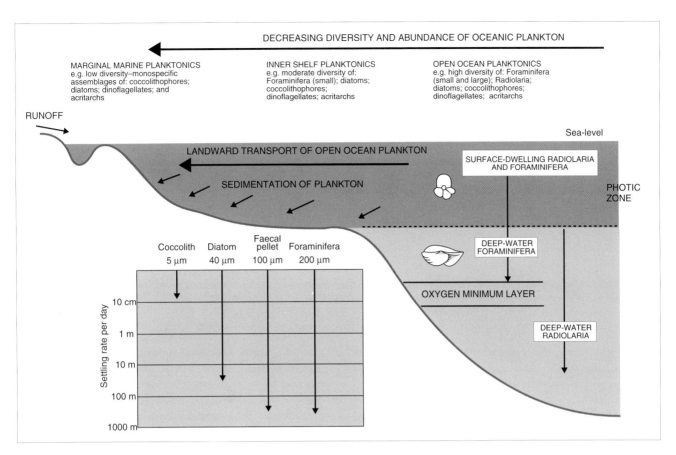

Fig. 6.6 Main environmental controls on the distribution of planktonic organisms

composition of a particular fossil assemblage in any biofacies is a function of palaeoenvironmental conditions, post-mortem redistribution by currents or mass gravity transport and diagenetic history. Most micro- and macrofossils species can be used to identify palaeoenvironment, however, the small size, preservability and widespread distribution of benthic foraminifera makes these organisms especially useful. The distribution of sediment type is only one of several environmental parameters controlling a biofacies. Consequently, there is no simple relationship between biofacies and sediment type, although in shallow water environments there is a closer relationship between biofacies and tidal or wave current energy and hence on sediment grain size.

In retrogradational and progradational depositional systems the environmental parameters controlling the distribution of the fossil assemblages and hence the biofacies migrate landwards and basinwards, respectively. Consequently, the vertical fossil record will record the bathymetric history of a basin, and inferences can be made as to whether the basin margin is prograding, retrograding or aggrading.

In progradational or retrogradational systems the boundaries between biofacies are diachronous surfaces (Armentrout, 1987). Consequently, the first and last fossil appearance datums that coincide with the environment changes are not necessarily evolutionary appearances or extinctions, but possibly diachronous biofacies boundaries related to progradation or retrogradation of the basin margin (Fig. 6.7).

Marine biofacies

Palaeoenvironments based on benthic and planktonic biofacies interpretations normally are related to present-day shelf and ocean bathymetry. Most modern biofacies, however, are only reliable analogues for late transgressive or early highstand systems tracts, when the shoreline is located a significant distance landward of the offlap break. During sea-level lowstands or when the coastline of a highstand systems tract has built out to the previous shelf break, the shelf and upper bathyal biofacies would have been more closely juxtaposed (Fig. 6.8). In such circumstances, proximal and distal biofacies mixing by currents, and subsequent downslope transport by gravity flows can complicate palaeoenvironmental determinations.

The identification of the deepest water palaeoenvironmental indicators in any assemblage can help to differentiate actual from apparent (derived) biofacies indicators. However, bathyal biofacies have a poorer bathymetric resolution by comparison with shelfal faunas. Consequently, changes in relative sea-level have only a limited bathymetric effect

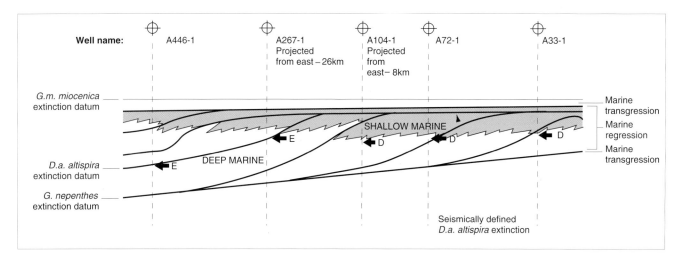

Fig. 6.7 Planktonic biofacies relationships in regressive and transgressive systems: an example of the effects of biofacies depression on the true extinction (tops) of planktonic foraminifera in the B High Island area of the Gulf of Mexico (after Armentrout, 1987). E, true extinction; D, ecologically depressed extinction

on bathyal biofacies. Glacial and interglacial alternations, however, influence the water mass properties of the oceans, such as oxygen availability, temperature and nutrient supply, and these have a pronounced effect on the composition of the bathyal biofacies.

Terrestrial biofacies

Fossil assemblages from terrestrial environments can provide a historic record of climatic conditions and a broad environmental setting of the land mass adjacent to a basin (Fig. 6.9). Microfloral assemblages indicative of warm, low latitude arid climates, infer low runoff and the potential to develop marine carbonate systems. Microfloral assemblages from humid environments infer greater clastic input to the basin and the development of fluvial and deltaic depositional systems. Humid environments also typically have lush vegetation cover, which binds or traps sediment, whereas more arid settings are prone to rapid sediment erosion and the redeposition of coarse clastics.

Land and freshwater fossil assemblages can be transported into adjacent marine environments, either by wind (particularly bisaccate pollen) or, more commonly, by fluvial systems (miospores, charophytes, ostracods and plant debris). In general terms abundance gradients of terrestrially derived fossils in the marine environment can be used to indicate the proximity of fluvial influx, and increasing proportions of smooth miospores and bisaccate pollen versus denser ornamented miospores and non-saccate pollen provide an indication of the proximity of land (Batten, 1974).

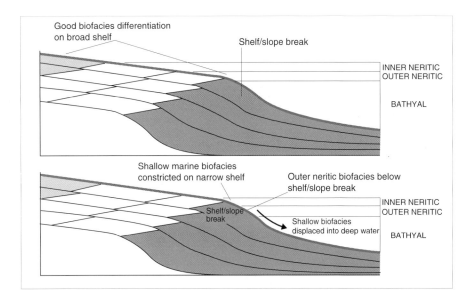

Fig. 6.8 Shelf width and proximity of the shore line to the deep basin during lowstand and early highstand and their effect on shelf biofacies. Flooded shelves are subject to strong or weak tidal energy depending on location, low wave energy and long transport distances to the deep basin, but during lowstand wave energy is at a maximum and transport distances are short

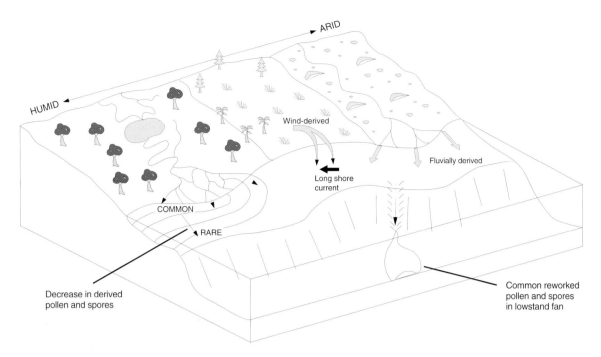

Fig. 6.9 Climatic controls on the terrestrial fossil signatures in marine deposits

6.4 Biostratigraphy and sequence stratigraphy

Our current understanding of this subject is limited by the small number of biostratigraphic studies that have been integrated fully with other sequence stratigraphic information from well and seismic data. The majority of our knowledge is derived from the Gulf of Mexico (Armentrout, 1987; Loutit *et al.*, 1988; Allen *et al.*, 1991; Armentrout and Clement, 1991; Armentrout *et al.*, 1991), although recent studies by McNeil *et al.* (1990) on the Beaufort–Mackenzie Basin, Jones *et al.* (1993) on the Barrow group, Northwest Shelf, Australia and Partington *et al.* (1993a) on the North Sea Jurassic, are improving this situation.

6.4.1 Sequence boundaries and their correlative conformities

A sequence boundary is a chronostratigraphically significant surface produced as a consequence of a fall in relative sea-level. If significant erosion has occurred at the sequence boundary, a biostratigraphic hiatus may be resolved by the juxtaposition of younger fossil assemblages on older ones and the negative evidence of absent fossil markers. The age and palaeoenvironmental contrast indicated by the fossil assemblages from above and below the boundary, are both a function of the magnitude of relative sea-level fall (McNeil *et al.*, 1990) and location within the basin. The magnitude of relative sea-level fall, as discussed in Chapter 2, may range from a dramatic, tectonically enhanced unconformity and associated erosion, to the more subtle changes in depositional facies experienced at a type 2 boundary. Irrespective of the magnitude of relative sea-level fall, however, the change in fossil composition across the boundary will vary along depositional dip. These variations may range from subtle changes in biofacies across the correlative conformity in deeper water settings, through shifts in biofacies accompanied by erosion marked by the absence of biochronostratigraphic markers on the upper slope, shelf and coastal plain, to more pronounced hiatuses landwards.

Major, tectonically enhanced sequence boundaries typically have rotated beds and widespread erosion of subaerially exposed strata. Together these result in significant widespread unconformities that are apparent in the fossil record both through the absence of age diagnostic species, and the abrupt superposition of markedly different biofacies. Examples include non-marine paralic deposits with pollen and spore assemblages overlying deep marine hemipelagics containing planktonic foraminifera, nannofossils and dinocysts.

The ability to resolve sequence boundaries, particularly smaller magnitude ones, using biostratigraphy, is limited by the resolution of marker fossils available. In the absence of sufficient fossil markers, Armentrout and Clement (1991) have proposed that minimum faunal abundances have the potential to be used to recognize periods of maximum regression and are therefore candidates for sequence boundaries. Gaskell (1991) has also demonstrated a possible correspondence between increased extinction rate in benthic foraminiferal faunas and rapid sea-level falls associated with type 1 sequence boundaries, which was not apparent for slower rates of sea-level fall.

The need to recognize reworked fossils is a common problem in biostratigraphy, and one that is associated closely with the erosion that occurs at sequence boundaries. Indeed reworked fossils are often the commonest palaeontologic component in rapidly deposited sediments. Their presence, together with the abrupt influx into deep marine basins of terrestrial fossils, such as pollen, spores and plant debris, may be used to help identify a sequence boundary (Fig. 6.13).

6.4.2 Lowstand systems tract

A significant fall in relative sea-level associated with the production of a type 1 sequence boundary, causes an abrupt basinward shift in facies that superimposes shallower or non-marine deposits on deeper marine ones. In essence the lowstand systems tract is recognized in the proximal fossil record by an underlying hiatus, a sudden shallowing up of biofacies or the superposition of non-marine assemblages on marine. In the deep basin it is recognized by increased rates of siliciclastic sediment supply and sediments that contain reworked fossils and a low abundance of indigenous fossils (Armentrout et al., 1991). Erosion below lowstand deposits is usually less widespread in the deep basin and often confined to channel systems or local slope instability. The bathyal benthos is also unlikely to be sufficiently sensitive enough to show a response to the bathymetric change involved in the lowstand (Armentrout et al., 1991).

The lowstand systems tract comprises two components; the lowstand fan, and the lowstand wedge. Lowstand fans (Fig. 6.10) are produced by gravity flow processes that result from fluvially supplied sediment bypassing the shelf and upper slope via incised valleys and canyons (Chapter 9). Consequently they are likely to contain derived terrestrial organisms and reworked fossil assemblages from the eroded shelf and slope (Van Gorsel, 1988), together with older, reworked fossils from the hinterland. Lowstand fan deposits can thus be recognized by the presence of exotic fossil assemblages encased within deeper marine shales containing indigenous assemblages.

Rapidly deposited fans are generally devoid of *in situ* deep-marine fossils (Armentrout et al., 1991), which poses a problem in placing the fan in its chronostratigraphic context. Stewart (1987), in an integrated bio- and sequence stratigraphic study of the central North Sea Palaeogene, noted that the microfossil assemblages in the Forties lowstand fan were impoverished and consisted predominantly of long ranging agglutinated foraminifera.

Rapidly deposited fans may contain rip-up clasts eroded from the slope during transport, which, if fossiliferous, may provide a maximum age for fan development. In the absence of indigenous fossils the age of the fan may be constrained by dating the encasing condensed shales immediately above and below the fan. Interfan lobes, however, may have indigenous fossil assemblages.

The reworked fossils provide information about the nature of the sediment provenance from which they were derived and, therefore, the probability of developing mud-rich, sand-rich or mixed sand–mud fans. Sand-rich fans are composed typically of a series of rapidly deposited, poorly fossiliferous, massive sand beds, which are difficult to date accurately. Lowstand fans that develop over a longer period are typically more mud prone and may have a greater proportion of indigenous basinal fossil assemblages.

The lowstand wedge is initiated as sea-level begins to rise following a rapid fall. Lowstand wedge deposits comprise progradational to aggradational parasequences (Fig. 6.11), and contain indigenous fossils with proximal to distal biofacies gradients. In a vertical succession through a prograding wedge the fossil assemblages indicate a shallowing-up signature from deeper marine, shallow marine, through marginal marine to non-marine biofacies. Aggradational wedges will not show such a pronounced shallowing-up succession, rather a thick accumulation of similar biofacies, either topset or slope. Consequently, lowstand wedge deposits have similar biostratigraphic characteristics to highstand shelf-edge prograding or aggrading systems tracts.

In nutrient starved basins, sediment bypass to the basin during lowstand may increase nutrient supply and lead to greater plankton productivity. If this occurs, the distal part of the lowstand wedge may be recognized by the occurrence of abundant planktonic fossils in condensed, hemipelagic shales overlying the basin-floor fan deposits. In the absence of fans, the fossil assemblages of the distal lowstand condensed shales will resemble those of the previous highstand.

Shelfal width is at a minimum during the lowstand systems tract, and wave energy impinging on the shelf is at a maximum. Such shelves are characterized by an epifaunal benthos and possibly a steeply declining planktonic gradient towards the land, depending on current distribution. Also the close proximity of land to the deep basin during the lowstand wedge is recognized by the presence of common plant material in basinal deposits.

The shelf-margin systems tract is associated with a type 2 sequence boundary (Chapter 2). The deposits of the shelf-margin systems tract are characterized by a progradational to aggradational parasequence stacking pattern. The fossil assemblages developed in the shelf-margin systems tract have similar proximal to distal biofacies relationships to the prograding and aggrading highstand systems tracts. The erosional and non-depositional hiatus developed up-dip of the coastal onlap point has insufficient magnitude to be distinguished in the fossil record (McNeil et al., 1990). Consequently shelf-margin systems tracts are poorly defined by their fossil assemblages and easily confused with highstand system tracts.

6.4.3 Transgressive surfaces

The transgressive surface separates the lowstand systems tract from the transgressive systems tract. It is characterized

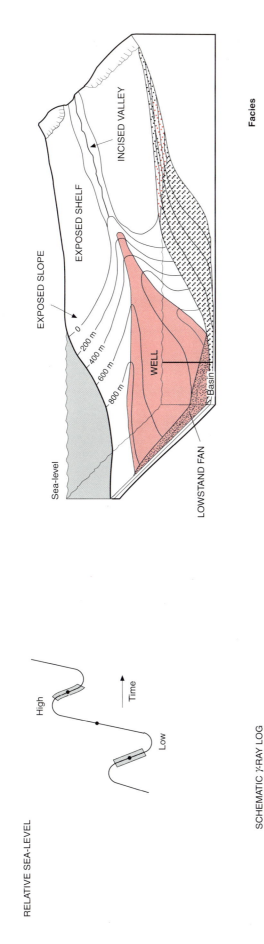
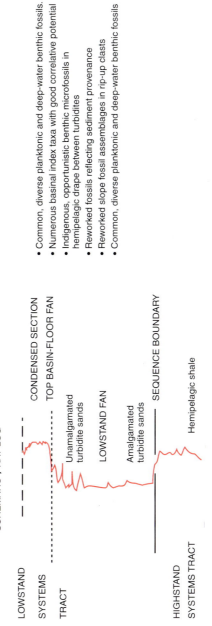

Fig. 6.10 Generalized fossil signature in a lowstand fan

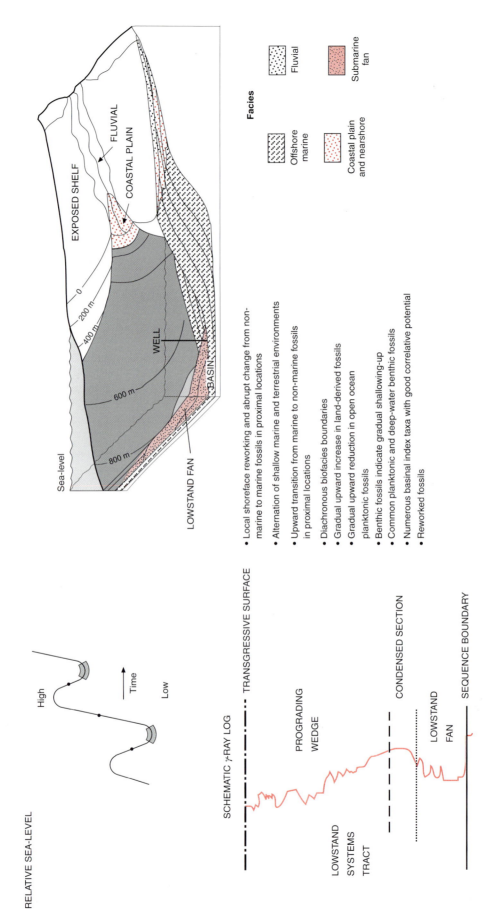

Fig. 6.11 Generalized fossil signature in a lowstand wedge. Note similarity with that for a highstand systems tract

by *in situ* reworking and winnowing of sediment, neither of which are conducive for the preservation of fossils. Hardgrounds and glauconite-rich deposits are also associated with the transgressive surface, and the diagenetic processes under which these are produced also may limit the preservation of fossils.

The presence of a transgressive surface may be inferred by the abrupt superposition of marine fossil assemblages upon marginal or non-marine ones. However, such an occurrence may easily be confused with a minor marine flooding surface, such as those associated with parasequence boundaries, although transgressive surfaces usually herald more prolonged and deeper water marine conditions. If the sediment supply to the drowned shelf is limited during the transgression, then the transgressive surface may be located within the condensed section containing the maximum flooding surface. Note that the transgressive surface represents a retrogradational biofacies boundary between terrestrial and marine environments, and is consequently diachronous.

6.4.4 Transgressive systems tracts

The transgressive systems tract comprises retrogradational parasequence sets, which have an overall deepening upward bathymetric signature in the fossil assemblages (Armentrout *et al.*, 1991), and usually the close superimposition of more distal fossil assemblages upon proximal ones. In a vertical succession the biofacies pass upwards, in a complete example, from terrestrial through brackish, to shallow marine and finally deep marine assemblages, which may be either fully open oceanic or a restricted basin depending on palaeogeography (Fig. 6.12).

Marine flooding during transgression creates new niches as facies belts step landward and are colonized by opportunistic species. The rapid rise in relative sea-level coupled with a low sediment input produces expanses of wetland, which may be evident in the terrestrial, floral fossil record. In warm climates, these are potential sites for coal swamp formation; thicker coals are progressively developed during the marine transgression as a result of higher rates of coastal plain aggradation (Chapter 11). Brackish environments developed in protected coastal plains during the transgressive systems tract are characterized by low-diversity assemblages of low salinity tolerant plant and animal species. These assemblages develop typically in low-energy conditions and are predominantly muddy. They represent a diachronous, retrogradational biofacies.

The shoreface deposits of the transgressive systems tract also comprise a diachronous, retrogradational biofacies. The marine flood events separating the parasequences contain more potentially datable marine fossils, although the periodicity of individual parasequence development is probably below biostratigraphic resolution in most cases.

As the rate of sediment supply to the shelf and basin is reduced during transgression, water turbidity decreases and clear-water marine microfaunas, including larger foraminifera and sea-grass species, become more frequent (Van Gorsel, 1988). Reduced sediment supply results in the widespread formation of condensed sections that contain abundant fossil assemblages, and include the presence of datable open marine planktonic markers. These sections diachronously onlap progressively younger marine deposits and reach their maximum development at the maximum flooding surface. Shaffer (1987) has used the presence of nannofossil abundances related to warm climatic periods to recognize the marine transgression of the previous shelf.

In the deep basin, open marine fossil assemblages in pelagic condensed sections are generally abundant, diverse and dominated by planktonics with cosmopolitan index taxa. The generation of submarine fans, due to slope regrading during marine transgression, as suggested by Galloway (1989), may be recognized by the presence of reworked or derived shallower marine microfossils within the deep basinal condensed shales.

6.4.5 Maximum flooding surfaces

The maximum flooding surface separates the transgressive systems tract from the highstand systems tract, and represents the maximum landward extent of marine conditions. The development of a widespread condensed section on the drowned shelf and deep basin may occur at this time, due to sediment starvation. These condensed sections typically comprise high gamma-ray, high sonic travel-time shales, associated with concentrations of uranium in organic-rich low-density sediments, and on seismic sections they form major downlap surfaces. Not all condensed sections, however, are indicative of maximum flooding surfaces. They may develop for a variety of reasons at any time within a sequence, for example on submarine highs, or due to delta lobe switching. Planktonic abundance events also may occur independently of condensed sections and be controlled by local climatic effects such as upwelling (Simmons and Williams, 1992).

The maximum flooding surface represents the most landward distribution of diverse, open marine, cosmopolitan, often abundant, plankton and deep water benthos (Loutit *et al.*, 1988; Allen *et al.*, 1991; Armentrout and Clement, 1991; Armentrout *et al.*, 1991) (Fig. 6.12). The condensed section associated with the maximum flooding surface comprises a biostratigraphically distinctive event usually with abundant planktonic fossils. It thus has the greatest potential for being dated and correlated across a basin (and possibly globally). It is, therefore, a more correlatable event than the sequence boundary, which sometimes is difficult to date or even recognize biostratigraphically. Partington *et al.* (1993b) have used palynomorph and microfossil assemblages from the maximum flooding surface associated with condensed sections, in the Jurassic to earliest Creataceous of the North Sea, to provide a robust and predictive biochronostratigraphic framework.

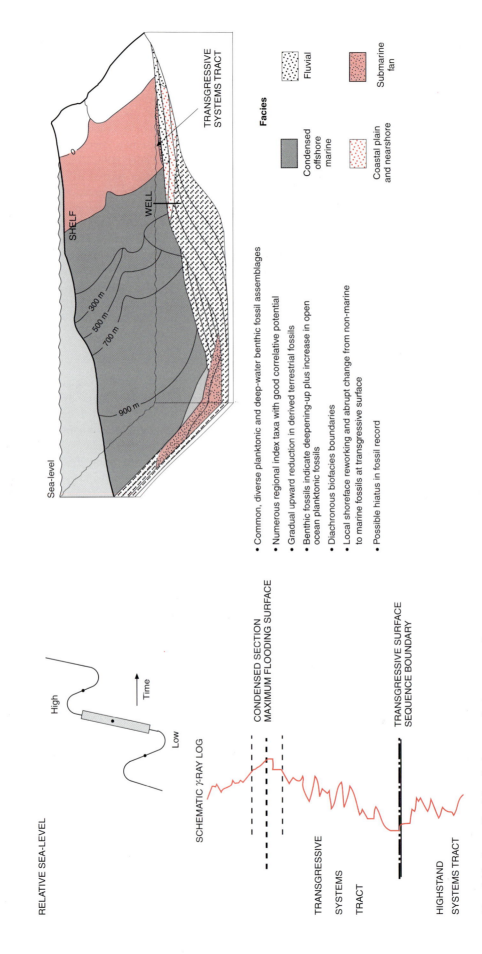

Fig. 6.12 Generalized fossil signature in a transgressive systems tract

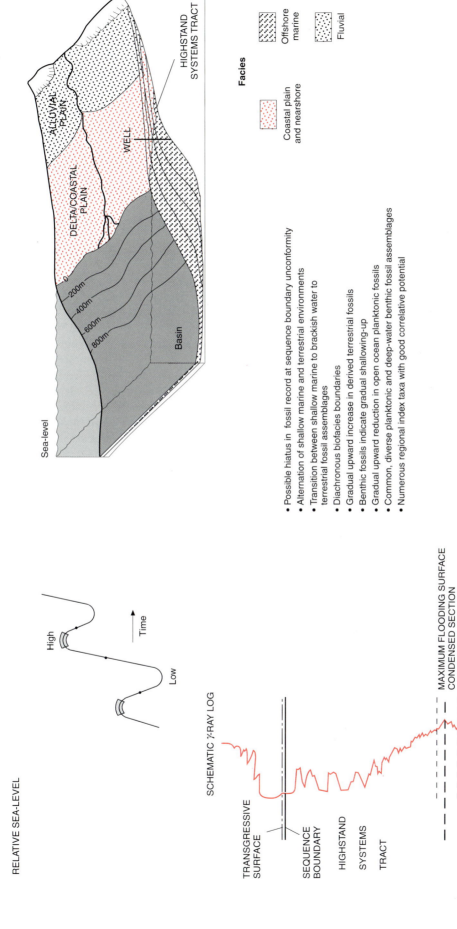

Fig. 6.13 Generalized fossil signature in a highstand systems tract

At the basin margin the maximum flooding surface of a condensed section is recognized by the sudden influx of low diversity, open marine plankton, sandwiched between shallower marine benthic or terrestrial fossil assemblages. On the shelf the maximum flooding surface is recognized by the presence of more diverse open marine plankton and possibly deeper water benthic fauna, whereas in a deep basin, sediment starvation may result in the development of highly fossiliferous deposits. In the case of extreme clastic starvation, pelagic carbonates, composed of the remains of calcareous microfossils, may develop. Alternatively, slow sedimentation may allow time for corrosive bottom-water to dissolve the calcareous fossils.

6.4.6 Highstand systems tracts

Aggrading highstand systems tracts occur when the rate of sediment supply is equal to the amount of accommodation space being created by rising relative sea-level. They are characterized by thick accumulations containing stacked shelfal or terrestrial fossil assemblages with no overall shallowing-up observed.

Progradational highstand system tracts occur when the rate of sediment supply exceeds the amount of accommodation space being created by rising relative sea-level. In essence they superimpose proximal fossil assemblages upon more distal ones (Fig. 6.13). In a vertical succession they pass upwards, in a complete example, from deep marine, shallow marine, through marginal marine to non-marine, possibly with an oscillatory character reflecting parasequences and minor flooding surfaces.

During early highstand, shelf deltas or the coastal margin must first advance across the drowned shelf of the underlying transgressive systems tract to the margin created by the previous lowstand wedge. At this time, shelf width is at a maximum and wave energy is at a minimum. In the absence of powerful tidal currents, muddy sediments tend to predominate. The fossil assemblages of these broad muddy shelves are dominated by distinctive, fine-sediment-dwelling, 'turbid water' benthic assemblages in shelf to slope environments. Tidally dominated shelves with coarse lag deposits predominantly have an attached epifaunal benthos and common derived planktonic elements owing to increased transportation across the shelf.

Shelfal fossil assemblages are strongly influenced by the presence of shelf deltas and associated rapid sedimentation, increased turbidity of the water and reduced salinity. In this nutrient-rich environment the benthic fossil assemblages are often abundant, diverse and dominated by infaunal species. Planktonics are generally scarce, although certain groups, such as dinocysts and acritarchs, have adapted to these conditions, and nannofossils, more easily transported in from the open ocean owing to their very small size, have important biostratigraphic correlation potential.

If sufficient time and sediment are available, the highstand progradation may be able to advance to the margin created by the previous lowstand wedge. These deltas now become shelf-edge deltas and capable of supplying sediment, and also terrestrial and shelfal organisms, directly to the deep basin.

Topset highstand deposits may comprise shelfal, paralic and fluvial deposits and associated shallow and non-marine fossil assemblages, with the proportion of each dependent upon the nature of the progradation. In extreme oblique progradation the highstand will comprise mostly slope and shelfal deposits, with significant sediment bypass and little evidence of deposition in paralic or fluvial environments. Accommodation space created during the highstand by a relative rise in sea-level will result in the accumulation of deposits with shallow marine and terrestrial fossil assemblages.

The prograding highstand slope comprises both gravity flow and hemipelagic deposits that are often eroded, slumped and contorted. Consequently they may contain mixed indigenous and derived fossil assemblages. High sedimentation rates can cause the dilution of indigenous fossil assemblages. A prograding highstand slope may be inferred in a vertical section, but not exclusively defined, on a gradual shallowing-up signature in the benthic fossil assemblages together with a decrease in the proportion of planktonics (Van Gorsel, 1988). The vertical passage through different, shallower biofacies as the highstand slope advances into the basin, produces apparent extinctions in the fossil record that may cause diachronous correlations (Armentrout, 1987).

Within the deep basin, slow sedimentation basinwards of the highstand toesets produces condensed sections that may contain abundant deep-marine fossil assemblages similar to those in the condensed sections of the transgressive systems tract and maximum flooding surface (Armentrout and Clement, 1991). More rapid sedimentation in the deep basin suggests erosion of the slope through mass wasting via slumping, debris and turbidity flows or possibly sediment bypass. These may introduce both slope and shelfal, or non-marine, fossil components to the deeper basin, diluting the indigenous fossil assemblages and inhibiting a true palaeoenvironmental determination. Turbidites are commonly barren of indigenous fossils (McNeil et al., 1990), and often contain only reworked fossils and faunas derived from further up the slope.

6.5 Conclusions

The biostratigraphic character of sedimentary sequences is controlled by the interplay of basin-specific environmental conditions, the evolution of organisms and the cyclic changes in depositional style related to base-level changes. Consequently, there are few concrete laws on the relationship between biostratigraphy and sequence stratigraphy. However, there are general observations and trends that do apply and are summarized as follows:

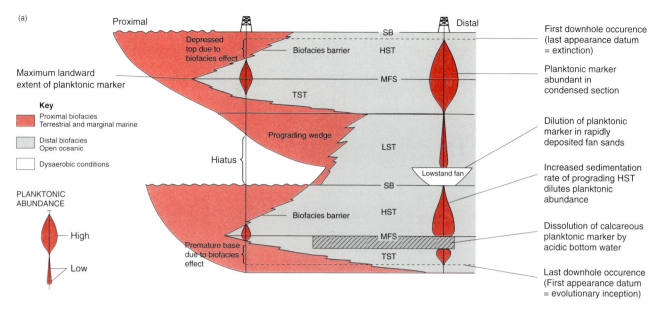

Fig. 6.14(a) The impact of biofacies, sedimentation dilution and fossil dissolution on planktonic fossil distribution and abundance

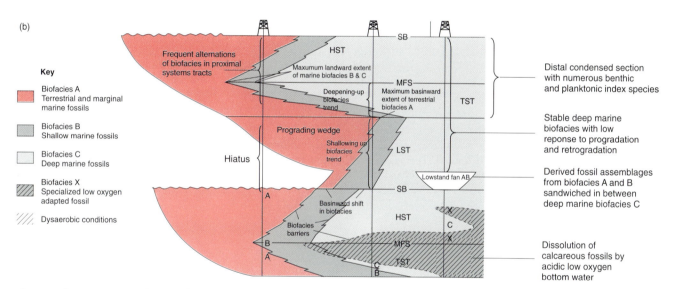

Fig. 6.14(b) Biostratigraphic resolution within different systems tracts

1 Individual fossil groups do not provide sufficient age and palaeoenvironmental determinations in all depositional environments and throughout Phanerozoic time. The combined use of different groups can complement each other and improve biostratigraphic resolution. Individual fossil groups used in isolation may result in incorrect palaeoenvironmental and age determination, with severe implications for geological modelling.

2 The late appearance or premature disappearance of a fossil from a stratigraphic section may be due to local environmental constraints, and relate to the biofacies of the fossil rather than true extinction (Fig. 6.14(a)). Correlations based on biofacies are generally diachronous and reflect progradation or retrogradation.

3 Fossil resolution is hampered by rapid sedimentation and by the degree of diagenesis (Fig. 6.14(b)). The highest resolution possible with fossil events (the smallest resolvable stratigraphic unit) may not be sufficient in all circumstances.

4 The ability to both identify and date sequence boundaries, transgressive surfaces or maximum flooding surfaces using biostratigraphy, depends upon the actual fossil resolution and the apparent resolution recorded by the sample programme. Sample spacing should be designed to solve the geological problem and, ideally, closely spaced samples

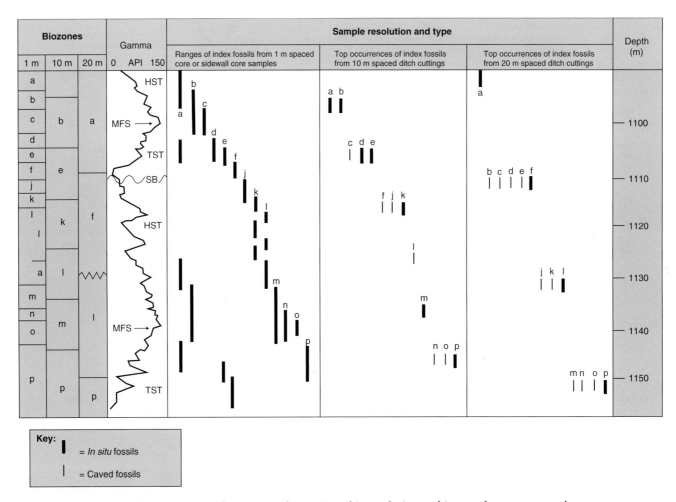

Fig. 6.15 The effects of sample type and spacing on biostratigraphic resolution and impact for sequence and systems tract recognition. HST, highstand systems tract; TST, transgressive systems tract; MFS, maximum flooding surface; SB, sequence boundary. Palaeoenvironmental indicators; j, terrestrial; a, d, e, f, l, p, shallow marine; b, m, deep marine; k, marginal marine; c, n, o, oceanic planktonics

are recommended in the vicinity of all potential boundaries. Figure 6.15 summarizes the limitations and advantages of different sample types. Note particularly the limited resolution of cuttings compared with recovered cores.

5 Caution is advised when tying fossil and seismic events, as both may have errors associated with depth conversion. It is particularly important to recognize that fossil and seismic ties in condensed sections may diverge significantly if correlated into a thicker sedimentary package, such as when correlating from condensed distal sediments into a prograding highstand systems tract.

6 Biostratigraphy and isotope stratigraphy are particularly useful for calibrating and correlating sequence boundaries and maximum flooding surfaces where there is poor seismic correlatability due to tectonic complexity (e.g. North Sea — pre-rift, Gulf of Mexico — growth faults) and poor or sparse data (e.g. Papua New Guinea — fold belt with karstified carbonates affecting seismic resolution, and tropical rainforest making seismic acquisition expensive).

7 Biofacies trends can be used to define progradational, aggradational and retrogradational trends and predict the timing of clastic accumulation on a shelf or bypass to a basin. An overall shallowing-up trend in biofacies applies equally to both highstand and lowstand progradation, whereas the transgressive systems tract is marked by an overall deepening-up trend in biofacies.

8 A maximum flooding surface is characterized by common and diverse planktonic assemblages at their widest distribution.

9 A sequence boundary is associated with erosion, a biostratigraphic hiatus and reworking.

10 The widespread distribution of planktonic markers

within the maximum flooding surface of a condensed section makes this the preferred correlative surface for biochronostratigraphic purposes.

11 The recognition of palaeoenvironments within systems tracts using fossil assemblages can provide a general indication of the potential type, distribution and sand content of reservoir facies.

Applications to Depositional Systems

CHAPTER SEVEN

Fluvial Systems

7.1 Introduction

7.2 **Fluvial processes and channel styles**
    7.2.1 Straight and anastomosing rivers
    7.2.2 High-sinuosity channel systems
    7.2.3 Low-sinuosity channel systems
    7.2.4 Classification of fluvial systems

7.3 The concept of the graded stream profile

7.4 **Fluvial architecture**
    7.4.1 Controls on fluvial architecture
    7.4.2 Sequence boundaries and lowstand systems tracts in alluvial strata
    7.4.3 Transgressive systems tracts and flooding surfaces in alluvial strata
    7.4.4 Highstand systems tracts in alluvial strata

7.5 **Reconstructing fluvial architecture**
    7.5.1 Eocene Castissent Formation, South Pyrenees, Spain
    7.5.2 Triassic, Ivishak Formation, Prudhoe Bay Field, North Slope, Alaska, USA

7.1 Introduction

The value of sequence stratigraphy as a predictive tool in the analysis of clastic shoreline and shallow marine systems is well documented in the literature and is reviewed in Chapter 8. Recent attempts to apply sequence stratigraphic concepts to fluvial systems has met with less success because the role of relative sea-level fluctuations in fashioning the fluvial stratigraphic record is less clear (Posamentier and Vail, 1988; Posamentier, 1993; Shanley, 1991; Shanley and McCabe, 1989, 1990, 1991, 1994; Westcott, 1993). Fluvial systems respond to a bewildering set of external (allocyclic) and internal (autocyclic) controls (Schumm, 1968, 1981; Schumm and Ethridge, 1991). The problem is compounded by rapid lateral facies changes and a lack of internal features in thick alluvial successions which allow them to be sub-divided into time stratigraphic units. As a result, the application of sequence stratigraphy to fluvial systems is still in its infancy, with concepts and models still the subject of lively debate (Galloway, 1981; Miall, 1986, 1991; Boyd et al., 1989; Walker, 1990; Posamentier and James, 1993; Schumm, 1993; Westcott, 1993; Koss et al., 1994; Shanley and McCabe, 1994).

The early sections of this chapter focus on depositional processes and fluvial channel patterns. The concept of the graded river profile is then introduced as the primary control on the development of accommodation in fluvial systems. Finally, the potential role of relative sea-level change in influencing the fluvial record is reviewed and illustrated with a number of examples.

7.2 Fluvial processes and channel styles

Fluvial systems are one of the best studied of all depositional environments. A detailed review of fluvial depositional environments is beyond the scope of this chapter and the reader is referred to a number of excellent summaries on this subject (Cant, 1978b; Miall, 1978, 1992; Ethridge and Flores, 1981; Collinson and Lewin, 1983; Collinson, 1986; Ethridge et al., 1987).

Four traditional styles of fluvial channel are commonly recognized; braided, meandering, anastomosing and straight (Leopold and Wolman, 1957; Leopold et al., 1964; Rust, 1978). These categories are useful but must be considered only as tendencies within a spectrum of channel types or classes (see discussion in Miall, 1992). River channel patterns are controlled by discharge, sediment supply and gradient (Bridge, 1985). As a result, changes between channel classes are gradational, with one or a number of different channel patterns containing similar morphological elements.

7.2.1 Straight and anastomosing rivers

Straight and anastomosing river systems are relatively rare in the recent and geological record. Straight rivers have well-defined single channel courses with stable banks flanked by levees. Anastomosing rivers form interconnected networks of low gradient, relatively deep and narrow channels of variable sinuosity, characterized by stable, vegetated banks composed of fine-grained silt or clay (Smith and Smith, 1980; Putman, 1983; Rust and Legun, 1983; Nanson et al., 1986; Fig. 7.1a). Lateral channel migration is limited by the development of fine grained bank material

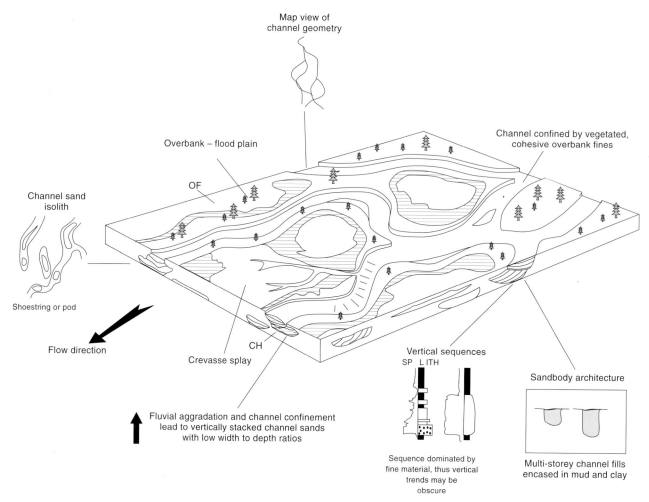

Fig. 7.1 (a) Block diagram of an anastomosing fluvial system illustrating the facies associations, channel belt and flood-plain subenvironments. Channel belts confined by fine-grained, vegetated overbank and flood-plain systems. Preserved channel belt systems form an interconnected network of low width-to-depth ratio sand bodies encased in overbank fines. Channel type common in low-gradient drainage systems and where increasing accommodation space leads to vertical rather than lateral accretion of alluvial deposits (after Galloway, 1981; Miall, 1992). OF, overbank fines; SB, sandy bedforms; CH, channels

and vegetation. Changes in channel course occur through avulsion (Smith, 1983); a process where successive major flooding events of the river result in progressive breaching and crevassing of the channel banks and alluvial ridge, leading to the formation of a new channel course along the lowest segment of the flood plain. Width to depth ratios of modern anastomosing channels are noticeably lower than for meandering systems (Smith and Smith, 1980; Smith and Putman, 1980; Smith, 1983; Tornqvist, 1993). The overbank and flood-plain areas separating channel courses consist of narrow natural levees, numerous crevasses, vegetated islands and wetlands (Smith and Smith, 1980).

Anastomosing channels have been documented in humid, tropical, semi-arid and arid climate settings, and appear to be typical of low-gradient downstream drainage systems dominated by cohesive sediments such as coastal plains and delta tops (Smith and Smith, 1980; Rust, 1981; Rust and Legun, 1983; Smith, 1983; Cairncross et al., 1988). This channel style may be developed preferentially in times of rapidly rising base level, where increasing accommodation space leads to a fluvial succession dominated by vertical rather than lateral accretion deposits (Smith and Smith, 1980; Smith, 1986; Kirschbaum and McCabe, 1992; Tornqvist, 1993). The depositional record of these systems is therefore dominated by isolated, shoestring-like sand bodies separated by levee–overbank, crevasse splay and flood-plain fines (Fig. 7.1a; Friend et al., 1978; Friend, 1983).

7.2.2 High-sinuosity channel systems

High-sinuosity channels develop in drainage areas dominated by low-gradient slopes, and where the river is commonly dominated by a high suspended load to bedload ratio (Leopold and Wolman, 1957; Allen, 1965; Schumm, 1971, 1977, 1981). They are typically organized into channel and overbank segments (Fig. 7.1b). Channels lie within a broad meander belt dominated by a complex

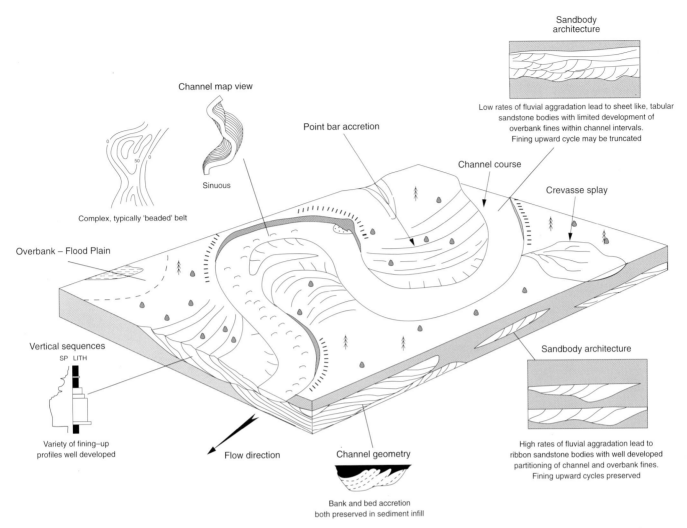

Fig. 7.1 (b) Block diagram of a high-sinuosity fluvial system illustrating the facies associations, channel belts and flood-plain subenvironments. The channel belt is confined within a raised alluvial ridge. The character of the channel-fill may be highly variable. Channels may migrate laterally to develop tabular to sheet-like sand bodies separated by fine-grained overbank and flood-plain sediments, or deposition may be confined by channel plugs, resulting in ribbon sand bodies. In these cases the width-to-depth ratios may therefore be variable. The stacking patterns of channel sand bodies will be affected significantly by rates of flood-plain subsidence (accommodation space). (Modified after Galloway 1981; Miall, 1992)

distribution of active and abandoned channels. Active deposition is confined largely to the channel belt, resulting in a well developed, raised alluvial ridge which stands above the general level of the flood plain. The distal margins of the alluvial ridge form overbank areas that interfinger with the adjacent flood plain. The course of the channel may be constrained by abandoned channel plugs or be relatively free to migrate laterally, developing point bars and associated lateral accretion deposits.

The sediment load of modern high-sinuosity channels is highly variable, ranging from systems dominated by a fine-grained suspended load (Jackson, 1976, 1978; Stewart, 1983) to coarse sand, gravel, pebble (Bernard and Major, 1963; McGowen and Garner, 1970; Levey, 1975; Arche, 1983) and gravel/pebble-rich systems (Gustavson, 1978; Ori, 1982; Forbes, 1983; Campbell and Hendry, 1987). A wide variety of facies and vertical facies successions may develop in the ancient record and facies successions may grade into the deposits of low-sinuosity systems (Puigdefabregas, 1973; Miall, 1983, 1987; Stewart, 1983).

Channel facies form tabular to sheet-like sand bodies separated by fine-grained overbank and flood-plain deposits (Fig. 7.1b; Friend, 1983). Areally restricted ribbon sand bodies may be more prevalent where channel belt migration is impeded (Puigdefabregas and Van Vliet, 1978). Associated overbank deposits comprise wedged shaped accumulations that thin and fine uniformly from the channel belt into the adjacent lacustrine or flood plain (Tornqvist, 1993).

7.2.3 Low-sinuosity channel systems

Low sinuosity or 'braided' channel systems occur where the coarser grain sizes, such as gravels and/or sands, form the dominant load of the system. In these cases, the lack of

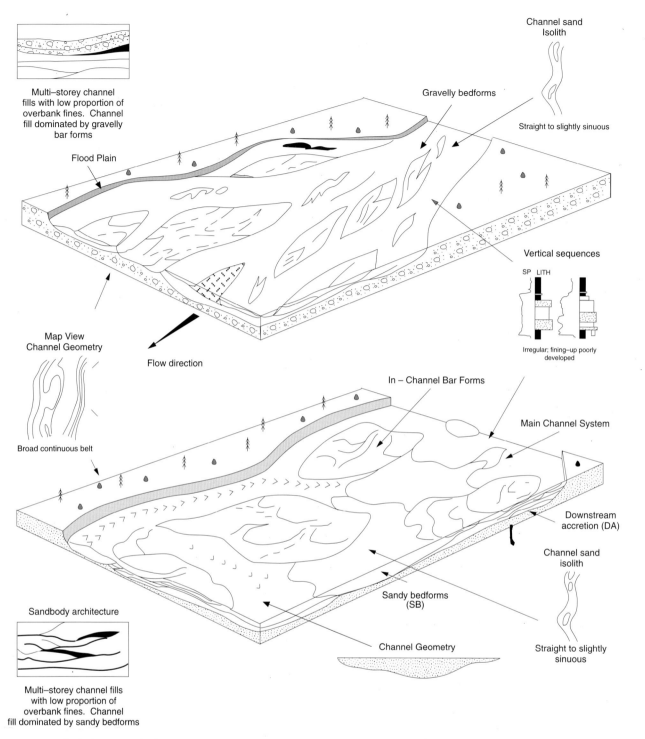

Fig. 7.1 (c) Block diagram of a low sinuosity (upper) gravelly and (lower) sandy fluvial system illustrating the facies associations, channel belts and flood-plain subenvironments. The lack of well-developed fine-grained overbank sediments within these systems leads to mobile channel courses and well-developed sheet sandstones (SB) and gravel bodies (GB) (modified after Galloway, 1981; Miall, 1992)

significant cohesive bank material leads to extremely mobile channel courses (Fig. 7.1c). Individual channels continually shift and bifurcate producing a rapidly evolving network of variable-scale braid channels with submerged in-channel migrating bedforms (Leopold and Wolman, 1957; Coleman, 1969; Collinson, 1970; Smith, 1974; Cant and Walker, 1976, 1978; Miall, 1977; Cant, 1978a,b).

The high bedload content of the low-sinuosity rivers, combined with the highly mobile nature of channel systems leads to a stratigraphic record dominated by lenticular, concave-upward sand bodies characterized by variable scale cross-stratification, lateral accretion deposits and a lack of channel-margin facies (Moody-Stuart, 1966, Campbell, 1976; Hazeldine, 1983; Bristow, 1987).

7.2.4 Classification of fluvial systems

The basic channel styles described above are often difficult to recognize in the geological record. The grain size of the fluvial system has been suggested as a parameter to subdivide fluvial systems because it can be measured in both the ancient and modern, at outcrop and in the subsurface. On this basis, river systems can be broadly subdivided into four principal types comprising; high-bedload, bedload, mixed-load and suspended-load dominated rivers (Schumm, 1977; Schumm and Brakenridge, 1987; Orton and Reading, 1993). Each of the four system-types displays characteristic channel-fill geometries, facies assemblages and vertical successions (Fig. 7.2 and Table 7.1).

7.3 The concept of the graded stream profile

In all depositional systems, deposition, burial and erosion are controlled by an equilibrium surface or base level which defines and influences the amount of accommodation space (see Chapter 2). The equilibrium surface separating erosion from deposition in fluvial systems may reflect the influence of numerous base levels (Miall, 1987, 1992; Posamentier, 1988; Westcott, 1993). These include lake-level, the level of trunk-stream drainage, the position of nick points, the position of ground-water tables and relative sea-level. This is in contrast to shoreline and shallow marine systems, where base level generally equates to sea-level. For this reason, the concept of a graded stream profile has been applied to fluvial systems to define an equilibrium surface that separates erosion from deposition and controls available sediment accommodation (Mackin, 1948; Sloss, 1962).

Mackin (1948) defined a graded stream or river as:
One in which, over a period of years, the slope is delicately adjusted to provide, with available discharge and with prevailing channel characteristics, just the velocity required for the transport of load supplied to the basin. The graded river is a system in equilibrium: its diagnostic characteristic is that any change in one of the controlling factors will be transmitted throughout the whole profile.

A graded profile can be considered as representing a balance between erosion and deposition; it will be graded such that the stream can transport its load along the profile without significant erosion or deposition. The slope at any point on the graded river profile will be a function of river discharge and sediment load (volume and capacity). Downstream reductions in slope relate to increasing discharge and down-valley decreases in grain size (Fig. 7.2). The overall shape of the slope will be modified through time to approximate a concave upward graded profile, flat at the river mouth and steepening towards the source.

Table 7.1 Summary characteristics of fluvial depositional systems according to grain size (from Orton and Reading, 1993). Facies codes after Miall (1978, 1985).

	1 Gravel	2 Gravel and sand	3 Fine sand	4 Mud/silt
Hinterland				
Catchment area	Small ($< 10^3$ km^2)	Intermediate ($< 10^5$ km^2)	Intermediate ($< 10^6$ km^2)	Large
Relief or topography	High	Intermediate	Intermediate	Low
Climate	Arid, arctic	Temperate	Temperate	Humid, tropical
Alluvial form				
Size of stream	Small	Intermediate	Intermediate	Large
Stream gradient	Very steep (> 5 m km^{-1})	Intermediate (> 0.5 m km^{-1})	Intermediate (> 0.05 m km^{-1})	Low
Flow velocity	High to very high	Intermediate	Intermediate	Low
Discharge	Low (< 100 m^3 s^{-1})	Intermediate ($< 10^3$ m^3 s^{-1})	Intermediate ($< 10^4$ m^3 s^{-1})	High
Discharge variability	Very irregular	Irregular–regular	Regular–irregular	Very regular
Sediment load	Low ($< 10^6$ tons year^{-1})	Intermediate ($< 10^7$ tons year^{-1})	Intermediate ($< 10^8$ tons year^{-1})	High ($< 10^{10}$ tons year^{-1})
Load : discharge ratio	High/very high	Intermediate	Intermediate	Low
Channel type	Bed-load	Bed-load	Mixed load	Suspended load
Channel pattern	Braided/absent	Braided	Meandering/braided	Straight/meandering
Bank strength	Moderate	Low–moderate	Low	High
Width : depth ratio	Intermediate	High–intermediate	High	Low
Channel mobility	Intermediate	High–intermediate	High	Low (fixed)
Architectural elements in deposits*	GB, SG	SB, FM, LA	LA, SB, FM, OF	OF, LA, SB

* GB, gravel bar/bedforms; SG, sediment gravity flows; SB, sandy bedforms; FM, foreset macroforms; LA, lateral accretion; OF, overbank fines

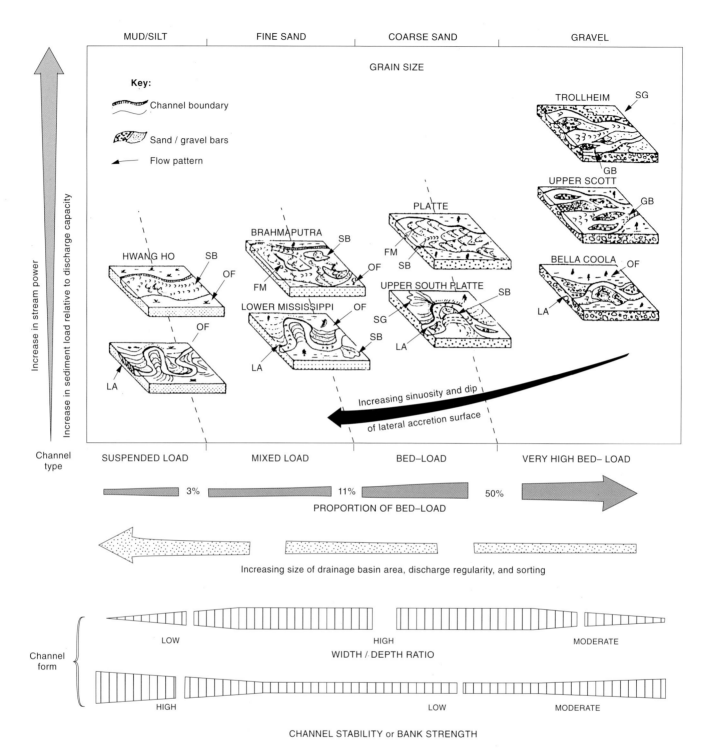

Fig. 7.2 Relationship between grain size and channel pattern (after Orton and Reading, 1993). Note general tendency of increasing sinuosity, confinement of channel belts and discharge regularity with decreasing grain size and low-gradient slopes. Position of fluvial examples approximate (based on Schumm, 1981; Ferguson, 1987; Miall, 1985) (Fluvial facies and architectural elements, after Miall, 1985). Architectural elements: GB, gravel bar/bedforms; SB, sandy bedforms; FM, foreset macroforms; LA, lateral accretion; SG, sediment gravity flows; and OF, overbank fines

Through time, rivers will attempt to establish a stable graded profile for a given discharge of water and sediment. Disturbances in this system, such as sea-level fluctuations, climatic change and tectonics, cause the stream to re-establish a new state of equilibrium by changing the external and internal attributes of the river (Schumm and Ethridge, 1991; Germanoski and Schumm, 1993; Schumm, 1993). This can be achieved by altering a river's characteristics such as (Table 7.2):

channel width

sediment calibre
velocity
boundary roughness
depth
sediment discharge
slope
planform.

Although autocyclic mechanisms may modify the short-term character of the fluvial system, only long-term allocyclic events may alter fundamentally one or a number of these variables, leading to major changes in fluvial architecture. Studies of modern river systems suggest that these adjustments may take considerable periods of time. Indeed many modern rivers exhibit graded profiles still adjusted to Quaternary glacial events (Wilcox, 1967; Church and Slaymaker, 1989). Recognition of these large-scale systematic changes, and their attendant stratigraphic boundaries, can provide a sequence stratigraphic framework for the evaluation of ancient fluvial deposits.

7.4 Fluvial architecture

Fluvial architecture can be defined as the three-dimensional geometry and interrelationships of the deposits of channel, levee, crevasse splay and flood-plain and other sub-environments of the fluvial depositional system (Miall, 1983, 1985). Fluvial architecture can be regarded as the result of the interaction between river process and long-term allocyclic controls. The term can be applied at varying scales from the deposits of a single river to the non-marine depositional record of a basin fill. Most recent studies suggest that all fluvial deposits can be subdivided into a number of key *architectural elements*, defined by overall shape, distinctive facies assemblages and internal geometries (Miall, 1985, 1988). These elements form the basic building blocks of all fluvial systems, yet their overall geometry and stacking patterns vary between different fluvial channel styles. Variations in the proportions of particular architectural elements can be used to interpret different fluvial styles in the geological record.

7.4.1 Controls on fluvial architecture

In modern systems, channel morphology evolves downstream in response to changes in valley slope, sediment load, bank materials, climate and tectonic regime. These same controls may change abruptly or evolve over time to cause significant or subtle morphological changes throughout the whole, or part, of the fluvial system (Table 7.2) (Burnett and Schumm, 1983; Schumm, 1993). For this reason, fluvial channel patterns in the geological record are unlikely to remain constant through a stratigraphic unit. To review how these changes take place, is it useful to divide the fluvial system into three geographical areas (Fig. 7.3). These comprise: (1) the *upstream* area forming the hinterland and source area for the river; (2) the *mid-stream* region forming much of the drainage basin over which

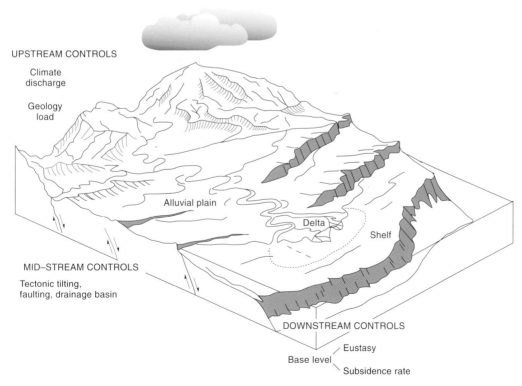

Fig. 7.3 Key controls influencing the graded profile within the upstream, mid-stream and downstream portions of the fluvial drainage system

Table 7.2 Channel adjustment to changes in hydrology and sediment load based upon empirical equations derived from Schumm (1968, 1977). The data illustrate how decreases and increases in annual discharge rates (Q) and percentage sediment load transported as bed-load (QS) influence channel depth (D), meander wavelength (l), sinuosity (P), slope (s) and width:depth ratio (F)

Annual discharge (Q)	Percentage bed-load transport (QS)	≈	Channel width (B)	Channel depth (D)	Meander wavelength (l)	Sinuosity† (P)	Slope (S)	Width-to-depth ratio (F)	Impact on fluvial system
Increase	Increase	≈	Increase	Increase/decrease	Increase	Decrease	Increase/decrease	Increase	Broadening of channels, decreasing sinuosity, increase in channel width and slope
Decrease	Decrease	≈	Decrease	Increase/decrease	Decrease	Increase	Increase/decrease	Decrease	Narrowing of channels, increase in channel depth, increasing sinuosity, decreasing slope
Increase	Decrease	≈	Increase/decrease	Increase	Increase/decrease	Increase	Decreases	Decrease	Narrowing and deepening of channels, increased sinuosity, decreased slope
Decrease	Increase	≈	Increase/decrease	Decrease	Increase/decrease	Decrease	Increases	Increases	Widening of channels, increase in slope, decrease in sinuosity

† Sinuosity (P) = channel length (l_c)/valley length (l_v) = valley slope (S_v)/channel slope (S_c)

sediment is transported; and (3) the *downstream* or lower reaches of the river in which sediment is deposited (Schumm, 1977). The controls influencing each of these regions of the system are discussed below.

Upstream controls

The graded profile within the upstream portion of the fluvial system is affected by tectonics, climate and bedrock geology. Tectonics plays a significant role in influencing the type of channel system, the nature of the sediment load and the calibre of sediment deposited (Cant, 1978b; Miall, 1981). For example, source areas undergoing rapid uplift may generate large volumes of coarse-grained clastic sediments and steep gradients. These conditions may favour the development of high-gradient braided or low-sinuosity channel systems within the upstream portions of the drainage basin. In contrast, lack of relief in the stable hinterland and river catchment areas may result in reduced regional gradients and the transport of finer grained clastic sediments by river systems. In these cases, meandering or anastomosing river systems may dominate the upstream region.

The impact of tectonics on the architecture of upstream fluvial systems is illustrated in studies of modern systems (Coleman, 1969; Alexander and Leeder, 1987), outcrop (Heward, 1978; Gloppen and Steel, 1981; Lawrence and Williams, 1987; Nichols, 1987; Jolley et al., 1990; Turner, 1992; Garcia-Gil, 1993) and in laboratory flume experiments (Burnett and Schumm, 1983; Ouchi, 1985; Fig. 7.4). Active tectonics in the upstream areas of the fluvial system may lead to stream rejuvenation, incision by nick-point retreat, river capture, and the erosion and cannibalization of basin-margin clastic sediments. These events may alter the discharge and load characteristics of the fluvial system, leading to changes in shape of the graded profile. Where these modifications are sustained, abrupt changes in channel pattern, style and architecture may be recorded in the stratigraphic record (Blakey and Gubitosa, 1984; Butler, 1984; Turner, 1992).

The upstream portion of the fluvial system is affected additionally by climate and bedrock geology (Schumm, 1977; Westcott, 1993). Climate influences the vegetation types, the degree of rainfall and run-off and the long-term discharge of the river (Knighton, 1984). Similarly, variations in bedrock geology control the type of material available and the load transported.

Mid-stream controls

Intrabasinal tectonics exert the dominant control on the mid-stream portion of the fluvial system by affecting the location and types of drainage patterns on the alluvial plain (Miall, 1981; Alexander and Leeder, 1987; Kraus and Middleton, 1987). Tectonic tilting associated with extension or thrust-related asymmetric subsidence superimposes a new slope on the pre-existing flood-plain topography (Alexander and Leeder, 1987; Wells and Dorr, 1987). These changes influence the graded profile of the river, leading to stream deflection and the shift of the channel course in the direction of maximum subsidence. Tilting

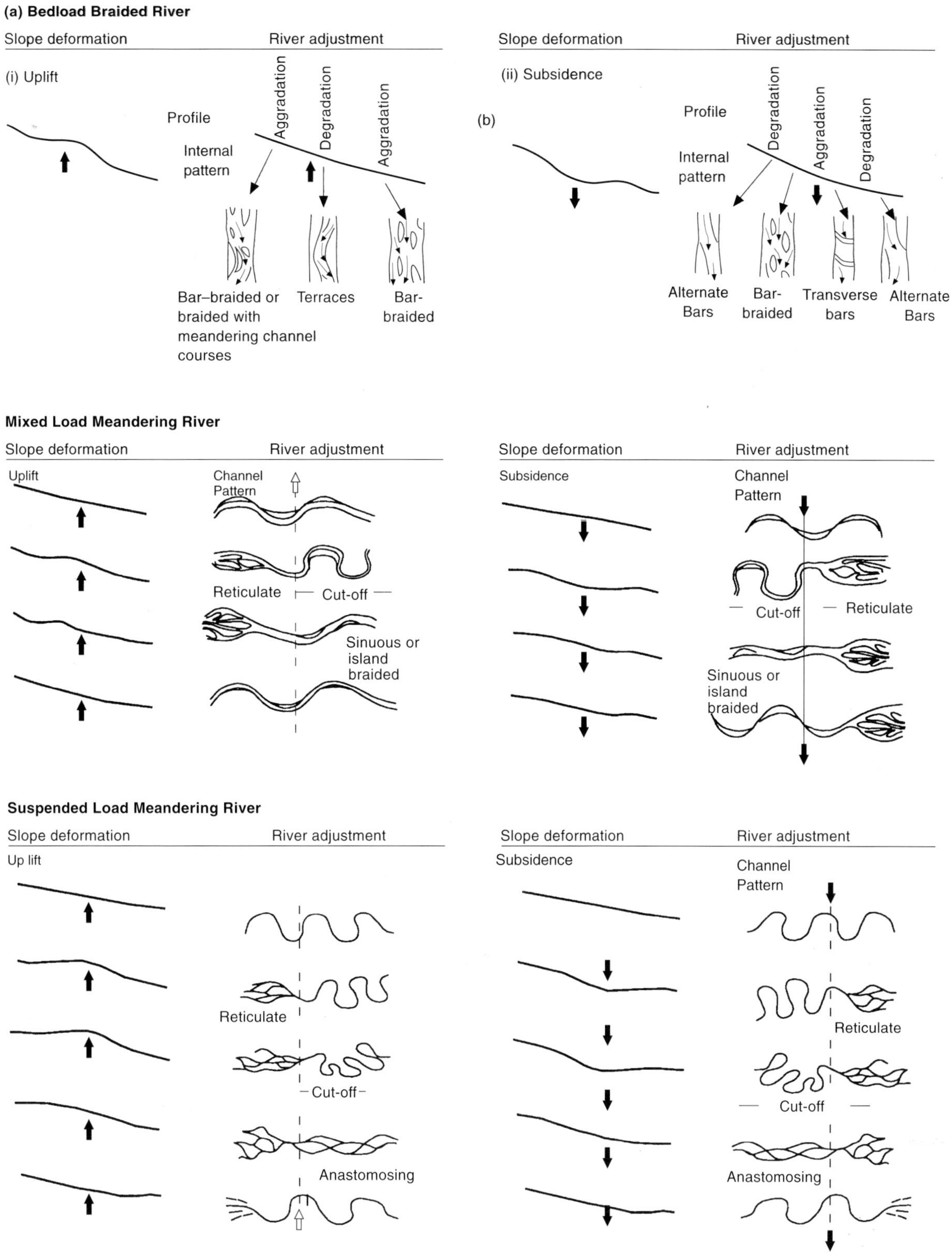

Fig. 7.4 Response of contrasting fluvial systems to uplift and subsidence from experimental flume studies by Ouchi (1985). Adjustment of (a) bed-load-dominated braided, (b) mixed-load meandering and (c) suspended load meandering river to (1) anticlinal uplift, (2) synclinal subsidence across it. The time-scale runs from top to bottom in each diagram (published with the permission of Geological Society of America)

Table 7.3 Characteristic fluvial depositional patterns of systems tracts and their stratal boundaries associated with relative sea-level changes (compiled from Nami and Leeder, 1978; Marzo et al., 1988; Van Wagoner et al., 1990; Kirshbaum and McCabe, 1992; Shanley and McCabe, 1993; Tornqvist, 1993; Westcott, 1993)

Sequence boundary	Lowstand systems tract	Transgressive systems tract	Maximum flooding surface	Highstand systems tract
Surface related to a relative sea-level fall. Boundary separates change from decreasing to increasing rate of generation of accommodation space. Associated with a basinward shift in facies and potentially accompanied by fluvial incision	Deposition on the climbing limb of the base level curve (cf. lowstand prograding wedge). Deposition propagates upstream depending on the rate of relative sea-level rise. Progressive reduction in stream gradient with increasing accommodation space	Coarse-grained braided or high-sinuosity channel deposits overlain by rapidly accreting flood plains with meandering or anastomosing channels. Channel patterns/styles reflecting increased rates of generation of accommodation space	Surface marking the most landward extent of marine inundation. Period prior to and following the maximum flooding event may be contiguous with the lowest fluvial gradients, reduced river discharges and stream power	Period characterized by a change from increasing to decreasing accommodation space
Relative sea-level fall causes rivers to straighten, deepen and widen their courses resulting in incised valleys. Width of channel belt significantly smaller than the extent of the valley incision	Braided or high-sinuosity fluvial deposits directly overlying marine shales or lower net : gross fluvial/deltaic facies. Channel sandstones multistorey and interconnected. Potentially coarsest grained systems developed during late LST and early TST	Increasing recognition of tidal influences with down-dip transition to estuarine systems. Dominance of mixed-load moderate/high-sinuosity channel systems, with variable net : gross characteristics. Upward decrease in sand-body amalgamation and continuity	Maximum flooding surface may be equivalent to the development of tidally influenced heterolithic facies in fluvial deposits, the development of poorly drained palaeosols, swamp peats or the formation of lacustrine carbonates	Early part of highstand characterized by poorly connected, vertically isolated sand bodies. Character of deposits may be broadly similar to the fluvial deposits of the late transgressive systems tract
Sequence boundary may define an abrupt change in facies, e.g. change from high- to low-sinuosity channel patterns, a change in grain size in response to increased stream power and sediment load, and/or a change in sediment composition	Limited accommodation space promotes flood-plain reworking during initially low rates of base level rise. Reworking by low- and high-sinuosity systems leads to amalgamated channel sandstone storeys	Base level rise may promote the preservation of complete fining upward channel units associated with more rapid flood-plain aggradation. Period associated with poorly drained soils, flood-plain peats and development of high water tables	Low net : gross channel sand bodies forming discontinuous, ribbon or shoestring geometries. Crevasse units may be more prevalent as the river system continually attempts to switch to areas of maximum gradient advantage on low-gradient flood plains	Late highstand fluvial deposits may be characterized by increasing amalgamation to form tabular channel sand bodies. Sand-body geometries reflect lateral accretion owing to low accommodation space
Sequence boundary may separate different degrees of channel sandstone amalgamation. Change in architecture reflecting change in accommodation space	Intermittent nature of base level fall marked by the preservation of coarse-grained terrace deposits and exposed interfluves with well-drained palaeosols	Transition from early to late transgressive systems tract associated with decreasing sand-body amalgamation. Upward change from amalgamated tabular sheets to isolated channel ribbons. Crevasse units may become more prominent in succession	Transition from 'maximum flooding surface' to highstand systems tract may be seen as subtle changes in channel stacking patterns, changing channel sinuosity and increasing sand-body amalgamation	Low rates of accommodation may lead to truncated or poorly preserved complete channel-fill successions
Fluvial systems linked down-dip to coarser grained lowstand shorelines or bypass shelf to feed submarine fans. High stream powers, discharge rates and sediment loads may lead to constructive fluvial-dominated deltas	Lowstand shoreline systems equivalent to up-dip coarse-grained, stacked fining upward channel storeys forming the incised-valley fill	Period of shoreline retrogradation and transgression. Marine processes may dominate at the shoreline leading to destructive wave or tidal-dominated deltas. However, constructive/destructive nature of deltas dependent on rates of sediment supply	Period of progressive shoreline transgression and ravinement	Period of shoreline progradation associated with increased rates of sediment supply and progressive reductions in accommodation space

Net : gross ratio, overall sand to shale proportions in fluvial channel and floodplain systems.

events may increase accommodation space and focus the position of channel belts, leading to increased lateral migration, greater channel deposit density and sand-body interconnectedness (Allen, 1978, 1979; Bridge and Leeder, 1979). Intrabasinal tectonics may additionally promote the formation of raised interfluve areas with more rapid soil development, or trap flood waters within the low points on the alluvial plain to form flood-plain lakes (Steel, 1974; Bown and Kraus, 1981; Alexander and Leeder, 1987; Kraus and Middleton, 1987).

Downstream controls

The downstream reaches of the fluvial system are affected most by changes in relative sea-level or lake-level. It is within this portion of the river system that the concepts of sequence stratigraphy may have greatest application in understanding the fluvial record (Shanley and McCabe, 1994). For river systems developed in close proximity to marine and lacustrine basins, changes in base-level may impose gradient changes in the lower reaches of the river that are propagated upstream. The river responds to these changes by aggradation, degradation or by altering its internal characteristics to re-establish grade. Downstream base-level changes will be felt only a finite distance upstream from the river mouth. In the case of the Mississippi River, the impact of the late Wisconsin lowstand propagated some 220 km upstream from the river's present mouth (see discussion in Shanley and McCabe, 1994).

Changes in architectural patterns and channel styles also may reflect the influence of upstream and mid-stream controls on the graded profile. For example, downstream aggradation may reflect increased sediment yield and discharge ratios associated with upstream climatic change. Similarly, upstream entrapment and the storage of bedload may lead to fluvial degradation.

The degree to which a fluvial system responds to a relative sea-level fall is controlled by a number of factors, including the gradient of the graded shelf and slope, the power of the river, the substrate and the rate and magnitude of the base level change. Differences between fluvial and shelf gradient exert a significant control on the response of the fluvial system to a base level fall (Schumm and Brakenridge, 1987; Miall, 1991; Schumm and Ethridge, 1991; Posamentier and Weimer, 1993; Schumm, 1993; Westcott, 1993).

Where the gradient of the shelf and the river system are equivalent, the fluvial system requires little or no adjustment to maintain equilibrium, and the graded profile may simply extend further across the exhumed shelf (Fig. 7.5b). Where the gradient of the shelf is less than the river, a base level fall may result in a reduction in stream power and transport capacity, leading to deposition and/or an evolution in channel style and channel-fill character, such as a change from bedload dominated braided systems to mixed-load or suspended-load dominated meandering systems (Fig. 7.5c).

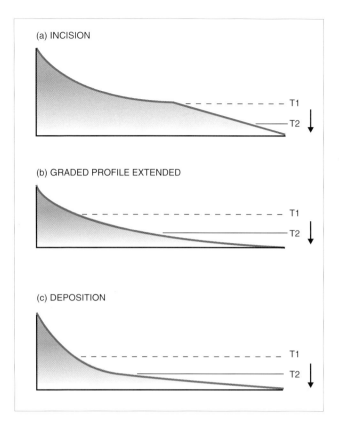

Fig. 7.5 Response of the fluvial system to base level falls; (a) where the gradient of the shelf > river, stream power and transport capacity may increase leading to erosion and fluvial incision — this particularly may be the case where base level falls below the shelf–slope break (type 1 sequence boundary); (b) where the gradients of the shelf and river system are equivalent, the stream requires little or no adjustment to maintain equilibrium and the stream extends its profile across the shelf; (c) where the gradient of the shelf < river, a reduction in slope may reduce stream power and transport capacity, leading to deposition. SL1, initial sea level; SL2, sea level datum at lowered base level

Commonly, the gradient of the lower reaches of the river is less than the gradient of the coeval shelf and a relative sea-level fall will impose a steeper gradient on the graded profile of the river. In such cases stream power and discharge may increase with a resultant increase in the potential for erosion and fluvial incision (Fig. 7.5a). The river system may attempt initially to reach a new equilibrium profile by changing its load characteristics, sinuosity and in-channel bedform styles. However, if the imposed gradient is greater than can be accommodated by an increase in channel width or by a changing channel pattern, the river will tend to incise in order to reduce the slope. Rivers of all scales commonly show a characteristic evolution in response to base level falls, in which they first narrow and straighten, then deepen and finally widen (Schumm, 1981). In theory the incision is propagated upstream as a nick point, and the whole river profile eventually lowers to reach an equilibrium with the new base level (Begin *et al.*, 1981; Posamentier and Vail, 1988).

It has been suggested that the gradients of the incised river are approximately double that prior to the base level fall (Salter, 1993). This change increases stream power, allowing the river to further incise unless the increased energy of the river can be dissipated by the erosion of channel banks, the widening of the river, and increasing roughness of the channel floor or changing sediment load characteristics, such as coarsening grain size. This perhaps explains why the width of incised valleys is often much greater than the river channels within the valley fills. It further suggests that wider channel belts with coarse bed-load will be a logical outcome of a base level fall. These characteristics have been noted in a number of Quaternary and ancient examples (Marzo *et al.*, 1988; Eschard, 1989; Van Wagoner *et al.*, 1990). The greater the slope created by the base level fall, the greater the slope imposed on the fluvial system and the greater the degree of incision. Incision is therefore particularly significant for type 1 sequence boundaries (Posamentier and Vail, 1988; Van Wagoner *et al.*, 1990; Wood, 1991; Wood *et al.*, 1991; Westcott, 1993).

In practice, rivers often appear not to fully equilibrate with the lowered base level, or the incision propagates only a finite distance upstream before a subsequent base level rise induces aggradation. In the absence of significant base level fall, rivers are more likely to adjust their profiles by changing their channel pattern, discharge and load characteristics rather than incising major valleys into the shelf (Suter and Berryhill, 1985; Blum, 1990; Autin *et al.*, 1991; Schumm and Ethridge, 1991; Westcott, 1993; Koss *et al.*, 1994). Sometimes base level fall may be marked not by erosion and valley incision but by a more subtle boundary separating fluvial deposits with distinctly different fluvial architectures (Shanley and McCabe, 1993; Westcott, 1993).

The impact of base level rise (relative sea-level rise at the shoreline) on the fluvial system is equally complex. Base level rise may impose a lower gradient on the lower reaches of the river system, reduce stream power and discharge and decrease the capacity of the river to maintain sediment transport. In these cases the river responds by depositing sediment. Many workers have suggested that the rate of base level rise is particularly important in controlling how fluvial systems respond (Posamentier and Vail, 1988; Posamentier *et al.*, 1988; Shanley and McCabe, 1993, and references therein). Rapid base level rise may be considered analogous to damming the downstream portions of the river valley (Fig. 7.6a). In these cases the rate of rise will be greater than the rate of deposition, and the lower course of the fluvial system will be flooded. Flood-plain aggradation during shoreline transgression is limited. Upstream of the flooded area, the effect on the river channel may be limited, with it still remaining at grade.

Significant fluvial aggradation will occur mainly as the shoreline progrades basinward by deltaic extension. Signifi-

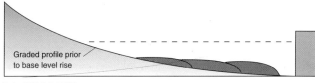

(a) STAGE ONE

- Initial rate of rise > rate of deposition, river course is flooded
- Upstream the river channel is essentially unaffected

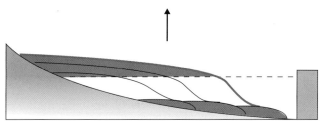

(b) STAGE TWO

- Rate of rise slows and deltas build out at the new (higher) base level
- Fluvial systems aggrade in response to imposed lower slope

Fig. 7.6 Fluvial system response to a base level rise. The response of the fluvial system will be affected by the rate of relative sea-level rise and the contrast in gradient between the shelf and the alluvial system. Response of the fluvial system to (a) rapid base-level rise and (b) slow base-level rise and/or high sediment supply rate (modified after Posamentier and Vail, 1988; published with the permission of the Society of Economic Paleontologists and Mineralogists)

cant fluvial aggradation will mainly occur as the shoreline progrades basinward by deltaic extension; *when the rate of sediment supply exceeds the rate of base level rise* (Fig. 7.6b; Posamentier and Vail, 1988; Shanley, 1991). In effect, the river system will deposit sediment in order to increase the slope and maintain grade with the upstream discharge and load. The net result may be a narrowing of the river channel belt, a change in channel style and a decrease in sand-body connectivity. Aggradation of the flood plain during rapid base level rise is therefore a function of the rate of rise and sediment supply.

Coastal and marginal marine processes (i.e. waves, tides and storms) may also affect the graded stream profile during transgression. In the Canterbury Plains of New Zealand, lowstand regional flood-plain deposits are presently being actively incised by braided river systems in an area currently undergoing transgression during a period of eustatic highstand (Leckie, 1994). Coastal erosion and cliff development during shoreline retreat has steepened fluvial gradients, resulting in incision. This example serves to further illustrate the potential difficulty in distinguishing cause and effect in ancient fluvial deposits. Changes in fluvial architecture related to downstream controls are summarised in Table 7.3 and discussed below in the context of the key stratal surfaces and systems tracts generated during a relative sea level cycle.

7.4.2 Sequence boundaries and lowstand systems tracts in alluvial strata

Sequence boundaries represent a surface along which sediment is bypassed during a relative sea-level fall. The boundary is coincident with subaerial exposure and erosion, and a basinward shift in facies belts and coastal onlap (Posamentier and Vail, 1988; Posamentier et al., 1988; Van Wagoner et al., 1990). In alluvial successions, the recognition of sequence boundaries is complicated by lateral facies changes and the difficulty of distinguishing major incision and abrupt basinward shifts in facies from localized channel scour (Posamentier, 1993; Posamentier and Weimer, 1993; Westcott, 1993). Where the gradient of the shelf exceeds the gradient of the fluvial system, significant incision may occur, leading to the formation of an incised valley (see also section 8.3). The fill of the incised valley will be a complex function of the rate of base level rise relative to sediment supply. Low rates of sedimentation are likely to lead to rapid drowning of the fluvial system and the formation of an estuary. High sediment supply rates will result in thicker fluvial, deltaic and tidal deposits (Eschard, 1989; Van Wagoner et al. 1990; Allen, 1991; Dalrymple et al., 1992; Shanley and McCabe, 1993; Shanley et al., 1993; Richards, 1994). Valleys abandoned by drainage capture may be filled entirely by marine mudstones (Wheeler et al., 1990). Within the more downstream portions of the fluvial systems, the abrupt vertical juxtaposition of coarse-grained fluvial deposits on marine or marginal marine strata may record the presence of a sequence boundary.

Some caution should be applied in interpreting all cases of stream rejuvenation as the product of relative sea-level fall (Posamentier, 1993). Valley incision additionally may occur in response to: (1) increased discharge and stream power, (2) decreases in stream load, or (3) from tectonic uplift within the mid-stream and upstream portion of the drainage basin. In all three cases fluvial incision results in sediment bypass and the development of localized unconformities.

Sequence boundaries developed within the interfluve areas adjacent to incised valleys are represented by a subaerial exposure surface recording periods of sediment bypass. They may be recorded by the juxtaposition of marine shales on thin emergent pedogenic horizons, or marine shales overlain by flood-plain and overbank fines. Associated pedogenic horizons and coals are developed but are often thin owing to the limited sediment accommodation available (Van Wagoner et al., 1990; and section 7.4.1).

Where valley incision is less clear, such as where the slope of the shelf is equal to or less than the fluvial gradient, the position of the sequence boundary is less easily interpreted. In these cases recognition of sequence boundaries must come from identifying systematic changes in channel stacking patterns; vertical variations in channel style and in the degree of sandstone amalgamation. These changes may be paralleled by variations in the grain size, abrupt changes in sediment composition and variations in the scale of sedimentary structures within the channel fills (Marzo et al., 1988; Eschard, 1989; Shanley, 1991).

The lowstand systems tract in the fluvial record may represent a period of renewed alluvial aggradation associated with the initial period of slow relative sea-level rise, following base level fall. Channel courses are confined initially to the central axis of the incised valley, resulting in repeated reworking of channel and flood-plain deposits (Fig. 7.7). At this stage, differences in channel styles may be less important in affecting the gross sand-body architecture of the lower valley fill than the dominant sediment load of the fluvial system. In effect, the depositional products of the bedload-dominated, low- and high-sinuosity channel systems may be identical because the lack of accommodation promotes reworking and bypass of fines and the limited differentiation between channel and flood-plain deposits. As a result, the lower valley fill may be dominated by high net:gross, multistorey–multilateral sand bodies.

Increasing rates of relative sea-level rise will promote increasing accommodation and an overall reduction in stream gradients. Low-sinuosity fluvial systems may respond by increasing their sinuosity to maintain grade, whereas high-sinuosity systems may become more effectively partitioned into channel and flood-plain components without a change in channel style.

The complex facies character and stratal patterns of lowstand fluvial deposits are illustrated in studies of the Blackhawk Formation of the Cretaceous Mesaverde Group, Book Cliffs, Utah (Van Wagoner et al., 1990). The formation represents a time interval of 6 million years and can be subdivided into three units; the Grassy, Castlegate and Desert Members. Four sequence boundaries are recognized within the formation, the first occurring at the top of the Grassy Member, a second and third within and bounding the Desert Member, and a fourth within the Castlegate Member (Fig. 7.8). The Desert Member lowstand systems tract consists of point-bar deposits with probable tidal influence (clay-draped lateral accretion surfaces), coals up to 30 cm thick and thin fluvial sheet sands. In places incision has resulted in coals lying directly on the Mancos Shale marine shelf facies, and meandering channel deposits erosively overlying lower shoreface sandstones. The succeeding Castlegate lowstand systems tract shows a downstream evolution in fluvial channel style, reflecting a basinward reduction in stream gradient following a relative sea-level fall. The most proximal portions of the system are represented by braided channel deposits, 50 m thick, which thin and pass down-dip into high-sinuosity channel and meandering point-bar systems down-dip. The downstream sinuous channel patterns can be traced up-system to a series of narrow separate valley fills with broad interfluves over a total distance of approximately 100 km.

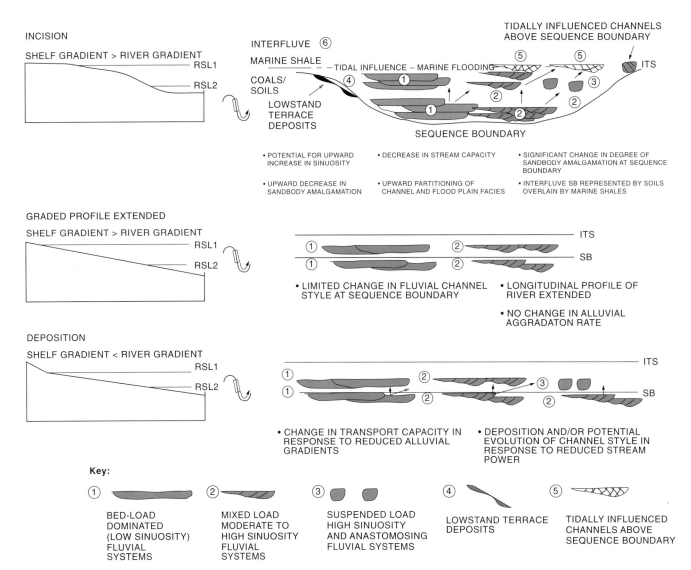

Fig. 7.7 Conceptual models illustrating the potential evolution of contrasting types of fluvial system in response to relative sea-level fall. In this example, the impact of reductions in accommodation space leads to changes in the character of the fluvial system, which may or may not include incision depending on the relationship between shelf and fluvial gradients. In all cases changes in architecture or stream type are assumed to reflect changes in accommodation space in response to relative sea-level fluctuations. No consideration has been given to the potential for stream capture or changes in upstream and mid-stream controls. ITS, initial transgressive surface or equivalent alluvial surface; SB, sequence boundary or equivalent alluvial surface

7.4.3 Transgressive systems tracts and flooding surfaces in alluvial strata

In shoreline and shelf successions, the transgressive systems tract occupies the interval between the transgressive surface and the maximum flooding surface (see Chapter 2 for definition). The identification of a transgressive systems tract in alluvial successions is less clear because neither of these stratal surfaces can be recognized in non-marine sections. Instead, the transgressive systems tract may be recognized only through changes in fluvial channel stacking pattern related to rapid base-level rise and shoreline transgression (Fig. 7.9).

Following a type I sequence boundary, the fluvial deposits of the lowstand and transgressive systems tracts can be distinguished by the identification of the first major marine flooding surface over the interfluves to the incised valleys. In other instances, the distinction between the fluvial deposits of the lowstand and transgressive systems tracts may not be clear. This may be the case where the initial flooding and marine onlap of the sequence boundary occurs further downstream of all, or part, of the major valley incision. Where shelf and fluvial gradients are similar, the initial deposits overlying the sequence boundary represent a depositional episode post-dating the initial transgressive surface, and true lowstand deposits will be absent. For this reason, Shanley (1991) coined the term *alluvial transgressive* deposits to encompass the complete fill of the

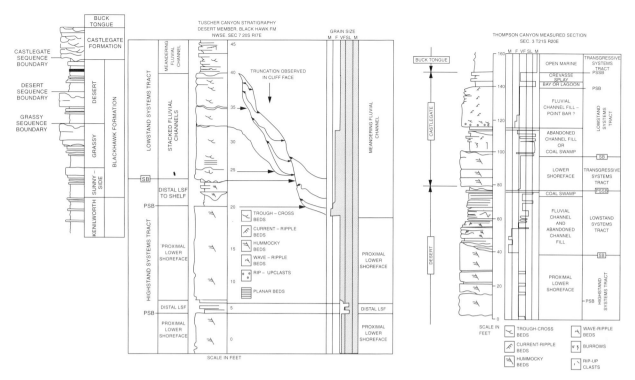

Fig. 7.8 Summary stratigraphy and selected vertical sections illustrating the range of facies juxtapositions associated with sequence boundaries in the Blackhawk Formation, Book Cliffs (after Van Wagoner, 1992). Inset stratigraphy illustrating sequence boundaries associated with the Castlegate, Desert and Grassy Members. Sections are 20 miles apart. Sequence boundaries represented by SB, possible sequence boundaries by PSB (published with the permission of the American Association of Petroleum Geologists)

incised valley. Use of this term can avoid misapplication of systems tract terminology in areas where regional stratal relationships are not apparent.

The fluvial deposits of the transgressive systems tract reflect the balance between sediment supply and the rate of relative sea-level rise. Where sediment supply is relatively low only a thin veneer of fluvial sediment may be preserved as the sea transgresses rapidly landwards. Where the rate of transgression is lower, the fluvial record of a transgressive systems tract may include the increased partitioning of channel and flood-plain deposits in vertical section, the decreasing connectivity of channel meander-belt sands, and the increasing influence of tidal processes in the lower reaches of the fluvial system (Allen, 1991; Shanley, 1991; Shanley and McCabe, 1993). Reduction of fluvial gradients may promote an evolution in channel styles to more mixed load, meandering or anastomosing rivers, and the increasing representation of crevasse deposits in the stratigraphic record as rivers attempt to maintain grade with rising base level (Ryseth, 1989; Kirschbaum and McCabe, 1992; Tornqvist, 1993).

The period of maximum flooding may be represented by evidence for subtle or pronounced tidal influence within the fluvial system (Shanley and McCabe, 1991). Poorly drained palaeosols or thick coals may occur. Relatively thick clastic and carbonate lacustrine facies reflecting rising water tables and incursions into poorly drained flood plains also may be developed (Ryer, 1981; Atkinson, 1986).

7.4.4 Highstand systems tracts in alluvial strata

The highstand systems tract represents the period of decreasing rates of relative sea-level rise. In shoreline and shelf strata the interval is bounded at its base by a maximum flooding surface and at its top by a sequence boundary. Early highstand fluvial deposits may be indistinguishable from the facies of the late transgressive systems tract. During the late highstand, reductions in flood basin accommodation may result in the increased potential for lateral rather than vertical accretion, and the formation of laterally interconnected and amalgamated channel and meander belt systems with poorly preserved flood-plain deposits (Shanley and McCabe, 1993; Fig. 7.10). In all cases the fluvial depositional record of the highstand will be affected significantly by the rate of sediment supply to the system, the character of the fluvial system during base level rise, and the extent to which the highstand deposits are eroded during a subsequent relative sea-level fall.

The broad links between a relative sea level cycle and its impact on fluvial and shoreface architectural patterns are schematically represented in Fig. 7.11 (from Shanley and McCabe, 1994).

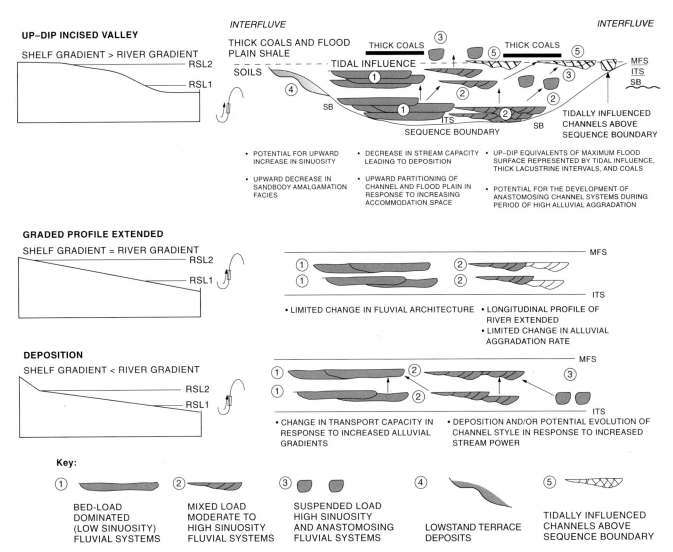

Fig. 7.9 Conceptual models illustrating the potential evolution of contrasting types of fluvial system in response to relative sea-level rise. In this example, the impact of increasing accommodation space leads to changes in the character of the fluvial system and the increased potential for alluvial aggradation, depending on the relationship between shelf and fluvial gradients. In all cases changes in architecture or stream type are assumed to reflect changes in accommodation space in response to relative sea-level fluctuations. Rates of sediment flux are assumed to balance rates of relative sea-level rise. MFS, maximum flooding surface; ITS, initial transgressive surface or equivalent alluvial surface

7.5 Reconstructing fluvial architecture

Reconstructing the architecture of alluvial successions requires extensive, three-dimensional outcrops and the ability to correlate information from widely spaced sample points (such as road side outcrops, cores and logs). Conventional methods of correlation are less easily applied to alluvial successions owing to lateral facies variability and the presence of multiple erosion surfaces. The lack of diagnostic fauna and flora within channel and flood-plain deposits, and the absence of anomalous facies transitions also hamper the definition of stratigraphic boundaries. Successful attempts at reconstructing alluvial architecture have been achieved with magneto-stratigraphy (Behrensmeyer and Tauxe, 1982; Johnson et al., 1985, 1988), the correlation of tuff layers (Allen and Williams, 1982), vertebrate taphonomy (Behrensmeyer, 1987) and the lithostratigraphic correlation of coals and palaeosols (Ryer et al., 1980; Ryer, 1981). The value of palaeosols as a correlation tool has long been recognized and is of particular interest because soil processes are intrinsic to fluvial environments (Allen, 1974; Johnson, 1977; Bown and Kraus, 1981; Kraus, 1987).

In the absence of these types of data, stratigraphic correlation in fluvial deposits should emphasize anomalous or systematic changes in facies, channel styles and sand-body stacking patterns in defining the position of key bounding surfaces. Once recognized, these surfaces can provide the means of reconstructing the three-dimensional architecture of the fluvial system, as illustrated in the following examples. It is often unclear whether changes in alluvial architecture reflect changes in local base level, tectonic subsidence or

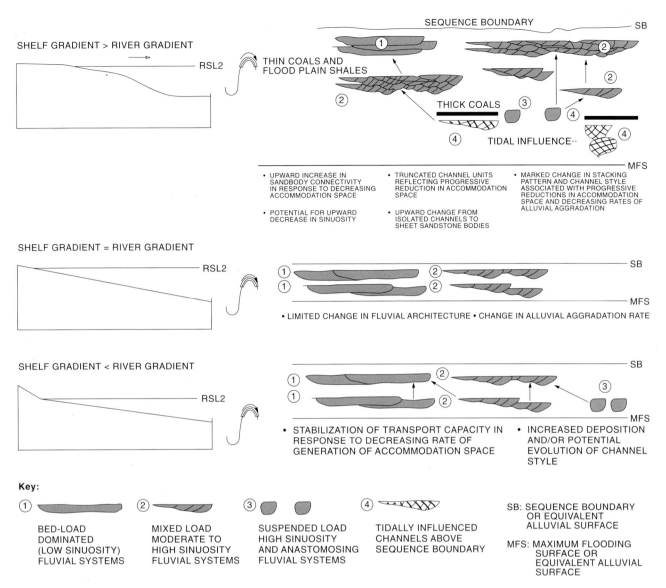

Fig. 7.10 Conceptual models of evolving alluvial styles during highstand. Period characterized by a reduction in the rate of creation of accommodation space. In this example, the impact of decreasing accommodation space leads to changes in the character of the fluvial system and the decreased potential for alluvial aggradation, depending on the relationship between shelf and fluvial gradients. Rates of sediment flux are assumed to exceed rates of relative sea-level rise during the late highstand. SB, sequence boundary or equivalent alluvial surface; MFS, maximum flooding surface or equivalent alluvial surface

sea-level. However, systematic changes in the architecture of alluvial successions clearly reflect changes in accommodation however formed, and the basic principles of sequence stratigraphic models still apply.

7.5.1 Eocene Castissent Formation, South Pyrenees, Spain

The Eocene Castissent Formation of the Tremp-Graus Basin illustrates fluvial and deltaic deposition in a small thrust-top basin in the Southern Pyrenees Foreland (Marzo et al., 1988). In this area, fluvial and deltaic deposition throughout the Eocene was strongly controlled by the emplacement of Pyrenean thrust sheets (Seguret, 1972; Nijman and Nio, 1975; Ori and Friend, 1984). The Castissent formation comprises a series of coarse-grained multi-lateral, multi-storey fluvial sandstones, sandwiched between two pronounced marine to brackish mudstone intervals of Ypresian age (Nijman and Nio, 1975). The contact between the fluvial sandstones and the underlying brackish-marine mudstones is unconformable and represents a sequence boundary (Marzo et al., 1988). Detailed mapping of the sequence boundary shows that the Castissent Formation forms an incised valley-fill and upper deltaic plain complex dominated by three multilateral and multistorey sandstone bodies (Marzo et al., 1988). These sand bodies are separated by thicker developments of finer grained flood-plain and brackish water deposits reflecting periods of enhanced flood-plain aggradation and sustained marine inundation (Fig. 7.12).

Lateral variations in sand-body architecture and fluvial

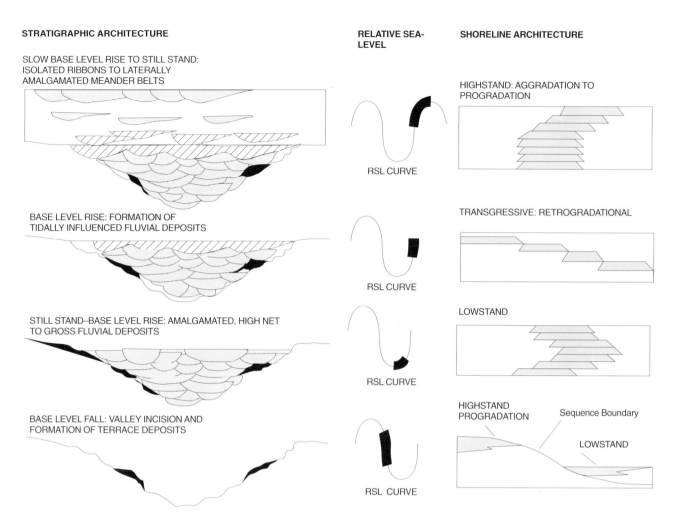

Fig. 7.11 Summary diagram illustrating the relationship between shoreface and fluvial architecture as a function of base level change (after Shanley and McCabe, 1993). Detailed correlations suggest the timing of the incised-valley fill occurred after the initial transgressive surface. For this reason Shanley and McCabe (1993) regard the valley-fill as 'alluvial-transgressive' deposits

style are best illustrated in the lowest sand-body complex (Complex A of Marzo et al., 1988). The upstream valley-fill is dominated by a braided network of channels showing a high degree of vertical and lateral amalgamation to form a sheet-sandstone complex. Correlative flood-plain deposits form widespread red beds with caliche soils, suggesting that most of the fines bypassed the alluvial area and the low rates of sedimentation on the adjacent flood plains promoted oxidation and soil formation. The valley-fill sandstones can be traced down-system into mixed-load meandering channel deposits. This down-dip change in channel style is accompanied by reduced incision, a decrease in channel interconnectedness and the increasing preservation of flood-plain deposits, due to high rates of flood-plain aggradation. Lateral changes in fluvial architecture in the Castissent Formation are therefore interpreted to reflect reduced fluvial gradients from the hinterland to the margins of the marine basin.

Systematic vertical changes in alluvial sand-body architecture also occur within the Castissent Formation, and are interpreted to reflect relative sea-level fluctuations (sand-bodies A1-A3, B1-B2 in Fig. 7.12). These changes are particularly well developed in the proximal valley-fill of the alluvial system, where they form repeated erosional–aggradational cycles (Marzo et al., 1988). Each cycle commences with a phase of valley incision into the underlying

Fig. 7.12 (*opposite*.) Summary diagram illustrating the geometry and correlation of sand bodies within the Castissent Formation, Eocene, Pyrenean Basin (after Marzo et al., 1988). Note the evolution of fluvial channel styles from coarse-grained sheet-sand bodies generated by low sinuosity systems to more tabular or ribbon sand bodies of meandering channel patterns. Individual channel cycles are separated by periods of flood-plain aggradation, palaeosol formation and down-dip equivalent marine incursions. Inset figure illustrates the main depositional facies and stratigraphic context of the Castissent Formation. Formation forms as incised-valley fill sandwiched between two marine flooding events and their up-dip equivalents. Key: 1, Castissent Formation (in black); 2, fluvial and upper delta plain; 3, lower delta plain; 4, fan-delta; 5, delta front; 6, slope and turbidites (Hecho Group)

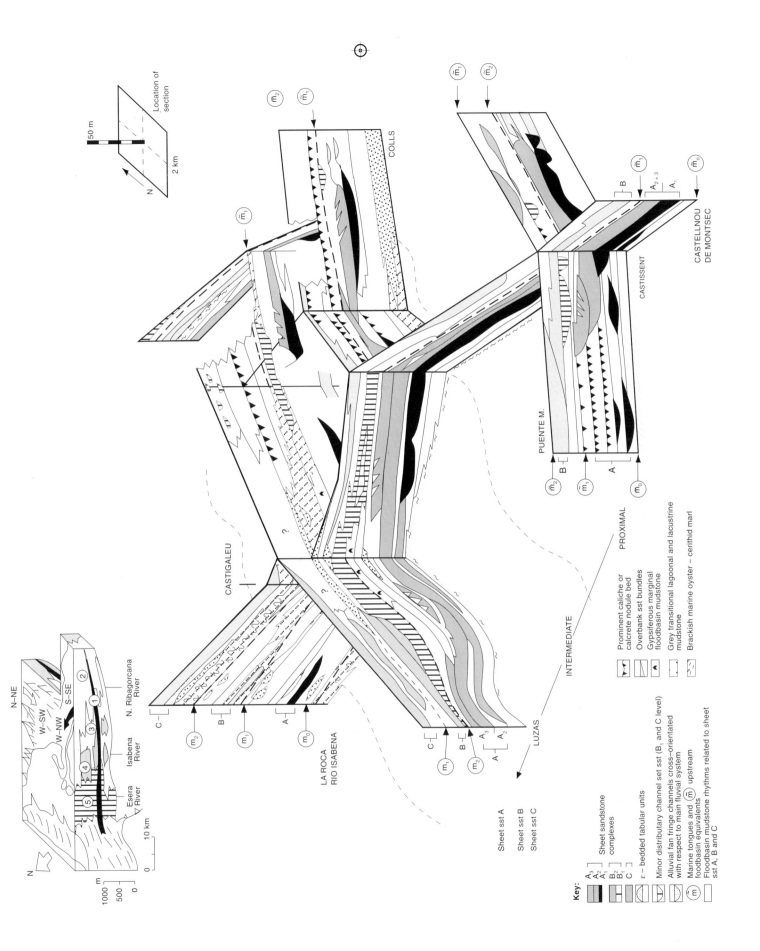

flood plain or channel deposits. Deposition within the valley is accomplished by vertical aggradation and lateral switching of braided channel networks resulting in highly connected sandstone bodies. The valley-fill deposits evolve from braided to more mixed-load, high-sinuosity channel systems and associated flood-plain deposits. This evolution in channel style is accompanied by a decrease in sand-body interconnectedness and increasing preservation of overbank and floodplain deposits in the succession. This erosional–aggradational cycle is repeated three times within each sandstone complex, with successively later phases of incision cutting down to the previous channel deposits to form progressively wider valley margins. Evidence for sustained, periodic flooding of the flood plain is manifested in the formation of hydromorphic soils and brackish water marls. Multiple phases of alluvial aggradation and degradation indicate the operation of higher frequency allocyclic controls. The upward change in alluvial architecture can be interpreted in terms of a lowstand incised-valley fill, and downstream flood plain, overlain by fluvial channel and floodplain deposits within transgressive and highstand system tracts.

7.5.2 Triassic, Ivishak Formation, Prudhoe Bay Field, North Slope, Alaska, USA

Coarse-grained alluvial and fluvio-deltaic deposits of the Triassic Ivishak Formation form the principal reservoir facies of the Prudhoe Bay Field, North Slope, Alaska. The field is the largest oil and gas field in the USA, with original in-place reserves of about 22 billion barrels of oil and 42 trillion cubic feet of gas (Morgridge and Smith, 1972; Jones and Speers, 1976; McGowen and Bloch, 1985; Atkinson *et al.*, 1988). The Ivishak Formation forms part of the Sadlerochit Group, a clastic wedge of Permo-Triassic age that unconformably overlies carbonates of the Mississippian and Pennsylvanian Lisburne Group. The Ivishak Formation is sandwiched between shelf mudstones of the Kavik Formation and shelfal and phosphatic carbonates of the Shublik Formation (Fig. 7.13). The contact with the Kavik Formation is considered conformable whereas the upper boundary with the Shublik Formation is sharp and associated with a phosphatic and pyritic pebbly lag. This boundary is interpreted as a disconformity developed as a result of shoreface reworking of the Ivishak Formation during transgression.

The Ivishak Formation varies from 400 to 700 ft thick across the oil-field and is generally subdivided into two large-scale depositional cycles; a lower coarsening upward, fluvio-deltaic and fluvial interval of largely progradational character, dominated by mudstones, siltstones, sandstones and conglomerates, and an upper aggradational to retrogradational cycle dominated by finer grained fluvial sandstones and shales (Atkinson *et al.*, 1988, 1990). Previous studies have suggested that the formation was deposited by a prograding fluvio-deltaic complex related to a braided-river-dominated fan delta or coastal braid plain (Jamieson *et al.*, 1980; Melvin and Knight, 1984; McGowen and Bloch, 1985; Lawton *et al.*, 1987).

Within the Prudoe Bay field the Ivishak reservoir has historically been subdivided into four zones on lithostratigraphic and petrophysical grounds (zones 1–4, Atkinson *et al.*, 1988). Recent studies demonstrate the complex stratigraphic nature of the Ivishak Formation (Richards *et al.*, 1994). The Sadlerochit Group can be subdivided into seven sequences based on detailed facies analysis of core data, sequence stratigraphic correlations of over 1300 wells and parallel evaluation of dip-meter and three-dimensional seismic data. Sequence boundaries are recognized by abrupt changes of facies associations in cores, systematic variations in the internal sedimentary structures, thickness and the degree of amalgamation of channel-fill deposits. Four orders of sequences are recognized and referred to here as first-order, second-order, etc. In all cases, sequence boundaries can be tied to the maximum progradation point of lower order sequences. Note that the term order is used here for ease of discussion, and no inferences should be made concerning the duration of the sequence as defined in section 2.2.4 (Fig. 7.13).

First-order sequence

At the largest scale, the Sadlerochit Group forms part of a first-order lowstand delta and fluvial braid-plain complex developed in response to a major relative sea-level fall. The base of the sequence conforms to the sequence boundary at the top of the Lisburne Group. The top of the sequence is represented by the major sequence boundary separating the Triassic Sag River Formation from the overlying Jurassic Kingak Formation (Hubbard *et al.*, 1987). The earliest fluvio-deltaic deposits are predominantly progradational in character, suggesting sediment supply exceeded the rate of creation of accommodation. Overlying alluvial deposits dominate the remainder of the Ivishak Formation to form a thick, aggradational interval of fluvial and flood-plain deposits. These topset deposits show an evolution with time from mixed-load fluvial and deltaic systems, to coarser conglomerate and sand-dominated fluvial systems. Gross changes in fluvial channel style and stacking pattern reflect an increasing rate of creation of accommodation over time consistent with a relative sea-level rise. The end of Ivishak deposition is marked by a gradual fining of alluvial deposits and the eventual reworking of fluvial sandstones during marine transgression, leading to eventual establishment of Shublik Formation carbonate and Sag River Formation clastic shelf facies (Fig. 7.13).

Second-order sequences

The first order sequence can be divided further into two second-order sequences by a major sequence boundary defined as the '6000' surface. The 6000 sequence boundary is a seismically mappable event (within the limits of seismic resolution), and marks a major change in the sand:shale

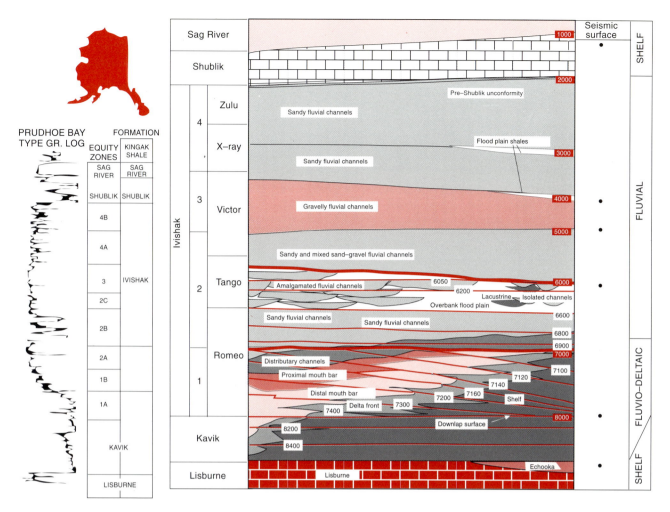

Fig. 7.13 Reservoir sedimentology, stratigraphy and equity subdivisions of the Ivishak Formation, Sadlerochit Group, Prudhoe Bay Field, North Slope. The stratigraphic subdivision recognizes a hierarchy of key stratal surfaces including (a) sequence boundaries (1000 series surfaces), (b) parasequence set boundaries or their up-dip equivalents (100 series surfaces), and (c) parasequence boundaries (intra-100-series surfaces) (the authors gratefully acknowledge the permission of the Prudhoe Bay Owners in the publication of this work by the Joint BPX–Arco–Exxon RAZOR team)

ratio of the succession. It also defines the upper boundary to a suite of early Triassic faults, which are confined to the base of the Prudhoe Bay reservoir. Palaeocurrent analysis from core-calibrated dip-meter data also shows that the sequence boundary marks a major switch in the palaeoflow direction of Ivishak river systems from NW–SE below the 6000 surface to N–S above. The lower second-order sequence displays a systematic change in architectural style, from predominantly progradational to aggradational to deltaic and fluvio-deltaic deposits consistent with increasing rates of relative sea-level rise. The succession above the 6000 surface is fluvially dominated, and it is unclear whether the initial deposits above the sequence boundary represent the depositional products of fluvial systems developed within lowstand or alluvial transgressive systems tracts. Succeeding fluvial deposits show an overall aggradational to retrogradational character consistent with constant or declining sediment flux associated with increasing accommodation.

Third-order sequences

Three third-order sequences are recognized within the succession based upon the recognition of basinward shifts in facies belts and changes in the stacking patterns in both marginal marine and fluvial deposits. Only the upper two sequences will be discussed here (Fig. 7.14).

Sequence 2

The sequence is defined at its base by the 8000 surface and at its top by the second-order 6000 sequence boundary. The earliest depositional units of the sequence comprise mixed-load to sandy deltaic systems (7400–7100), which prograded from northwest to southeast into the Kavik sea. The earliest deltaic lobes are progradational in character, whereas later deltaic episodes show increased evidence of aggradation and the increasing preservation of fluvial topset deposits. These changes in deltaic style, combined with the increasing thickness and continuity of deltaic and flood-plain shales favour increasing sediment accommodation

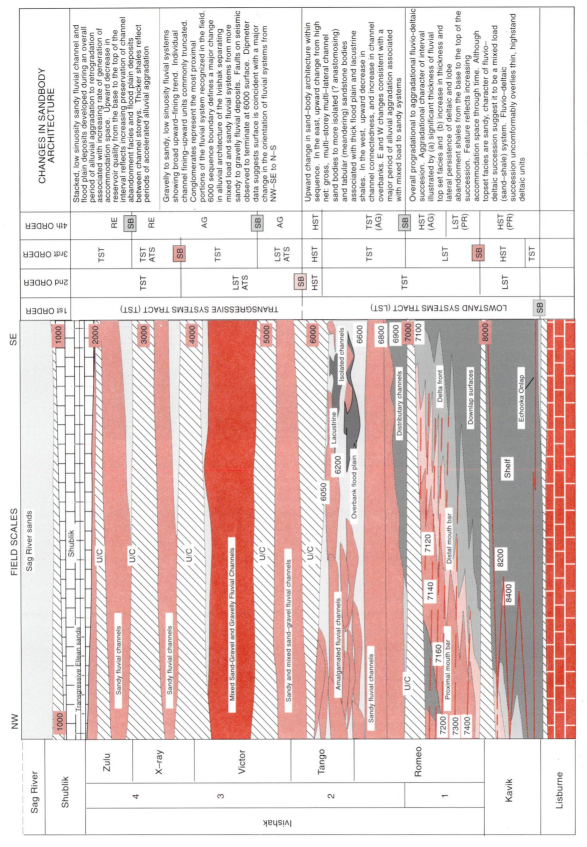

Fig. 7.14 Summary sequence stratigraphy of the Triassic Ivishak Formation and relationship to fluvial style and stacking pattern (the authors gratefully acknowledge the permission of the Prudhoe Bay Owners in the publication of this work by the Joint BPX–Arco–Exxon RAZOR Team). LST, lowstand systems tract; TST, transgressive systems tract; SB, sequence boundary; ATS, alluvial transgressive tract; HST, highstand systems tract; PR, progradational stacking pattern; AG, aggradational stacking pattern; RE, retrogradational stacking pattern

following a period of lowstand. The maximum progradation point of the sequence corresponds to a third-order sequence boundary (7000 surface) separating fluvio-deltaic from alluvial deposits. The boundary is recognized in core data within the southern portion of the field by the abrupt vertical juxtaposition of distributary channel facies on distal mouth-bar and lower delta-front deposits. In the northwest of the field the same boundary is defined by a change in channel stacking pattern from distributary channel and delta-plain channels to sandy fluvial systems. The boundary can be traced throughout the field and corresponds to a shift in gamma-ray log response and a change in petrophysical properties.

In the eastern portion of the field, relatively thick floodplain and lacustrine shale complexes dominate the upper part of the sequence, where they form laterally extensive intra-reservoir seals. These shales contain poor to moderately well-developed pedogenic horizons, which can be correlated between wells (Atkinson et al., 1988). Within the northeastern parts of the field, channel-fill deposits associated with lacustrine shales form isolated fining upward units with low width-to-depth ratios ($<1:200$). These channel types evolved from amalgamated tabular sand bodies within lower parts of the sequence. The overall width to depth characteristics of these channel systems together with their close association with thick lacustrine and flood-plain deposits favours an overall evolution from moderate sinuosity to anastomosing channel styles associated with very low alluvial gradients and periodic floodplain flooding. Towards the southeast and central portions of the field these isolated channel sand bodies give way to more tabular, amalgamated channel storeys with higher width-to-depth ratios ($>1:200$) associated with floodplain rather than lacustrine shales. These facies compare favourably with the deposits of amalgamated sandy to mixed-load, moderate-sinuosity channel systems developed within the central part of the sequence. Both types of channel deposits can be traced to the northwest, where they give way to multistorey fluvial channel sand bodies. The increasing development of thick lacustrine and floodplain shales in the east of the field is paralleled by a decrease in the degree of channel sand-body amalgamation in the west. This may reflect increasing rates of creation of accommodation through time. The scale and extent of the shales towards the top of sequence 2 may further suggest that they represent the fluvial equivalents of a maximum flooding surface in marine strata.

Rates of alluvial aggradation decreased during the final stages of sequence development, resulting, in the west of the field, in an upward increase in the connectivity and amalgamation of channel sand bodies to form welldeveloped fluvial sand-sheets. This change in architectural style is less transitional in the eastern part of the field, and isolated channel sand-bodies are abruptly overlain by stacked, multistorey channel sand-bodies dominated by incomplete fining upward intervals and a lack of floodplain-abandonment fines (Atkinson et al., 1988; Richards et al., 1994). The major change in stacking pattern and degree of sandstone amalgamation marks the 6000 sequence boundary.

Sequence 3

The final third-order sequence commences with the 6000 sequence boundary and terminates in close proximity to the Pre-Shublik unconformity. The exact position of the upper boundary of the sequence is unclear in view of the fluvially dominated nature of the succession and the evidence for additional sequence boundaries within the succession. The sequence exhibits an overall progradational to aggradational style. Sandy and mixed sand and gravelly fluvial deposits dominate the base of the sequence, where they form a well-developed sheet sandstone covering the entire extent of the field. The deposits give way abruptly to coarse-grained fluvial channel conglomerates and conglomeratic sandstones. The junction between the conglomerates and underlying sandstones has been interpreted traditionally as a sequence boundary based on the abrupt change in grain size (Atkinson et al., 1988). Although this clearly represents a period of fluvial incision, the significance of the surface may have been overstated in the past because similar coarse-grained facies occur within fluvial deposits underlying the boundary. This suggests that the magnitude of facies dislocation is limited and that it represents a higher order boundary. The coarse-grained fluvial deposits are dominated by erosive-based, multistorey conglomerate bodies up to 10 ft thick, characterized by normally graded, horizontally bedded and cross-stratified conglomerates. Overbank fines are rarely preserved within the section.

The upper part of the sequence is dominated by finer grained, fluvial-channel sandstones, which form a broadly upward-fining succession related to increasing sediment accommodation. These deposits reflect the products of sandy, distal coastal plain environments (Atkinson et al., 1988). Within the western and eastern part of the field, relatively thick shales occur as fine-grained flood plain and lacustrine intervals of similar character to the shales of sequence 2. The widespread nature and thickness of these deposits may reflect periods of reduced alluvial gradients and enhanced flood-plain aggradation associated with rising base level. The top of the sequence may pass abruptly into the carbonates of the Shublik Formation or be overlain by thin, transgressive shelf sands of the Eileen Formation.

In summary, the stratigraphic record of the Ivishak Formation illustrates the complex response of alluvial and fluvio-deltaic systems to changes in sediment accommodation. Recognition of sequence boundaries and the updip equivalents of flooding surfaces is complicated by the lateral and vertical variability of fluvial and deltaic deposits within the succession. Changes in sediment accommodation are reflected at a variety of scales, from the evolving patterns of channel sand-body amalgamation and shale development to changes in channel pattern and style as the fluvial depositional systems attempted to achieve equilibrium.

CHAPTER EIGHT

Paralic Successions

8.1 Introduction

8.2 Paralic depositional systems
    8.2.1 Deltas
    8.2.2 Delta physiography
    8.2.3 Deltas and sedimentary processes
    8.2.4 Coastal plain to shoreline–shelf systems
    8.2.5 Estuaries

8.3 Sequences in paralic successions
    8.3.1 Sequence boundaries and valley incision
    8.3.2 Interfluve sequence boundaries
    8.3.3 The transgressive surface
    8.3.4 Forced regressions
    8.3.5 The maximum flooding surface

8.4 Parasequences in paralic successions
    8.4.1 Sedimentary processes
    8.4.2 Accommodation space
    8.4.3 Grain size
    8.4.4 Climate

8.5 The sequence stratigraphy of distinct paralic systems
    8.5.1 The stratigraphy of coastal plain to shoreline–shelf systems
    8.5.2 The stratigraphy of deltaic systems
    8.5.3 The stratigraphy of estuarine systems

8.6 Correlation procedure
    8.6.1 Depositional environments
    8.6.2 Parasequence correlation
    8.6.3 Progradation, aggradation, and retrogradation
    8.6.4 Sequence boundaries

8.7 An example: the Viking Formation, Western Canadian Basin

8.8 Reservoirs in paralic successions
    8.8.1 Stratigraphic traps
    8.8.2 Seals
    8.8.3 Analogues and flow units

8.9 Paralic systems at a seismic scale
    8.9.1 Seismic-scale models
    8.9.2 Seismic resolution
    8.9.3 Seismic facies

8.10 Variations in paralic systems within a sea-level cycle
    8.10.1 Shelf processes
    8.10.2 Spatial changes in processes
    8.10.3 Sequence stratigraphic framework

8.11 Summary

8.1 Introduction

Paralic successions include a wide range of environments — deltas, coastal plains, shoreline–shelf systems and estuaries — each deposited at or near to sea-level (Table 8.1). As such they are extremely sensitive to changes in relative sea-level and are therefore particularly suitable for high-resolution sequence stratigraphic analysis.

The chapter commences with a summary of different paralic depositional systems, and a discussion of high-resolution sequence stratigraphy. It proceeds by showing (i) how stratigraphic signatures vary in distinct paralic environments; and (ii) how sequence stratigraphy can be applied to paralic petroleum reservoirs. The chapter concludes with an analysis of the seismic expression of paralic successions, and how paralic successions may vary during a sea-level cycle.

8.2 Paralic depositional systems

8.2.1 Deltas

A delta is the prism of sediment that accumulates where a river enters a standing body of water. Deltas comprise a

Table 8.1 Environments and subenvironments encountered in paralic successions

Environments
Deltas
Coastal plain to shoreline–shelf systems
Estuaries
Incised valleys

Subenvironments
Distributary channels
Distributary mouth bars
Crevasse channels
Crevasse splays
Levees
Lagoons
Lakes
Cheniers
Beaches
Shoreline–shelf complexes
Tidal ridges
Tidal channels
Flood and ebb tidal deltas
Barrier islands
Bay-head deltas
Tidal flats, etc.

subaerial portion and a subaqueous portion. The subaerial portion, the delta plain, can be commonly divided into two parts: the upper delta plain, which is fluvially dominated, and the lower delta plain, which is marine influenced, particularly by tides (Fig. 8.1). The following two sections discuss variations in delta physiography, and the influence of sedimentary processes on deltaic systems.

8.2.2 Delta physiography

Shelf-edge deltas (or deep-water deltas)

Shelf-edge deltas are located at the shelf–slope break. They pass directly into an extensive coeval slope (with an angle of 2–5°) and deep-water sedimentation system (Fig. 8.2a). Shelf-edge deltas commonly exhibit a range of syn-sedimentary deformational features, such as growth faults, slides and mud diapirs. These features are gravity driven,

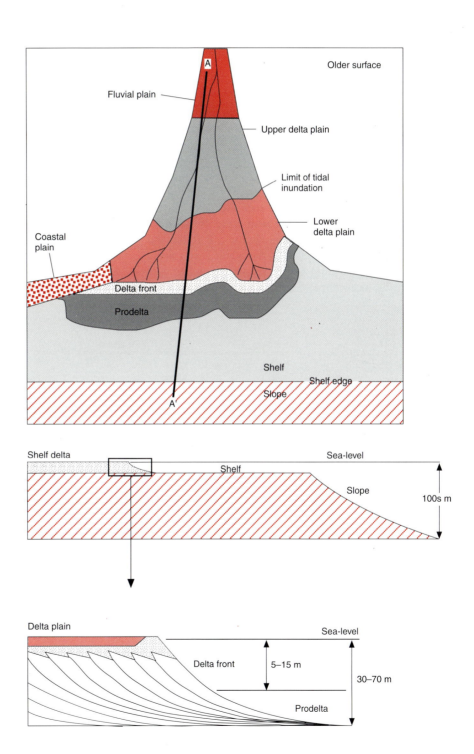

Fig. 8.1 Delta terminology (after Coleman and Prior, 1982)

and result from the rapid accumulation of a thick pile of soft sediment fronted by an unconstrained slope.

Shelf deltas (or shoal-water deltas)

Shelf deltas develop in shallow water depths, commonly 30–70 m, landwards of the shelf–slope break. The subaqueous portion of shelf deltas comprises a relatively coarse-grained, steep part (angles of 1–2°), known as the delta front, and a fringing muddy, lower angle section (angles < 0.5°), known as the prodelta (Fig. 8.2b). Shelf deltas lack large coeval slope and deep water systems, and, in general, are devoid of large-scale soft-sediment deformation features.

Gilbert deltas

Gilbert deltas are coarse-grained fan-deltas characterized by steep delta foresets (> 20°) that are dominated by sediment gravity flow processes (Fig. 8.2c; Colella, 1988; Braga *et al.*, 1990). Foresets range from subseismic to seismic in scale (up to 700 m; Ori, 1987). Gilbert deltas are common in rift and strike-slip settings, where subsidence and uplift combine to generate the requisite basin-margin water depth and alluvial fan catchment areas. If the water depth is too shallow, rapid progradation will result and steep foresets may not develop. If the water is too deep, sediment may not accumulate as a subaerial fan, but pass directly to the basin floor to accumulate as a submarine deposit. Fault-induced slumping may characterize the early portions of Gilbert deltas while silty foresets up to 10 m thick tend to increase in thickness and abundance basinwards. The foresets are commonly inversely graded.

8.2.3 Deltas and sedimentary processes

Fluvially dominated deltas

Deltas with a strong fluvial signature in the delta front result from high sediment input and relatively low-energy shelf processes. They are commonly characterized by a sheet-like sediment body comprising coalesced mouth bars. The core of each mouth bar, adjacent to the distributary channel, is likely to be sandy. However, depending upon the calibre of the supplied load, sediment deposited away from the channel axis is likely to be increasingly mud-prone. As a result, a finger-like sandstone isopach develops in the lower delta plain and shallow delta front (Coleman and Prior, 1982). The delta plain of fluvially dominated deltas takes on the character of the fluvial system. Three main types are recognized: rivers, braidplains and alluvial fans, which feed river deltas, braid deltas and fan deltas respectively (Orton, 1988).

Wave- and storm-dominated deltas

Deltas characterized by strong wave energy have relatively straight coastlines, and shore-parallel sandstone isopachs

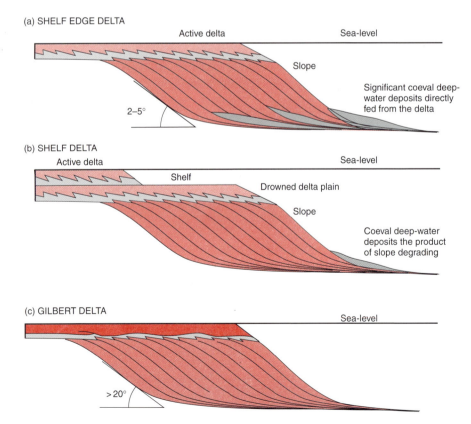

Fig. 8.2 Delta types: (a) shelf edge, (b) shelf, and (c) Gilbert deltas

skewed in the direction of prevailing longshore currents. Fairweather waves move sediment onshore and provide an effective barrier to its offshore transport. Storms redistribute the sand on to the inner shelf to increase the shore-perpendicular width of the coastal sands.

Tide-dominated deltas

Strong tidal currents produce numerous coeval channels at the delta front and result in a finger-like irregular sandstone isopach (Coleman and Prior, 1982). Depending on tidal strength, a tidal signature may penetrate deep into the delta plain, favouring the development of tidally influenced lagoons, tidal flats and creeks. The strongest tidal currents are likely to be in the upper portion of the delta front and in the lowermost delta plain, where daily reworking can produce clean-swept sandstones. The subaqueous delta may be extensive and grade into adjacent tidally influenced shelfal areas, providing one of the few fairweather mechanisms for offshore sand transport.

8.2.4 Coastal plain to shoreline–shelf systems

Coastal plain to shoreline–shelf systems lack major rivers. They receive the majority of their coarse sediment through longshore and along-shelf transport, largely from coeval deltas. Shelves are shallow, gently sloping, open-marine areas characterized by wave, storm, tidal and, rarely, oceanic currents. Shelves pass landwards into the shoreline, which comprises the shoreface and the beach. The shoreface is a steep, narrow zone (commonly less than 1 km wide), characterized by structures produced by shoaling waves (Shepard, 1960; Bernard and LeBlanc, 1965; Friedman and Sanders, 1978). The beach extends from the mean low-tide mark to a supratidal zone that is only periodically influenced by marine processes, for example during storms. The beach may be backed by aeolian dunes. Shoreline–shelf systems pass landwards into the coastal plain, which is commonly fronted by a lagoon. Small rivers that debouch into the lagoons form bay-head deltas.

8.2.5 Estuaries

Estuaries are drowned river valleys (Dalrymple *et al.*, 1992). They are characterized by sediment input from both fluvial and marine sources. At their heads, estuaries are characterized by fluvial input. Where fluvial input is strong, bay-head deltas may develop. Where fluvial input is weak and tidal currents relatively strong, fluvial channels become progressively tidally influenced and merge downstream into tidal channels. Estuary mouths range from tidal to wave-dominated, allowing estuaries to be divided into two broad groups (Dalrymple *et al.*, 1992).

In wave-dominated estuaries, along-shore and onshore sand transport results in the accumulation of a sandy plug at the head of the estuary (Fig. 8.3a). The sand plug commonly comprises two elements, (i) a barrier with associated washover deposits, and (ii) a tidal inlet that breaches the barrier allowing tidal exchange with the estuary and forming a flood tidal delta. Landwards of the sand plug, the central portion of the estuary is a low-energy zone characterized by muddy facies. Together, the sand plug, the central-basin muds and the bay-head delta form a tripartite subdivision of wave-dominated estuaries.

In contrast, strong tides ensure an active exchange between the estuary and the open sea, preventing the development of bay-head deltas, a muddy central basin or a discrete sand plug (Fig. 8.3b). Instead, in tide-dominated estuaries, a series of tidal sand bars ornamented by dunes are likely to pass up-system into sandflats, and then to tide-influenced and finally fully fluvial channels. Although a muddy central basin is not developed, an analogous zone of relatively low energy is recognized (Dalrymple *et al.*, 1992). It occurs landwards of the strongly flaring tide-dominated sandy estuary and is characterized by a systematic change in channel types 'straight–meandering–straight', reflecting changes in the balance between fluvial and tidal process. At the seaward end an outer straight section is tide dominated, with net headward sediment transport. At the landward end an inner straight portion is fluvially dominated with net seaward transport. In between, a zone of net bed-load convergence is characterized by tight meanders and fine grain sizes.

8.3 Sequences in paralic successions

8.3.1 Sequence boundaries and valley incision

In paralic successions sequence boundaries are recorded by basinwards or 'downwards' shifts in facies belts. Two important signatures are associated with base level fall: (i) valley incision and (ii) forced regression.

Valley incision

The response of fluvial systems to relative sea-level fall is discussed in detail in Chapter 7. If the new river course is steeper than the equilibrium river profile, the river will firstly straighten its coarse and then incise to form a valley. Incised valleys are common features in paralic successions (Van Wagoner *et al.*, 1990). They are important because they represent unequivocal evidence of a sequence boundary and also because they can form stratigraphic traps for hydrocarbons.

The recognition of incised valleys

With limited well or outcrop data it is often difficult to distinguish amongst (i) individual channels (e.g. fluvial or tidal distributary channels); (ii) multistorey channel sandstone bodies, and (iii) incised valleys. Both valleys and channels are sharp based and commonly filled by fining

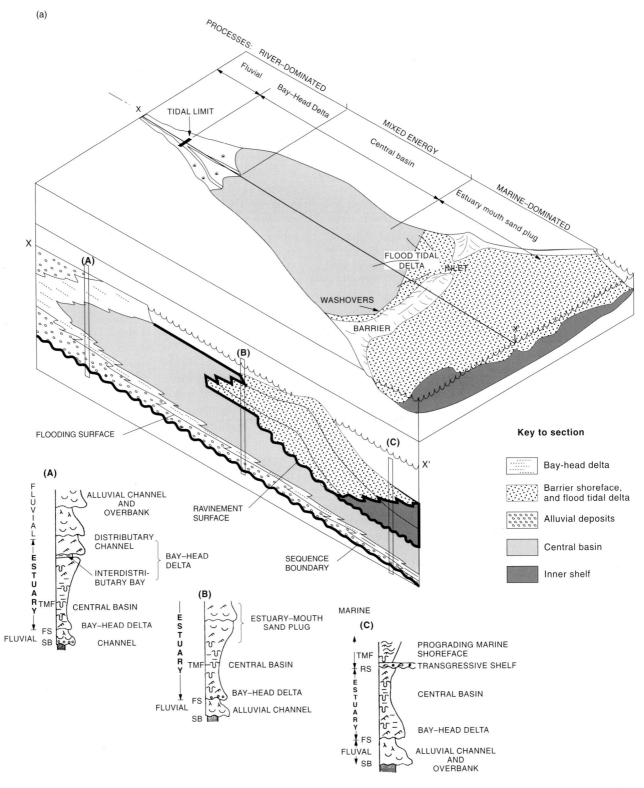

Fig. 8.3 (a) Wave-dominated estuary

upwards successions. Similarly, both valleys and multi-storey channel sandstones may comprise two or more stacked-channel sandstone bodies. However, some key features can allow these distinct sediment bodies to be distinguished:

1 Valley fills are commonly wider and thicker than channels. On average, valleys have width:thickness ratios in the order of 1:1000 or more, whereas channels are narrower, with width:thickness ratios of 1:100 (Reynolds, 1994b; Table 8.2).

2 Valleys can be inferred where deep incisions (deeper than the thickness of individual channels) can be recognized.

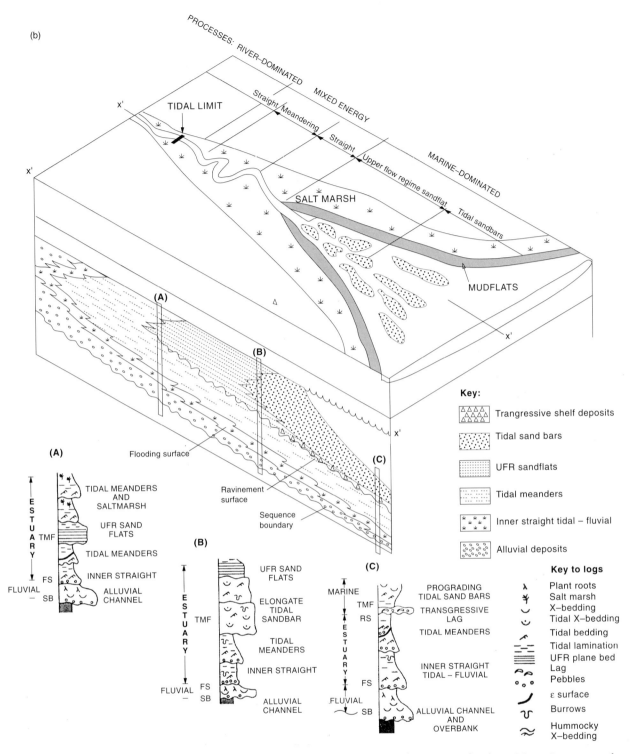

Fig. 8.3 (b) Tide-dominated estuary (after Dalrymple *et al.*, 1992). TMF, time of maximum flooding; RS, ravinement surface; SB, sequence boundary; FS, flooding surface; UFR, upper flow regime

Commonly, these incisions take the form of terraces.
3 Valley incision is often heralded by the generation of accommodation space at progressively slower rates, and followed by a phase during which accommodation is generated progressively more rapidly. Such progressive changes are likely to be recorded in the rock record, leaving signals that indicate proximity to a sequence boundary. For example, where the rate of generation of accommodation space is decreasing, the thickness of parasequences and the volume of crevasse splay deposits are likely to progressively decrease, whereas channel connectivity is likely to increase.
4 The facies that fill incised valleys record a basinwards shift in facies belts. Two types of valley fill are common in paralic successions: estuarine valley-fills, and valley-fills

Table 8.2 Summary statistics of paralic sandstones (after Reynolds, 1994b)

	Widths (m)		Lengths (m)		Thicknesses (m)		Number of data points
	Mean	Range	Mean	Range	Mean	Range	
Incised valleys	9850	63 000–500			30.3	2–152	91
Fluvial channels	750	1400–60			9	24–2.5	6
Distributary channels	520	5900–20			7.8	40–1	268
Crevasse channels	60	400–5			2.4	17–0.2	44
Shoreline–shelf sandstones							
Undifferentiated	25 000	106 000–1600	93 200	190 000–47 000	19.1	49–2.7	67
Highstand systems tract	16 400	43 000–16 000					36
Transgressive systems tract	7200	20 000–3300					5
Distributary mouth bars	2900	14 000–1100	6500	9600–2400	9.7	42–1.2	26
Tidal ridges	6400	29 300–750	17 500	61 000–3250	11.2	22.9–3	32
Crevasse splays	800	7700–20	5600	11 700–160	1.4	12–0.3	84
Flood tidal delta complex	6200	13 700–1700	12 300	25 700–2900	6.7	23–1.8	13
Lower tidal flat	1000	1550–400			4.6	9–2.0	14
Tidal creeks	810	1550–160			5.2	18–1	15
Tidal inlet	1850	2550–700	4300	4300	4.8	3–7	3
Estuary mouth shoal	2400	2900–1700	3750	4700–2200	10	10–35	4
Chenier	3650	6400–900	21 800	38 600–49 000	5.8	7–4.6	2
All sands	5050	106 000–5	35 300	190 000–160	?	?	669

that evolve from fluvial to estuarine fills. Marine mudstones may also fill incised valleys (Fig. 8.4).

5 As a result of river rejuvenation, incised valleys commonly contain the coarsest sediment available locally.

6 Haq *et al.* (1988) suggest that phases of eustatic sea-level fall are unlikely to exceed 100 m in magnitude. Therefore, valleys produced by eustatic sea-level fall are unlikely to be deeper than 100 m. This may help to distinguish valleys from slope canyons and gullies.

Incised-valley patterns

Incised valleys in the rock record have highly variable patterns. Regional studies typically reveal patterns comparable to modern tributive drainage patterns (e.g. Dolson *et al.*, 1991). By contrast, detailed mapping over relatively restricted areas commonly shows complex geometries characterized by short cuts between adjacent valleys, varying valley widths and a great range in the spacing between adjacent valleys (e.g. Jennette *et al.*, 1992). As with all fluvial systems, valleys are, in general, aligned down the palaeoslope. Where the basin floor is exposed during sea-level fall, the orientation of incised valleys may be perpendicular to fluvial and distributary systems in the preceding highstand systems tract.

8.3.2 Interfluve sequence boundaries

Valleys are restricted features separated by interfluves (Fig. 8.4). During falling sea-level the interfluves are exposed subaerially and subject to erosion and pedogenesis. With relative sea-level rise the valley aggrades and the interfluves are progressively onlapped. In paralic successions, valleys are incised into the pre-existing coastal or delta plain, and, therefore, the interfluves are likely to be flat topped. When a valley has filled to the level of the old coastal plain, the next increment of relative sea-level rise will flood the interfluves generating a proportionately large volume of accommodation space, and causing the shoreline to transgress rapidly. As a result, many interfluves are characterized by a sharp erosion surface, which is the product of ravinement, and are overlain by a thin transgressive lag.

In many ways this signature resembles a simple marine flooding surface. However, interfluve sequence boundaries may: (i) be underlain by well-developed palaeosols indicating prolonged exposure; and (ii) be capped by anomalously coarse lags, comprising clasts that were introduced to the basin by rivers during lowstand, but are not present in the underlying highstand successions.

8.3.3 The transgressive surface

The transgressive surface is the first significant marine flooding surface across the shelf within a sequence (Van Wagoner *et al.*, 1988). The surface defines the top of the lowstand systems tract and the base of the transgressive systems tract. Commonly, the transgressive surface is coincident with the interfluve sequence boundary described above. It will also cap valley-fill deposits (Fig. 8.4).

In delta plain and coastal plain successions it is often difficult to recognize the landwards equivalent of the transgressive surface. As a result, although valley fills and

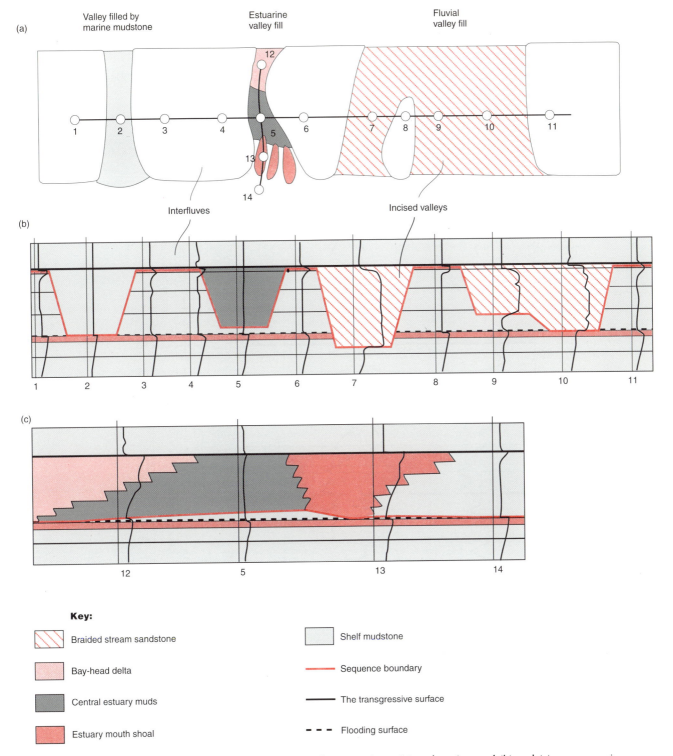

Fig. 8.4 Variable expression of sequence boundaries in paralic successions: (a) a plan view and (b) and (c) representative cross-sections (after Van Wagoner *et al.*, 1990)

maximum flooding surfaces can be recognized, it may be impossible to divide intervening successions into lowstand and transgressive systems tracts (Fig. 8.5).

8.3.4 Forced regressions

Forced regressions are basinward movements of the shoreline, caused by relative sea-level fall. They are independent of sediment supply (Posamentier *et al.*, 1992;

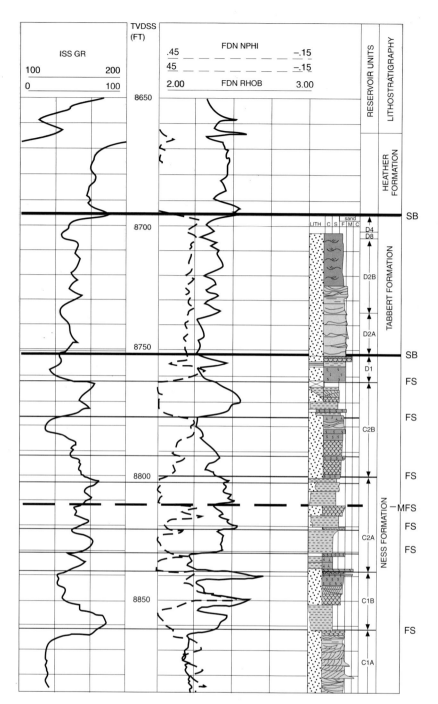

Fig. 8.5 (*continued opposite*) Wireline and associated graphic log of the Brent Formation in well 211/18a-7 in the Thistle field, North Sea. A sequence boundary is inferred at the base of the Etive Formation, where a multistorey succession of fluvial channel sands lie above upper shoreface sands, and a maximum flooding surface is recognized in the middle of the Ness Formation, between retrogradational and forestepping sets of bay-head deltas. However, there is no clear transgressive surface. As a result, the intervening succession cannot be divided into lowstand and transgressive tracts (from Reynolds, 1995). Lithology: C, clay; S, silt; F, fine; M, medium; C, coarse

Posamentier and James, 1993) and are expressed by basinwards or 'downwards' shifts in facies belts. Typically, shoreline sands sharply overlie outer shelf mudstones. The sharp contact reflects the re-establishment of a sea-floor profile in equilibrium with shoreface and inner shelf processes, prior to sand deposition and renewed shoreline progradation. Two types of lowstand shoreline are recognized: (i) attached shorefaces that overlie sands of the underlying highstand systems tract, and (ii) detached shorefaces that are isolated in offshore shales (Fig. 8.6; Ainsworth

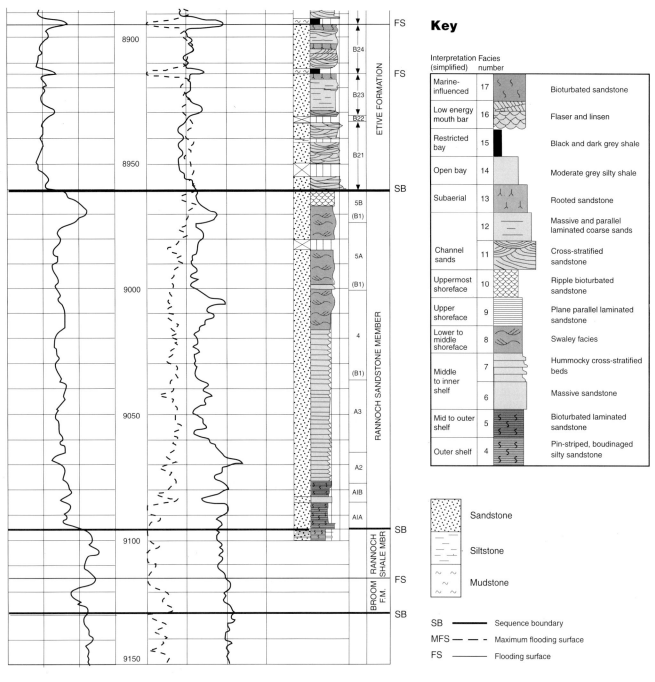

Fig. 8.5 (*continued*)

and Pattison, 1994; Fitzsimmons, 1994). In addition, two expressions of the sequence boundary are recognized. In the first the sequence boundary rapidly dies out basinwards into a correlative conformity (e.g. Plint, 1988; Posamentier and Chamberlain, 1993). In the second the sequence boundary remains a sharp erosive surface for 10–20 km basinwards of the lowstand shoreline, and is always overlain by a distinctive, commonly gutter cast, sand (Fitzsimmons, 1994).

To date, forced regressions have been recognized largely in wave- and storm-dominated successions, where they produce shore-parallel sand isopachs (e.g. Bergman and Walker, 1988). However, sea-level fall also occurs in successions dominated by tides or semi-permanent currents. Reynolds (1994a) argues that a series of extensive undulating erosion surfaces in the Viking Formation of Alberta, Canada, were cut subaqueously by tidal scour following sea-level fall (section 8.6).

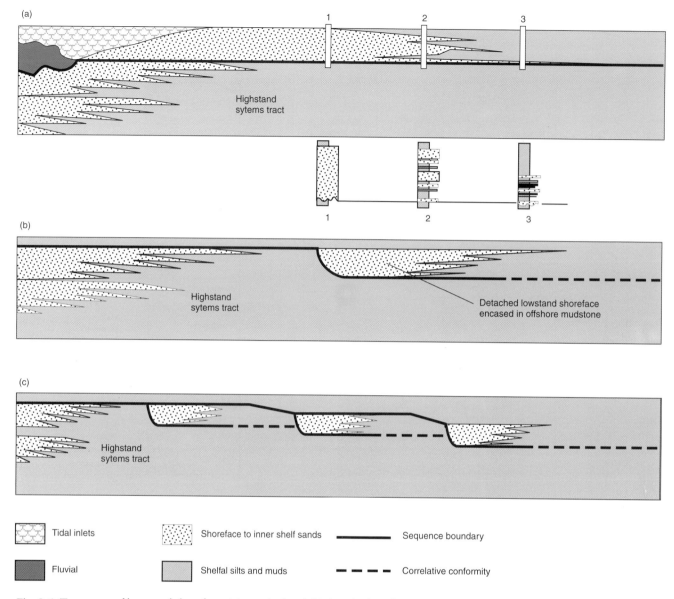

Fig. 8.6 Two types of lowstand shoreface: (a) attached and (b) detached, and (c) a set of lowstand shorefaces, or 'forced regressive set'. Note that this is only one example of a forced regressive set. The sand bodies could be spaced more closely or connected (after Ainsworth and Pattison, 1994; Fitzsimmons, 1994)

8.3.5 The maximum flooding surface

In paralic successions, the maximum flooding surface is coeval with the most landwards position of the shoreline. Commonly, the maximum flooding surface is underlain by a retrogradational parasequence set and overlain by a progradational parasequence set. In many cases, however, it is difficult to recognize a single, discrete surface. Instead, a 'maximum flooding zone' is recognized. Within this zone, two or three discrete surfaces may be candidates for a 'maximum flooding surface'.

Basinwards of the shoreline, maximum flooding zones are represented by outershelf silts and mudstones. Candidates for the maximum flooding surface are marked: (i) by the finest grained deposits; (ii) by evidence for condensation, such as firmgrounds; or (iii) by outer shelf carbonates (Fig. 4.3). High gamma-ray values, and high organic matter content reflect anoxic bottom water. These are common during transgression and in the lower portion of maximum flooding zones.

On the delta plain, maximum flooding zones are characterized by tidal influence in distributary channels (Shanley and McCabe, 1993), pronounced channel crevassing, lake expansion (Atkinson, 1983) and generally wetter palaeosols. In lagoonal successions, discrete maximum flooding surfaces may be recognizable between backstepping and forestepping bay-head delta parasequences (Fig. 8.5).

8.3.6 Parasequence sets

Parasequence sets record the net movement of the shoreline over two or more parasequences. Their wireline log expression, and relation to key surfaces and systems tracts, have been illustrated in previous chapters. In core and in outcrop, progradational and retrogradational parasequence sets are expressed as systematic changes in sedimentary structures, ichnofacies and grain size (Figs 4.3 and 8.7).

8.4 Parasequences in paralic successions

The detailed character of parasequences is extremely varied, and is controlled by a range of interconnected factors: (i) sedimentary process; (ii) the calibre of the supplied sediment; (iii) accommodation space; and (iv) climate. Each of these factors is discussed below. The following section, section 8.5, shows how they affect the detailed stratigraphy of coastal plain to shoreline shelf systems, deltas and estuaries.

8.4.1 Sedimentary processes

In general, as fluvial, wave and storm-dominated shorelines prograde they each produce a coarsening succession. However, the lateral and dip extent of the associated sand bodies, and the detailed sedimentary structures, differ depending on the dominant process (Fig. 8.8; Table 8.2). Shoaling waves redistribute sediment alongshore and straighten the plan-form of shorelines. As a result, wave- and storm-dominated shorelines produce tabular, shore-parallel sandstone bodies with a spatially uniform stratigraphic signature. Shoaling waves also generate an onshore-directed current that prevents offshore sand transport. By contrast, storms can transport sand to the inner shelf, so that storm-dominated sand bodies can extend further offshore than wave-dominated sand bodies (Fig. 8.8a,b). In fluvial- and tide-dominated settings the sand : shale ratio, and the grain size will vary across the shoreline portion of a parasequence, reflecting its composition by a number of coalesced mouth bars (Fig. 8.8c).

8.4.2 Accommodation space

Two models relate parasequences in shoreline successions to delta plain and coastal plain successions. Van Wagoner et al. (1990) show coastal plain mudstones developing contemporaneously with shoreline progradation (Fig. 8.9). By contrast, Devine (1991) shows strandplain and lagoonal successions developing, respectively, during (i) progradation, and (ii) aggradation and transgression (Fig. 8.10).

Both models are supported by detailed field-work, they each reflect distinct rates of generation of accommodation space. In Devine's model no accommodation space is generated during regression, resulting in toplap, whereas significant accommodation space is generated during barrier island aggradation and transgression, resulting in a dramatic expansion of the lagoon. By contrast, accommodation space is generated throughout strandplain regression in the model of Van Wagoner et al. (1990), allowing the synchronous deposition of coastal plain deposits.

Partitioning of sand-body types into systems tracts

Analysis of coastal plain and deltaic sand bodies shows (Table 8.3) that the overwhelming majority of sand bodies deposited landwards of the shoreline (distributary channels, crevasse splays, tidal creeks, and flood tidal deltas) occur in transgressive systems tracts (Reynolds, 1994b). The data indicate that accommodation space in the delta plain and coastal plain is generated largely during transgression, and that certain sand-body types are favoured when base level is rising rapidly. For example, flood tidal deltas develop in lagoons, which are favoured during transgression. Similarly, base level rise may also favour crevassing.

Sand-body dimensions

Recent studies have recognized the cyclic stacking of sequences into sequence sets and composite sequences (Mitchum and Van Wagoner, 1991; Jones and Milton, 1994). The low-frequency changes in accommodation space and sediment supply that generate these patterns also

Table 8.3 Data to show the preferential partitioning of sand bodies into particular systems tracts. LST, lowstand systems tract; TST, transgressive systems tract; HST, highstand systems tract; SMST, shelf-margin systems tract; %, the percentage of a given sand-body type in a given systems tract; N, the total number of sandstone bodies of a given type for which the systems tract is known

	Systems tract	%	N
Valleys	LST	100	47
Tidal creeks	TST	100	13
Tidal flats	TST	100	14
Shoreline–shelf	HST	75	48
	TST	17	
	LST	0	
	SMST	4	
Distributary channels	HST	22	78
	TST	60	
	LST	18	
Crevasse splays	TST	100	75
Flood tidal deltas	TST	92	12
	LST	8	

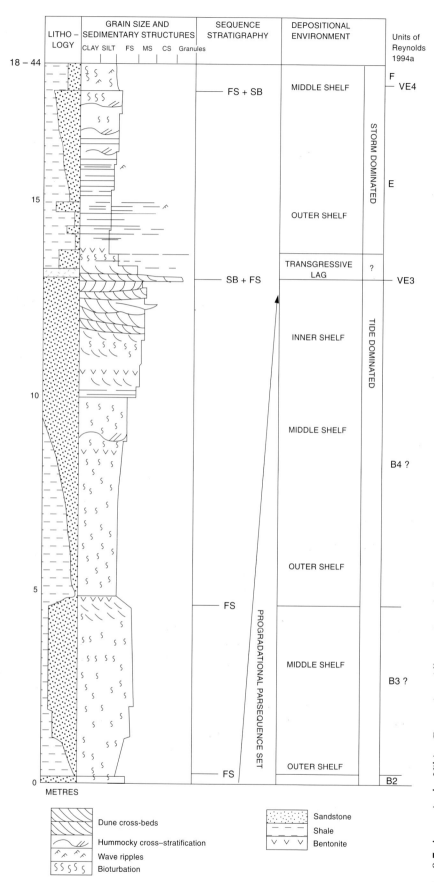

Fig. 8.7 A core in the Viking Formation, Alberta, Canada shows two complete parasequences, deposited in a tidal sand-sheet setting. The base of the core penetrates a strongly bioturbated, fine-grained sandstone with rare ripple cross-lamination. An abrupt decrease in sandstone content marks the first flooding surface. The flooding surface is bioturbated and characterized by scattered medium sand grains. Upwards there is a steady increase in sandstone content and a change from horizontal to vertical burrow forms. These changes, together with an increase in the preservation of unburrowed cross-beds, indicate shallowing and progradation within a parasequence. A second flooding surface is marked by a second decrease in sandstone content, and by more scattered medium-grained sands. At its base the overlying succession is similar to the first parasequence, but towards its top dune-scale cross-bedding (*sensu* Ashley, 1990) predominates, indicating daily current activity. The cross-beds indicate that the second parasequence prograded further than the first, forming a progradational parasequence set. The second parasequence is capped by a third flooding surface, marked by a thick lag and by a dramatic change in grain size and sedimentary structures. The thickness of the lag and the change in sedimentary processes indicate that the third flooding surface may be an interfluve sequence boundary — regional mapping supports this interpretation (Well 6-11-48-21-W4, based on Reynolds, 1994a). FS, flooding surface; SB, sequence boundary; MFS, maximum flooding surface

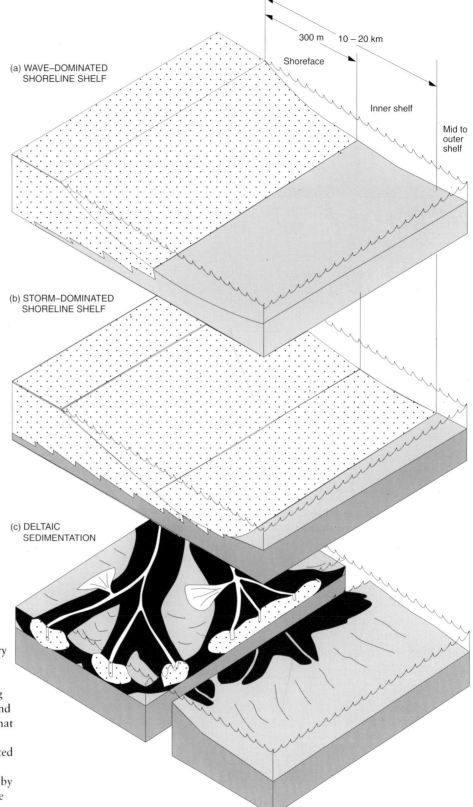

Fig. 8.8 Variations in sand distribution within parasequences characterized by distinct sedimentary processes. Wave-dominated (a) and storm-dominated (b) parasequences are characterized by simple cleaning and coarsening signatures that extend in three dimensions. They differ in that the sand belt extends much further from the shoreline in storm-dominated successions. By contrast, deltaic parasequences (c) are characterized by marked spatial changes in sandstone content

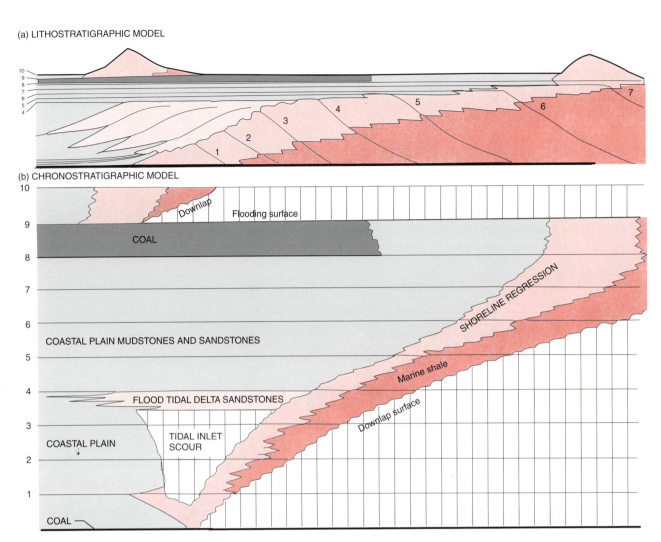

Fig. 8.9 Lithostratigraphic and chronostratigraphic models of parasequence development based on the work of Van Wagoner et al. (1990). Coastal plain mudstones and sandstones are shown to be coeval with shoreline progradation. For discussion see text

influence sandstone body dimensions on a parasequence scale. For example, the thickness of shoreline shelf sands decreases upwards through highstand sequence sets. Similarly, the dip extent of shoreline–shelf sands tends to be twice as extensive during highstand systems tracts as in transgressive systems tracts (Table 8.2).

8.4.3 Grain size

Coarse sediment (gravel and sand) favours: (i) low-sinuosity channels in the delta and coastal plain; (ii) well-drained delta and coastal plains that lack lakes; (iii) the rapid deposition of river load at river mouths; (iv) a steep 'reflective' shoreline that receives the full effect of wave energy; and (v) a steep delta front characterized by mass flow processes (up to around 25° in Gilbert deltas).

By contrast, fine sediment (silt and mud) favours: (i) high-sinuosity channels; (ii) poorly drained delta and coastal plains characterized by lakes; (iii) sediment dispersal beyond river mouths as a result of buoyancy processes; (iv) a low-angle shoreline that attenuates and dissipates wave energy, and forms a sandy shoreline and inner shelf zone; and (v) a low-angle delta front (around 1°).

The dominant grain size of deltas reflects the nature of the catchment area: the area and physiogeography of the catchment; the prevailing climate; and the bedrock lithology. These factors, in combination with tectonic activity and relative sea-level fluctuations, may change through time to generate fluctuations in sediment flux and calibre. For further discussion on the influence of grain-size, see Orton (1988) and Orton and Reading (1993).

8.4.4 Climate

On the delta and coastal plain climate influences the nature of lacustrine deposits, palaeosols and the development

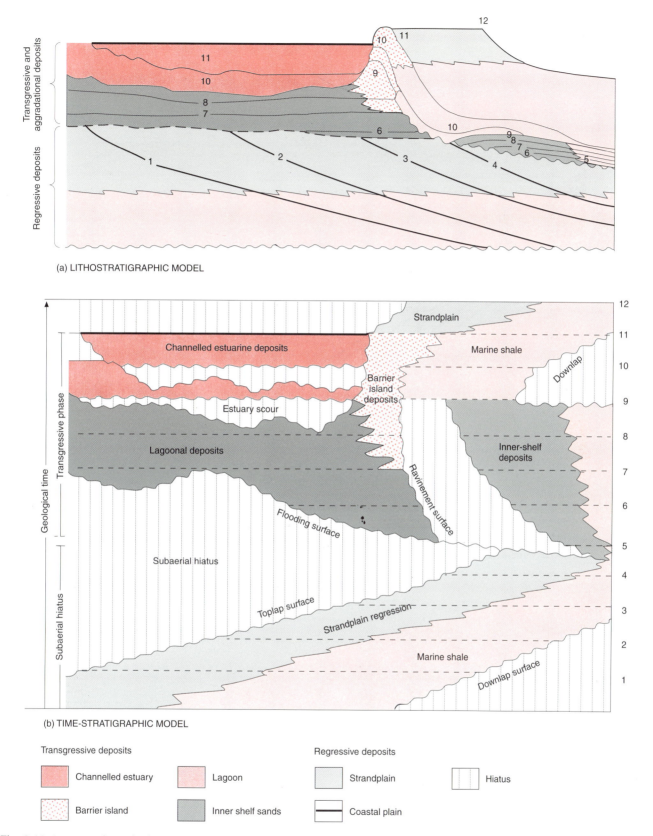

Fig. 8.10 (*continued overleaf*) Lithostratigraphic and chronostratigraphic models of parasequence development based on the work of Devine (1991). Coastal plain mudstones and sandstones are shown to be coeval with transgression. For discussion see text

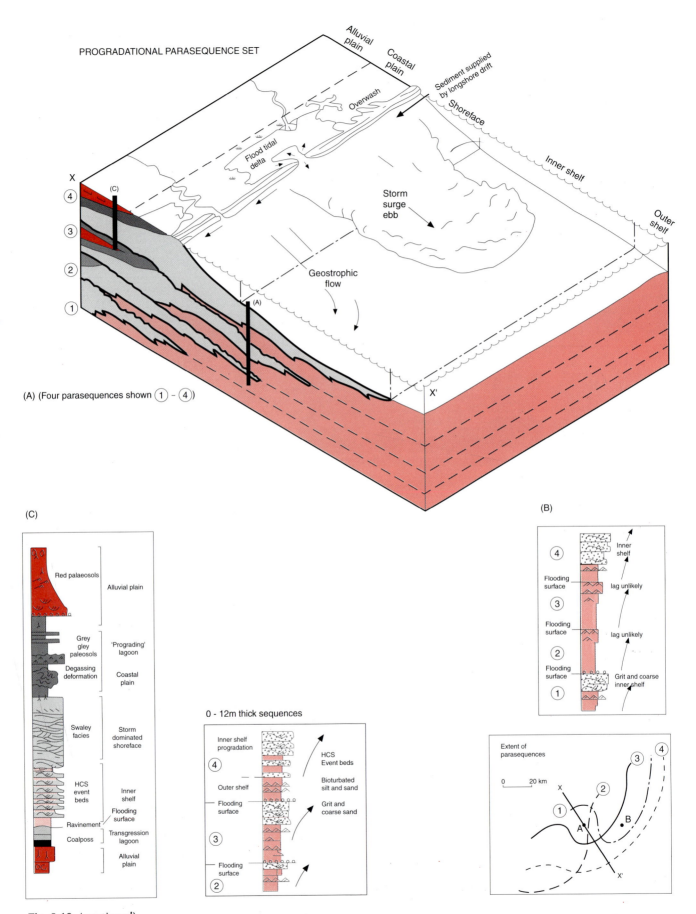

Fig. 8.10 (*continued*)

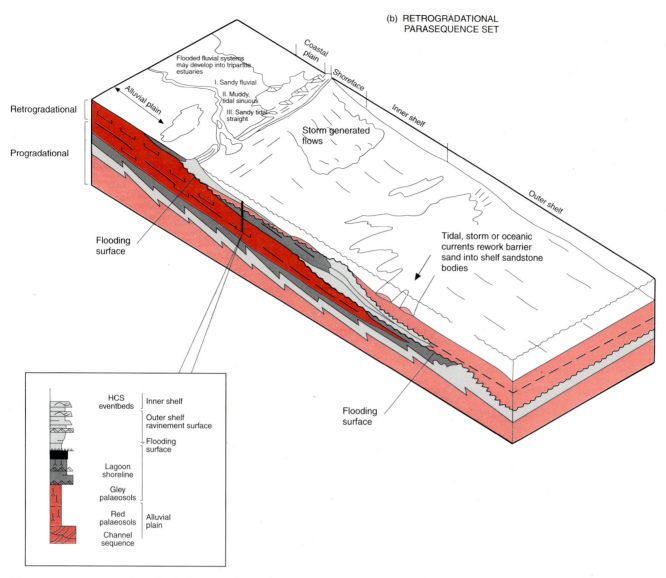

Fig. 8.11 (a) (*continued overleaf*) A progradational parasequence set in a storm-dominated shoreline–shelf succession. The work of Devine (1991) and Van Wagoner *et al.* (1990) has been combined to show coastal plain and lagoonal successions developing during progradation and transgression. The map shows the extent of shoreface sands in each parasequence 1–4. The short logs (A) and (B) illustrate spatial variations in parasequence set signature that result from lobate parasequences. A flooding surface lag is more likely if the underlying shoreface has prograded past a given location, as only then can shoreface erosion winnow and concentrate coarse material (based on Hamblin and Walker, 1979; McCrory and Walker, 1986; Plint and Walker, 1987; Eyles and Walker, 1988; Plint *et al.*, 1988). (b) A retrogradational parasequence set in a storm-dominated shoreline–shelf succession. During transgression, the shoreline may aggrade and keep pace with relative sea-level rise, so that significant accommodation space is generated in the coastal and alluvial plain. With an increase in the rate of relative sea-level rise, or a drop in sediment supply, the shoreline will jump landwards to a new position where these two variables are in balance (for a review see Elliott, 1986b)

of coals, evaporites and lacustrine carbonates. The wind regime has an important effect on the efficacy of storm processes, while the prevailing winds determine the extent of aeolian (coastal) dune fields, the direction of longshore sediment transport, and the presence of semi-permanent shelf currents. Climate also influences the development of shelf carbonates.

8.5 The sequence stratigraphy of distinct paralic systems

The fundamental controls outlined above affect the development of all paralic successions. However, they combine in different ways in different paralic systems. In many ways coastal plain to shoreline–shelf systems are the simplest. These are discussed and illustrated in detail below. Subsequent sections show how stratigraphy differs in deltaic and estuarine successions.

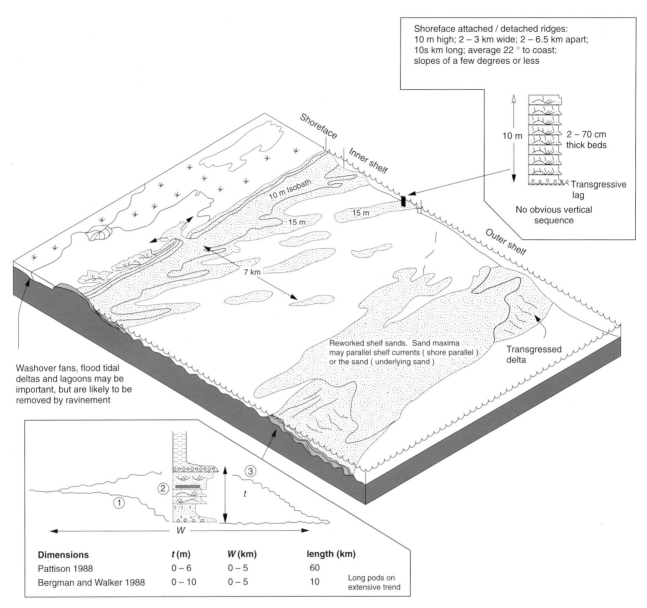

Fig. 8.11 (c) A transgressive parasequence set in a storm-dominated shoreline–shelf succession. Surface 1 is cut by shoreface erosion. The sand body, 2, is the product of short-lived progradation. All evidence for subaerial exposure is removed by ravinement, producing surface 3. Ancient examples suggest that the preserved sediment bodies are linear and shore-parallel (Bergman and Walker, 1988; Pattison, 1988). Modern shelves suggest the presence of numerous shore-oblique ridges (e.g. Rine et al., 1986) and ragged patches of sand above transgressed sandy successions. W, width; t, thickness

8.5.1 The stratigraphy of coastal plain to shoreline–shelf systems

Storm-dominated parasequences

In general, the lower part of storm-dominated parasequences (shelf and lower shoreface) comprises a series of sharp-based beds 5–30 cm thick that become thicker and progressively amalgamated upwards (Fig. 8.11a). The beds are often hummocky cross-stratified, with bioturbated, wave-rippled tops. The upper shoreface may be: (i) storm dominated and characterized by swaley facies (McCrory and Walker, 1986); (ii) wave dominated, with or without bars; or (iii) cut by cross-bedded tidal channel units. Barred shorefaces are characterized by erosively based, trough cross-beds, deposited in longshore troughs and/or rip channels. Non-barred shorefaces are characterized by wave ripples, onshore and offshore directed cross-beds, and planar laminae (for a review see Elliott, 1986b). It is critical, but not always easy, to distinguish channels generated by tidal inlets and longshore troughs, from incised-valley fills. Coastal plain successions are strongly influenced by climate, and are characterized by small fluvial systems. Lagoonal deposits are generally shale prone, but may also comprise sandy facies deposited in storm washovers, flood tidal deltas, or bay-head deltas (Plint and Walker, 1987; Devine, 1991).

In shelfal locations the stratigraphic signature of *progradational, storm-dominated parasequence sets* is generally simple, comprising a stack of cleaning and coarsening parasequences that become coarser and sandier upwards reflecting net progradation of the shoreline (Fig. 8.11a). An exception is suggested by detailed mapping in the Cardium Formation (Eyles and Walker, 1988), which has revealed a lobate form to individual parasequences that produces an overall progradational, but variable, parasequence-set signature. In proximal locations the balance between sediment supply and relative sea-level rise will control both the parasequence and the parasequence-set signature (section 8.4.2).

The nature of *retrogradational, storm dominated parasequence sets* depends on the balance between sediment supply and relative sea-level rise. Figure 8.11b shows one combination of these variables — the shoreline–shelf elements of the parasequences could be stacked more closely or separated entirely.

Transgressive, storm-dominated parasequence sets comprise a series of back-stepping parasequences, the sandy portions of which are neither connected nor overlain (Fig. 8.11c). Many transgressive parasequence sets are characterized by ravinement, removing all evidence for subaerial exposure, and by shelfal processes reworking the stranded shoreline deposits. Transgressive parasequence sets are extreme forms of retrogradational parasequence sets.

Forced regressive, storm-dominated sets are generated when sea-level falls in a series of more or less discrete steps (Fig. 8.6c). As discussed in section 8.3.4, the base of each sand package is characterized by a downwards shift in facies that records sea-level fall. If sea-level fall is associated with river incision, then each sand package is a high-frequency sequence. The sand bodies themselves are characterized by a progradational, coarsening signature. Forced regressive sets can have a geometry closely similar to transgressive parasequence sets (compare Fig. 8.6c with Fig. 8.11c). The shoreline position is likely to be fixed by shelf irregularities, generated, for example, by deep-seated faults.

Wave-dominated parasequences

The stratigraphic architecture of wave-dominated parasequences is closely comparable to that of storm-dominated parasequences. A key difference is that the predominance of shoaling waves commonly limits the distance from the beach to the edge of the sand belt to less than 1 km. As a result, dip-extensive wave-dominated sands can be generated only by progradation. As in storm-dominated parasequences, wave-dominated shoreline successions generally coarsen, and increase in bed thickness upwards. Again, the upper shoreface may be barred, non-barred, or cut by tidal channels. Conglomeratic wave-dominated shorelines are steeper and have their own distinctive suite of structures (Fig. 8.12; Massari and Parea, 1988; Hart and Plint, 1989).

Tide-dominated parasequences

Tide-dominated shoreline–shelf systems pass laterally into estuaries and deltas, and landwards into tidal flats (Fig. 8.13). If the tidal flats are richly fed with sediment by alongshore and along-shelf tidal currents, they can form the central portion of prograding coastal plain to shoreline–shelf successions. Subtidal flats and the lower portions of intertidal flats tend to be sandy, and to pass landwards into muddy and then vegetated, intertidal and supratidal flats. Consequently, prograding tidal flats generate fining upwards successions. These in turn may be cut by fining upwards channel fills, the deposits of complex channel systems which dissect the flats (for a review see Elliott, 1986a,b).

Progradational, tidal successions are illustrated in Fig. 8.13. Analogous transgressive tidal deposits are thinner, developing during periods when rivers are drowned, forming estuaries, and little sediment is transported to the shelf. In such cases sandy shelf deposits tend to be sourced by tidal scour and reworked from the underlying succession. The combined effects of tidal scour and shoreface erosion may generate complex erosional topography on the flooding surface. Deposits that overlie the surface may include sand sheets and sand ridges. Sand sheets may fine or coarsen upwards depending on sand-sheet movement. The internal structure of tidal sand ridges is poorly known, but is thought to be dominated by dune cross-bedding. For a detailed discussion see Stride (1982).

8.5.2 The stratigraphy of deltaic systems

Much of the stratigraphic variability observed in deltas is controlled by interplay amongst: sedimentary processes, accommodation space, sediment supply, climate and grain size. Accommodation space and sediment supply play a role similar in deltaic systems to that discussed for coastal plain to shoreline–shelf systems. Following Orton (1988) and Orton and Reading (1993) the importance of grain size and sedimentary process can be shown by extending the ternary process classification of Galloway (1975) into a prismatic form (Fig. 8.14). Their model can be refined by differentiating amongst shelf deltas, shelf-edge deltas and Gilbert deltas. Space precludes illustration of this extra diversity. For examples and discussion of the stratigraphy of Gilbert and shelf-edge deltas the interested reader is referred to Colella (1988), Braga *et al.* (1990), Ethridge and Westcott (1984), Rossi and Rogledi (1988), Collinson (1986), Pulham (1989) and (Elliott 1986a, 1989).

In high-resolution sequence stratigraphic studies, four features distinguish the stratigraphy of deltas from that of coastal plain to shoreline–shelf systems: delta lobe switching, lakes, major distributary channels, and growth faulting.

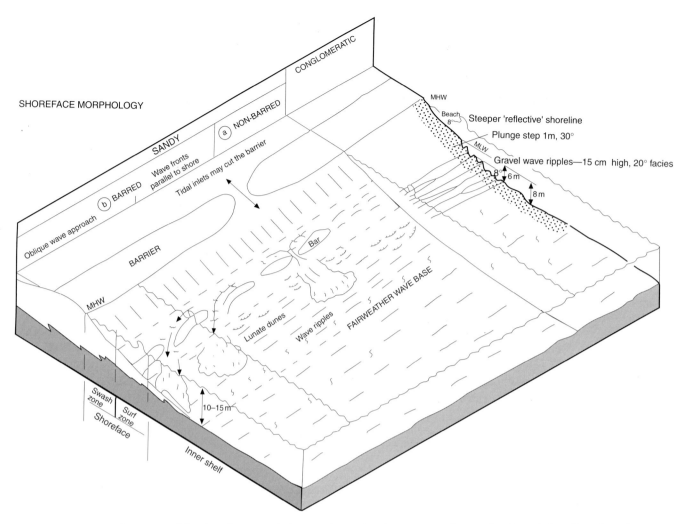

Fig. 8.12 (*continued opposite*) The shoreface morphology and associated structures of wave-dominated shorelines (based on Clifton *et al.*, 1971; Davidson-Arnott and Greenwood, 1976; Hunter *et al.*, 1979; Massari and Parea, 1988; Hart and Plint, 1989). MHW, mean high water; MLW, mean low water

Delta lobe switching is an autocyclic process that results in the deposition of local parasequences (Fig. 8.15). The process is initiated by river avulsion, which causes the abandonment of previously active delta lobes. An abandoned lobe subsides and is transgressed, generating a local flooding surface. When the river switches back, a new lobe is formed through the progradation of a deltaic shoreline. A second phase of abandonment generates a second flooding surface, and completes the deposition of a local parasequence areally restricted to the abandoned delta lobe. Lobe switching can occur at a variety of scales, from the large-scale switching of major trunk rivers, to the local switching of small-scale distributaries. As a result a complex hierarchy of parasequences may develop.

The stratigraphic signature of *lakes* that develop on the delta plain is commonly similar to that of interdistributary bays (Fig. 8.15). Phases of lake expansion generate lacustrine flooding surfaces, whereas distributary and crevasse channels feed a variety of prograding lacustrine shorelines, resulting in a 'parasequence' signature. Lakes, however, have no marine connection and the relationship between lacustrine flooding surfaces and marine flooding surfaces may not be clear.

Major distributary channels are a key element of both the delta plain and the delta front. In the delta front they commonly truncate cleaning and coarsening shoreface deposits, and with limited data they can be difficult to distinguish from incised-valley-fill deposits (section 8.3.1; Fig. 8.15).

Growth faulting is common in thick deltaic successions. The subsidence that occurs in the hanging wall of such faults, (i) enhances the effect of relative sea-level rise, resulting in the enhanced preservation of flooding surfaces in the hanging wall, and (ii) dampens the effect of relative sea-level fall, so that sequence boundaries in the footwall may pass into correlative conformities in the hanging wall (Fig. 8.15).

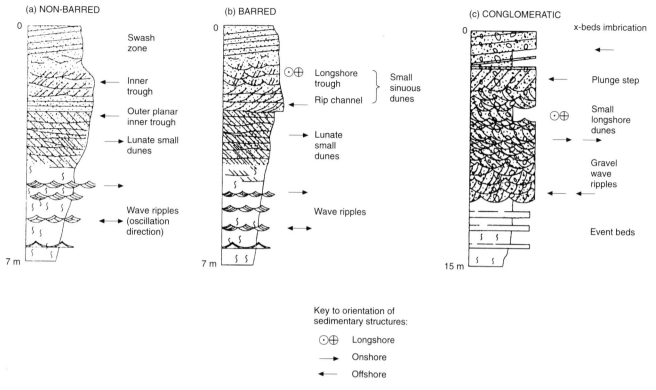

Fig. 8.12 (*continued*)

8.5.3 The stratigraphy of estuarine systems

Estuarine successions are valley fills that develop during relative sea-level rise. Variations in valley shape, sediment supply, sedimentary processes, and accommodation space result in a variety of stratigraphic signatures. However, many estuarine successions record steady transgression.

When initial rates of relative sea-level rise are low, the basal portions of valley fills comprise lowstand fluvial deposits (Dalrymple *et al.*, 1992; Allen and Posamentier, 1993). Where relative sea-level rises more rapidly, lowstand fluvial incision and bypass may be immediately followed by estuarine sedimentation (Wood and Hopkins, 1989). As estuarine systems backstep, fining upwards successions are generated in their mixed-energy and river-dominated portions (Fig. 8.3; Pattison, 1992). These may be truncated by transgressive erosion surfaces formed either by shoreface erosion (Fig. 8.3a) or by tidal scour (Fig. 8.3b). If the rate of sea-level rise decreases or sediment supply increases, estuarine facies belts can aggrade or prograde. In the latter case a flooding surface can be recognized between the backstepping and forestepping facies belts. However, parasequences are not generally described within estuarine successions. Rather, the high-frequency signature of estuaries is dominated by repeated relative sea-level fall, within overall transgression (e.g. Eschard *et al.*, 1991; Wood and Hopkins, 1989).

8.6 Correlation procedure

It is recognized increasingly that high-resolution sequence stratigraphy provides an excellent means of correlating within paralic successions. However, the ease with which paralic successions can be correlated is highly variable, being strongly dependent upon both the depositional environment and the mechanisms that generate the relative sea-level changes.

8.6.1 Depositional environments

Paralic subenvironments generate sand bodies with a range of scales and geometries (Table 8.2). These dimensions can be used to sense-check the detailed correlations required for oilfield reservoir zonation. The impact of this data is reinforced by considering the relative scale of different sand bodies and typical paralic oil and gas fields (Fig. 8.16). Most paralic oil and gas fields are small. For example, the vast majority of Indonesian paralic fields have a productive area of less than 10 km^2 (M. Eller, pers. comm., 1993), whereas even giant fields, such as Ninian in the North Sea (1045 × 10^6 barrels of recoverable oil and an area of 89 km^2; Abbots, 1991) commonly have areas less than 100 km^2. Comparing these sizes with the mean dimensions of paralic sandstones, it can be seen that individual

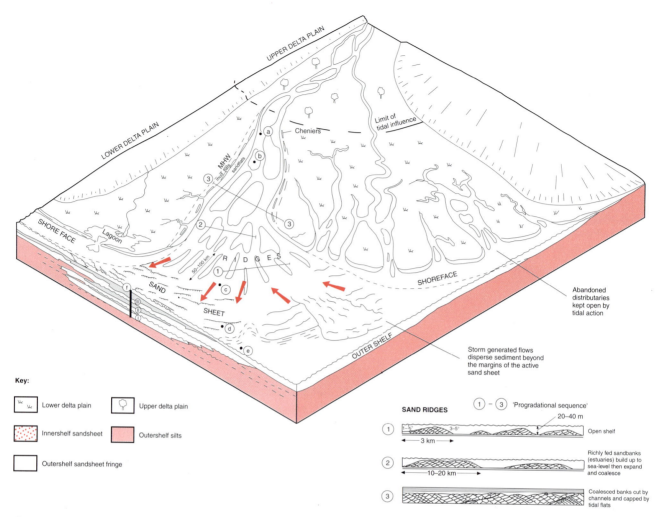

Fig. 8.13 (*continued opposite*) The stratigraphic signature of tide-dominated systems is poorly known, but likely to be highly variable. In this model an active distributary channel supplies sand to a prograding, sandy, tide-dominated delta. The delta passes gradationally into a prograding coastal plain to shoreline–shelf system that comprises tidal flats, and a sandy tide-dominated shelf. In its tide-influenced portion the active distributary comprises a series of mixed fluvial and tidal channels (log a; Meckel, 1975; Dalrymple *et al.*, 1990, 1992) separated by intertidal, and subtidal bars (log b; Mutti *et al.*, 1985a). Where they are active, the channels dominate the stratigraphic record. The bars pass down system into a series of ridges that straddle the transition from the delta to the inner shelf (Smith, 1966; Johnson and Baldwin, 1986; Elliott, 1989). The ridges are likely to be characterized by dune-scale cross-bedding, and with a plentiful supply of sand, the ridges will build to sea-level, and may grow laterally and coalesce (see inset sections 1–3; Stride, 1982). The shelf ridges pass into sheet sands characterized by downstream changes in bedform type: proximal portions have stronger tides and are dominated by large- and small-scale dunes, distal portions are characterized by small dunes, ripples, and storm reworking (c–e). Sand-sheet progradation results in vertical profiles closely similar to those seen in ancient tidal ridges (f; FS — flooding surface). Abandoned distributaries may develop an estuarine-like sedimentary fill, with a sand plug at their mouths fed by onshore transport of coarse sediment, passing landwards into a complex series of muddy tidal creeks

storm-dominated shoreline–shelf sands, and incised valleys occur on the same scale as all but the largest fields. Mouth bars and tidal ridges have areas comparable to small fields, or segments of large fields. By contrast, individual fluvial channels, distributary channels, and crevasse splays are small areally.

In addition to sandstone-body scale and geometry, the orientation and stacking of individual sandstones can have a critical influence on the ease of correlation, and on sand body continuity. For example, an incised valley 10 km wide would cover the mapped closure of the Statfjord field if it was parallel to the axis of the field, but only a small portion of the field if it was aligned perpendicular to the field axis (Fig. 8.16). Similarly, although distributary channels are relatively narrow they may be stacked, or connected laterally, resulting in discrete, field-wide reservoir zones.

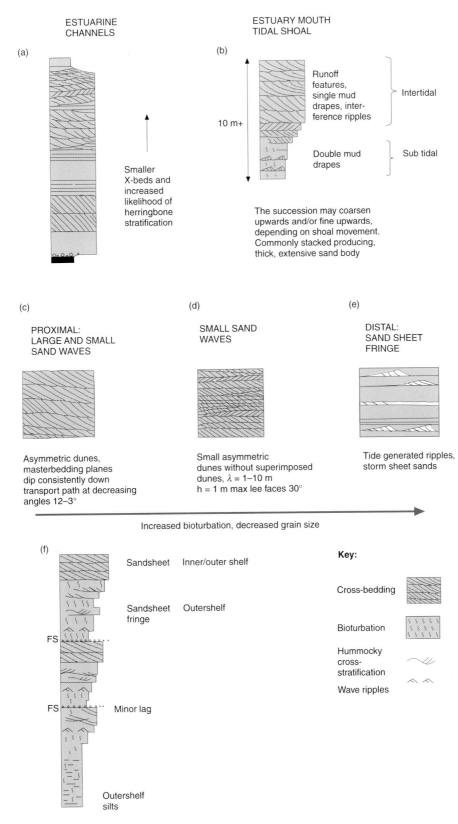

Fig. 8.13 (*continued*)

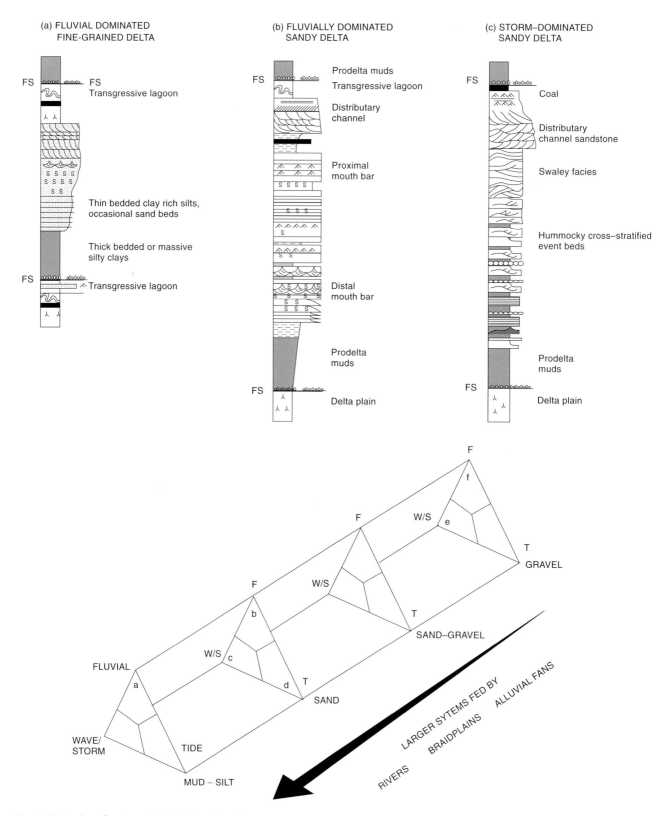

Fig. 8.14 A classification of shelf deltas based on grain size and sedimentary process. In general, the progradation of all deltaic shorelines results in a cleaning and coarsening upwards signature, which may or may not be cut by a fining upwards distributary channel succession. In detail, each sedimentary process and grain size combine to produce a distinctive suite of sedimentary structures and stratigraphic signature (after Orton and Reading, 1993). (a) Fluvially dominated, fine-grained shelf deltas are characterized by a semi-continuous delta-front sand sheet that results from a distributive channel pattern and the consequent

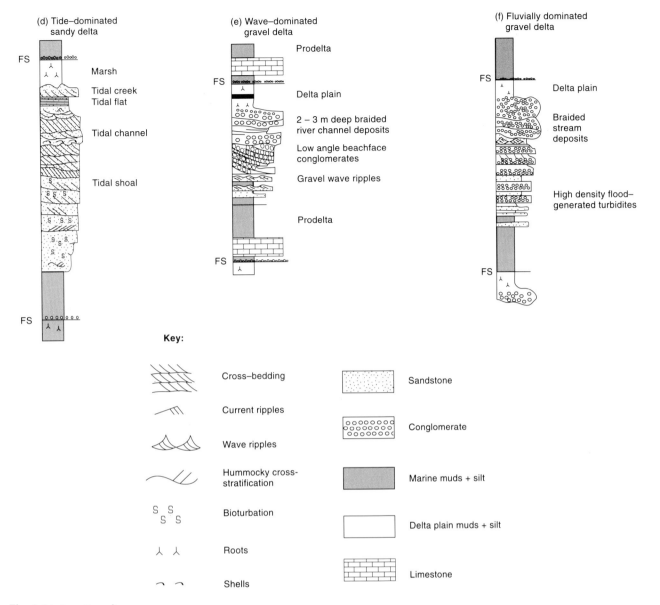

Fig. 8.14 (*continued*)

coalescence of mouth bars. The fine grain size favours the development of lakes and lagoons on the delta plain. Syn-sedimentary deformation is rare by comparison with fine-grained shelf-edge deltas (Gould, 1970; Edwards, 1981; Coleman and Prior 1982; Elliott, 1986a). (b) Fluvially dominated, sandy shelf deltas are characterized by the rapid deposition of coarse sediment in friction dominated, middle ground bars. Sharp-based, plane laminated, and climbing ripple sands are common (Wright, 1977; Elliott, 1986a; Orton, 1988). (c) Storm-dominated deltas are characterized by hummocky cross-stratified sands. The stratigraphic signature is comparable to storm-dominated shorelines (Fig. 8.11a) but can be distinguished by the presence of major distributary channels (Johnson and Stewart, 1985; McCrory and Walker, 1986). (d) In tide-dominated deltas a tidal signature is likely to be preserved in the delta front, in distributary channels and deep into the delta plain (see Fig. 8.13). (e and f) Gravel shelf deltas are typically medium-scale fan-deltas (9–70 km radius) with a relatively steep delta front and well-drained delta plains characterized by low-sinuosity rivers. Where wave-dominated they lack distinct delta lobes and are characterized by a cleaning and coarsening upwards signature, resulting from the progradation of a steep reflective shoreline. Where they are fluvially dominated, delta lobes occur that may be transgressed generating local flooding surfaces, and mouth bars dominate the shoreline succession (Dixon, 1979; Dutton, 1982; Colella, 1988; Braga *et al.*, 1990). Given suitable climatic conditions, abandoned delta lobes may be characterized by patch reefs

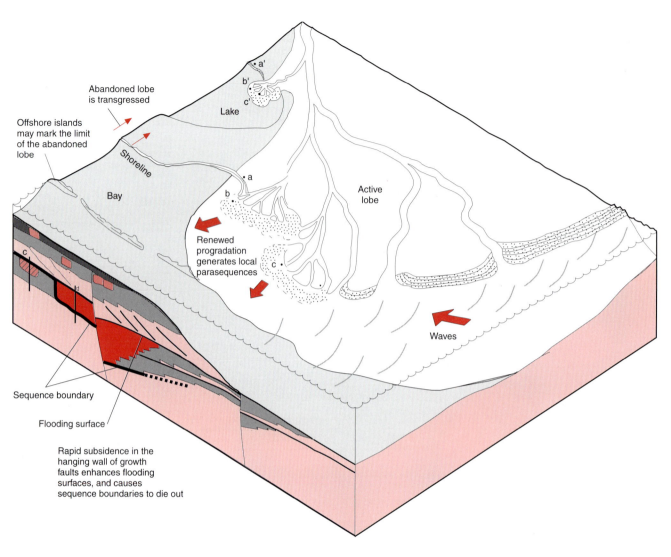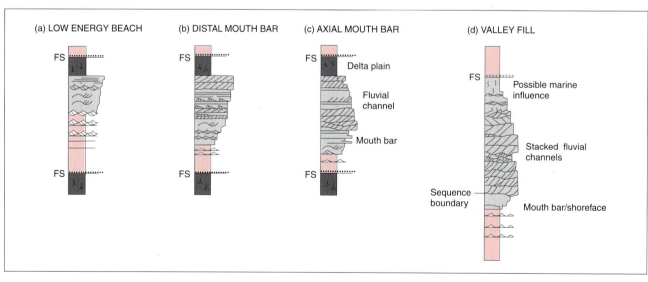

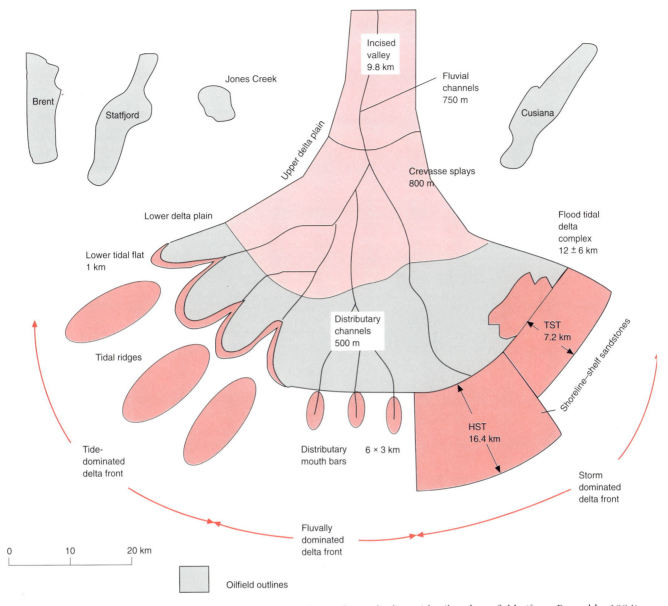

Fig. 8.16 Schematic comparison of the areal extent of paralic sandstone bodies with oil and gas fields (from Reynolds, 1994). TST, transgressive systems tract; HST, highstands systems tract

8.6.2 Parasequence correlation

The area over which parasequences can be correlated depends on the dominant process involved in creating the bounding flooding surfaces and their correlative surfaces: predominantly eustasy, tectonics and delta-lobe switching. For the petroleum geologist it is again instructive to compare the spatial extent of these mechanisms with the size of typical paralic oil and gas fields.

Flooding surfaces generated by eustatic sea-level rise may be recognizable globally, whereas variations in tectonic subsidence will, in general, occur on a basin or sub-basin scale. The smallest scale parasequences are likely to be generated by lobe switching, and will be as extensive as the

Fig. 8.15 (*opposite*) An abandoned delta lobe (shaded) subsides and undergoes transgression, creating a local flooding surface. When the river switches back to the abandoned lobe renewed progradation creates a cleaning and coarsening signature. Repetition of this process, known as delta-lobe switching, creates local parasequences. The stratigraphic signature of the parasequences varies spatially, depending on the size and location of distributary channels and the strength of shelf processes (logs a–c). Similar signatures may develop in lakes on the delta plain, positions a′–c′. Major distributary channels (hatched, log c) create stratigraphic signatures similar to valley fills (red, log d). For further discussion see text

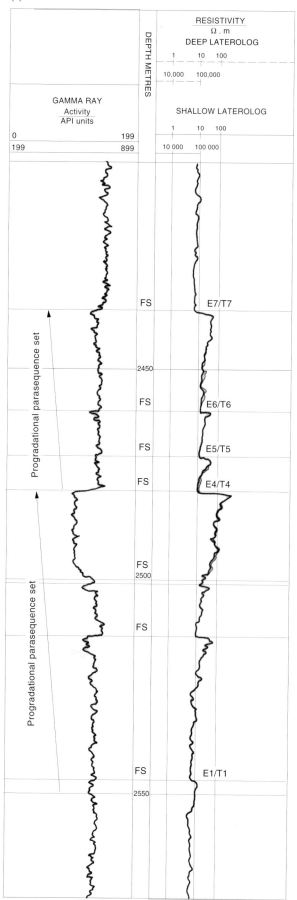

flooded delta lobe. The scale of delta lobes is variable, and related directly to the scale of the delta, which, in turn, is reflected commonly in the grain size of the delta (Table 8.4). In sandy and fine-grained deltas, delta lobes are likely to be larger than most paralic oil and gas fields. In coarser grained systems, delta lobes scale more closely with oilfields. In summary, the majority of parasequences are likely to have areal extents greater than most oil and gas fields.

Table 8.4 Estimates of delta areas and areas of delta lobes. Data from Orton and Reading (1993) and Frasier and Osanik (1967)

Grain size	Delta area (km²)	Delta-lobe area (km²)
Gravel	< 50	
Gravel and sand	< 1000	
Fine sand	< 50 000	
Mud/silt	< 500 000	777–7770

8.6.3 Progradation, aggradation, and retrogradation

In general, the most easily recognized stratigraphic signatures in paralic successions are intervals ranging from 50 to 150 m in thickness and characterized by overall progradation, aggradation, or retrogradation.

Such signatures may reflect parasequence sets or sequence sets. Commonly they can be traced for tens of thousands of square kilometres, forming excellent primary correlation markers at the sub-basin or basin scale. For example, Plint *et al.* (1986, 1888) have traced sequence sets in the Cardium Formation over an area in excess of 50 000 km², and Bhattacharya (1988, 1991) has mapped a parasequence set in the Dunvegan Formation over an area of some 16 000 km². Such areally extensive signatures are extremely useful in making preliminary regional correlations. It may, however, require extensive subsurface data sets before a parasequence signature and a sequence set signature can be distinguished (Fig. 8.17).

Fig. 8.17 (a) Wireline log suite of well 7-1-62-6W6 in the Kakwa Member of the Cardium Formation, Alberta (from Plint, 1988). The well is characterized by a number of cleaning and coarsening upwards successions bounded by marine flooding surfaces (FS). From well-logs alone it is tempting to interpret these cycles as parasequences deposited in a shoreline–shelf setting. Plint *et al.* (1986) confirm the environmental interpretation, but demonstrate that the surfaces, which they have termed E/T surfaces, are associated with dramatic erosion and conglomeratic lowstand deposits (Fig. 8.17b). The example demonstrates the difficulty in distinguishing between marine flooding surfaces and transgressive surfaces where well data is limited

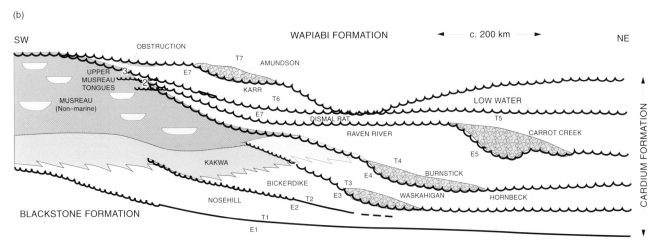

Fig. 8.17 (b) Generalized cross-section through the Cardium Formation, showing the architecture of sequence boundaries (E), conglomeratic lowstand deposits (pebble symbol) and transgressive surfaces (T). The formation is about 100 m thick, and the panel some 200 km in length. The Kakwa and overlying Musreau Members represent a storm-dominated shoreline–shelf and alluvial plain succession that prograded at least 150 km into the basin. Non-marine sediments of the Musreau and Obstruction Members were deposited directly on marine shelfal deposits of the Raven River, Dismal Rat and Karr members at times of low relative sea-level. At such times gravel was introduced to the basin by rejuvenated rivers, and reworked subsequently during ensuing transgressive phases, leaving coarse conglomeratic bodies far out into the basin (Waskahigan, Burnstick, Carrot Creek and Amundsen Members) (from Plint *et al.*, 1988)

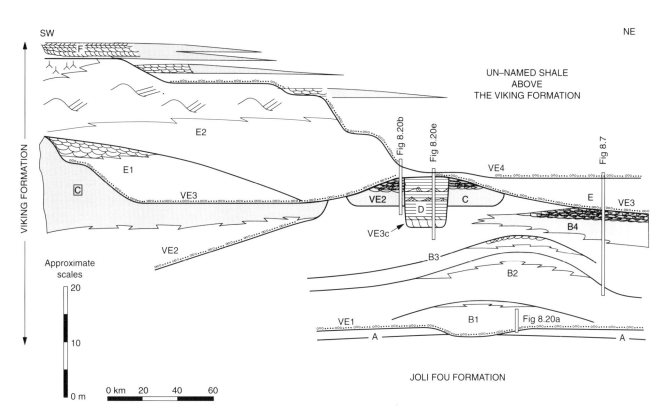

Fig. 8.18 The stratigraphy of the Viking Formation, Alberta, Canada. The 50-m-thick succession is divided into six units, A–F, by five erosion surfaces VE1–VE4 and VE3c. The lower three units are the product of prograding tidal sand-sheets. Their bases are characterized by dramatic increases in sand content that reflect forced regression. Unit D is an estuarine valley-fill succession — the climax of a period of overall falling relative sea-level. VE3 is a transgressive surface that establishes a storm-dominated shelf recorded by two parasequences E1 and E2. VE4 is a further regional erosion surface capped by a conglomerate and onlapping sands. The coarse deposits were introduced to the basin by a further fall in relative sea-level. The vertical bars show the approximate positions of core photographs and detailed graphic logs (after Reynolds, 1994a)

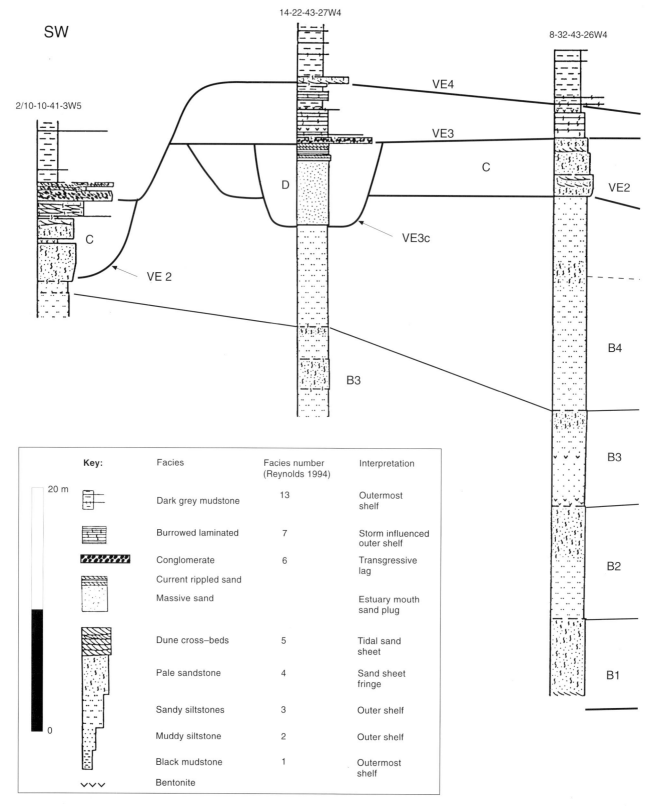

Fig. 8.19 (a) (*continued opposite*) Core and wireline log correlations of surfaces and facies in the Viking Formation, Chigwell area, Southern Alberta. (a) Core correlation

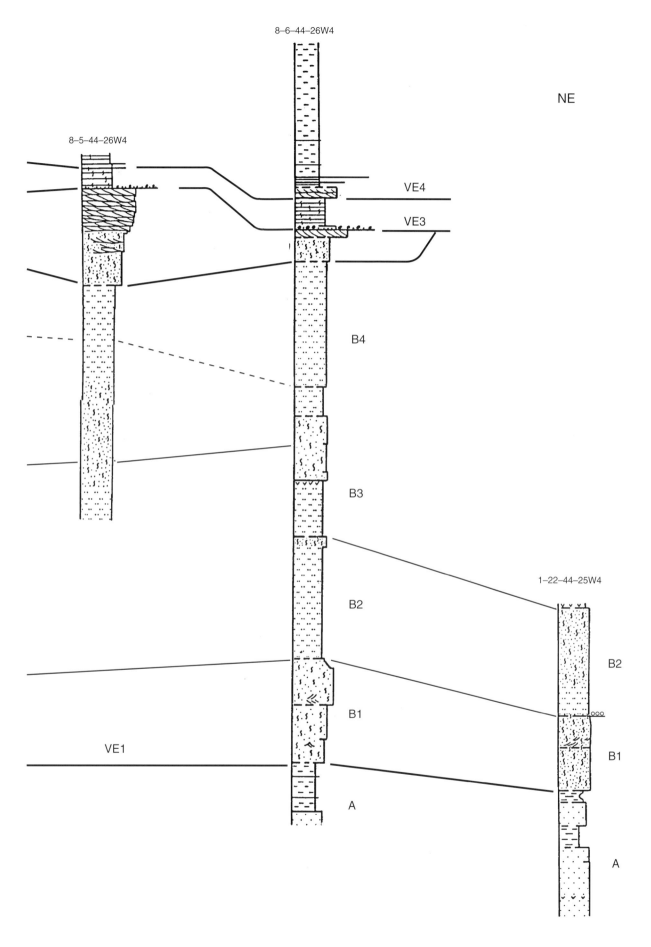

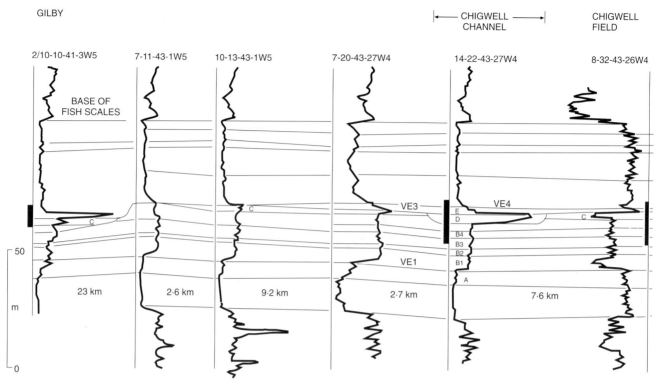

Fig. 8.19 (b) (*continued opposite*) Wireline log correlation. The wirelines are microresistivity curves (to the right) and gamma-ray curves (to the left). The cores are hung from a regional marker, the base of the Fish Scales Zone, some 50 m above the Viking Formation. Note that the datum parallels a suite of other marker picks both above and below the Viking Formation, suggesting that the topography on the surfaces VE2, VE3 and VE3c is erosional

8.6.4 Sequence boundaries

Single outcrops, or even a single cored well can reveal the presence of a sequence boundary. However, they are commonly difficult to recognize unequivocally in paralic successions. This may be because either they have variable expression over a given study area, or their signature is subtle within the study area. The tracing of a sequence boundary through a data set may be the last act of correlation, even though its position is locally well known.

8.7 An example: the Viking Formation, Western Canadian Basin

The Viking Formation is an extensive, predominantly shallow-marine, sandstone succession sandwiched between two marine shales, the Joli Fou Formation below and an unnamed shale above (Fig. 8.18). The formation was deposited on a ramp in the Canadian portion of the Rocky Mountain foreland basin. An extensive subsurface data base of wireline logs and cores allows the recognition of five erosional surfaces within the Viking Formation in south-central Alberta (Reynolds, 1994a). Four of the surfaces, VE1–4, are regionally extensive, one surface, VE3c, is of local extent. The surfaces divide the formation into six non-genetic 'units' A–F. In addition, units B and E are subdivided by subregional discontinuity surfaces that lack evidence for clear erosion. This section documents and interprets this stratigraphy, with the aim of illustrating a range of stratigraphic signatures and key surfaces.

The lowermost unit, unit A, comprises silty sandstones, siltstones and mudstones (Figs 8.19a and 8.20a) and is expressed as two subtle peaks on resistivity logs (Fig. 8.19b). The base of the unit has not been cored, but wireline logs show that it is planar and lacks evidence for erosion. However, it is always marked by an abrupt increase in sandstone content, and it may be an expression of a forced regression.

Unit B is characterized by a progradational parasequence set deposited in a tidal sand-sheet setting. The systems tract is interpreted to have been orientated NNW–SSE, with progradation towards the south-southeast. In proximal locations, and towards the top of the parasequence set, the parasequences are characterized by dune cross-bedding and clearly defined flooding surfaces (Fig. 8.7). By contrast, the distal and lower parasequences are thoroughly bioturbated, they lack clear flooding surfaces, and have a poorly developed cleaning and coarsening signature (Fig. 8.19a). The base of unit B is marked by a dramatic increase in sandstone content (Fig. 8.20a) and, on regional correlation panels, by subtle truncation of unit A below (Reynolds, 1994a). The increase in sandstone content

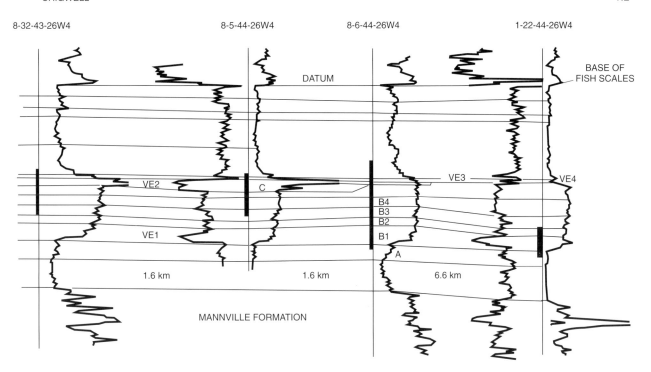

Fig. 8.19 (b) (*continued*)

reflects a downward shift in facies belts as a result of forced regression. The erosion dies out towards the southwest (Fig. 8.18) and is interpreted to be the result of tidal scour.

Unit C comprises a single cleaning and coarsening succession that records the progradation of a tidal sand sheet (Fig. 8.20b). At the base of unit C is a sharp, bioturbated surface, VE2, marked by a dramatic increase in grain-size (Fig. 8.20c) and clear topography on regional sections (Fig. 8.19b). VE2 is interpreted as a subaqueous scour surface reflecting base level fall on a tide dominated shelf.

Surface VE3c, the base of unit D, is a sharp surface that incises through unit C into unit B (Figs 8.19 and 8.20e). It cannot be traced regionally, but has been recognized in several areas (Reinson *et al.*, 1988; Boreen and Walker, 1991; Reynolds, 1994a). At each location the facies, such as massive, planar and ripple-laminated sandstones, with associated bioturbation and mud drapes, suggest deposition in an estuarine setting. The estuarine facies represent a basinwards shift in facies belts with respect to unit C below. This, together with the topography and restricted extent of VE3c suggests that unit D is a valley-fill deposit. Surface VE3c is therefore a sequence boundary, the culmination of a phase of overall base level fall that led to forced regressions at the base of units A, B and C.

The surface that caps units D and C, VE3, is overlain by a bioturbated pebble conglomerate interpreted as a transgressive lag (Fig. 8.20d). Surface VE3 is the *transgressive surface*, the first flooding surface across the old shelf.

Where it caps unit C it also represents an interfluve sequence boundary. No evidence for subaerial exposure is preserved at the boundary — presumably it was removed by ravinement.

Unit E comprises a burrowed succession of wave-rippled and low-angle cross-laminated event beds that punctuate a silty background. They record a middle shelf setting on a storm dominated shelf, and represent a dramatic change from the tide-dominated setting of units A–D (Fig. 8.20b). Such changes in sedimentary process may be common across sequence boundaries. Unit E is capped by a regional erosion surface, VE4, which is marked by a conglomerate layer (Fig. 8.20b). The conglomerate is interpreted as a transgressive lag, and is considered to be evidence for a further drop in relative sea-level subsequent to the deposition of unit E, i.e. VE4 is also a transgressive surface.

8.8 Reservoirs in paralic successions

Oil and gas fields in paralic successions commonly comprise a stacked series of reservoirs (e.g. Verdier *et al.*, 1980; Jev *et al.*, 1993). Each reservoir may reflect a distinct depositional environment or range of subenvironments. Sequence stratigraphy can help to unravel this complexity by: (i) elucidating the geometry of stratigraphic traps; (ii) outlining seal architecture; (iii) refining the choice of analogue data for input into stochastic reservoir models; and (iv) delineating flow units.

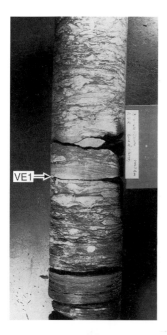

Fig. 8.20 Cores illustrating the surfaces and stratigraphic signature of the Viking Formation in the Chigwell area of Southern Alberta. (a) The VE1 surface in well 1-22-44-26W4 is marked by a dramatic increase in sandstone content, and a downward shift in facies with bioturbated mid-shelf sands overlying laminated outer shelf mudstones. The facies dislocation and grain-size increase are interpreted to reflect forced regression. The contact is strongly bioturbated by robust trace fossils which form a firmground or *Glossifungites* ichnofacies. The ichnofacies is characteristic of many paralic marine erosion surfaces (MacEachern *et al.*, 1992; Taylor and Gawthorpe, 1993). (*continued*)

Fig. 8.20 (b) Well 2/10-10-41-3W5 shows the cleaning and coarsening succession of unit C. Its base and top surfaces, VE2 and VE3, are detailed in c and d. Burrowed laminated sands and siltstones comprise unit E. The base of the overlying conglomerate marks the VE4 erosion surface. The top of the core is to the right. (*continued*)

8.8.1 Stratigraphic traps

Stratigraphic traps are common in paralic systems. Some are the result of the pinchout of individual sands. Many others reflect the erosion, topography, and facies dislocations associated with sequence boundaries. Valleys and onlap pinch-outs are key elements (Fig. 8.21.).

Many valleys incise into marine mudstones. Often they are filled at their base by coarse sands and capped by transgressive shales (Fig. 8.4). The valley walls provide lateral seals, and the transgressive shales a top seal. In such a situation, the key risk on trap integrity is the up-dip seal of the coarse sands. Fluvial sands tend to be deposited continuously during valley-fill, providing an effective thief sand for migrating hydrocarbons (Fig. 8.3). By contrast, central-basin mudstones can provide an up-dip seal for sands deposited as estuary mouth shoals (Zaitlin and Shultz, 1990; Fig. 8.4c). Of course not all valleys are orientated parallel to the dip direction. Sinuous valleys may form a string of reservoirs located in meanders that point up-dip. Up-dip seals can also occur where valleys terminate (Fig. 8.22).

Some valleys are filled with mud: most commonly transgressive marine mudstones, or the central-basin facies of estuaries. Such valleys can provide lateral seals to petroleum trapped in a range of other paralic sandstones (e.g. Wood and Hopkins, 1989).

Forced regressions also can produce stratigraphic traps, with incised lowstand shoreface sands underlain by highstand and capped by transgressive shales (Fig. 8.6). Similarly, onlap of transgressive sandstones against a sequence boundary can generate stratigraphic traps (e.g. Pattison, 1988).

Fig. 8.20 (c) The VE2 surface in well 2/10-10-41-3W5. Bioturbated sands interpreted to reflect the fringes of a tidal sand sheet sharply overlie outer shelf muddy siltstones. A *Glossifungites* ichnofacies marks the contact. (d) The VE3 surface in well 2/10-10-10-41-3W5 has a well developed *Glossifungites* ichnofacies. *Arenicolites* burrows (two vertical pipes joined to form a stretched U-shape) penetrate the contact and have been filled by coarse material from above VE3. The muds below the contact are also strongly sideritized, a feature typical of many key surfaces in the Alberta subsurface. (*continued*)

(c)   (d)

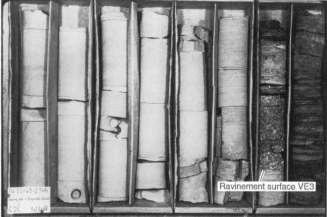

Fig. 8.20 (e) Well 14-22-43-27W4 shows the sharp base to unit D, VE3c, overlain by massive clean sandstones. The massive sandstones are interpreted to have been deposited in an estuarine setting, and VE3c is interpreted as a sequence boundary. The conglomerate near the top of the core is a transgressive lag, the base of which marks a ravinement surface, VE3

8.8.2 Seals

Paralic successions are characterized by numerous intraformational seals. Many of these seals are formed by shales that overlie marine flooding surfaces. Their extent reflects the mechanism that generated the flooding surface, for example, delta-lobe abandonment or eustatic sea-level rise. As paralic oil and gas fields are small with respect to these features, well-developed marine flooding surfaces are likely to be field-wide in all but the largest fields (Table 8.4). Maximum flooding surfaces have great extent. They form regional seals that control the long-distance migration of petroleum in the subsurface, and form major pressure barriers within fields.

8.8.3 Analogues and flow units

Sand-body dimensions vary with sequence stratigraphic setting (8.4.2). Therefore, sequence stratigraphy is helpful in choosing the correct analogue sand-body dimensions with which to populate stochastic reservoir models.

For modelling purposes reservoirs are commonly divided into 'flow units', i.e. units that have a distinct and internally consistent effect on fluid flow (Ebanks, 1987). In many cases flow units are bounded by sequence stratigraphic surfaces. For example, intraformational seals above flooding surfaces mark flow unit boundaries. In addition, the grain-size and facies changes that occur across sequence boundaries typically result in permeability contrasts, and therefore, in flow unit boundaries (Fig. 8.23).

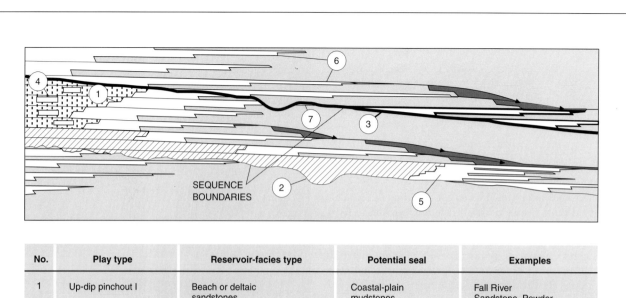

Fig. 8.21 Many stratigraphic traps in paralic successions are associated with sequence stratigraphic surfaces (after Van Wagoner *et al.*, 1990)

(a) VALLEY CHANGES DIRECTION

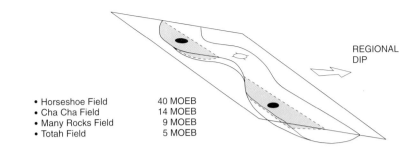

- Horseshoe Field — 40 MOEB
- Cha Cha Field — 14 MOEB
- Many Rocks Field — 9 MOEB
- Totah Field — 5 MOEB

(b) VALLEY TRUNCATED BY YOUNGER SEQUENCE BOUNDARY

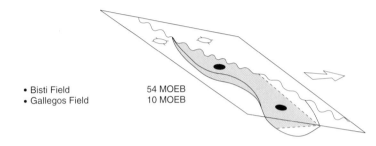

- Bisti Field — 54 MOEB
- Gallegos Field — 10 MOEB

(c) VALLEY TERMINATION

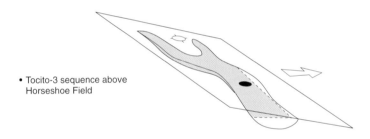

- Tocito-3 sequence above Horseshoe Field

Fig. 8.22 Incised-valley-fill trapping mechanisms (from Jennette *et al.*, 1992)

8.9 Paralic systems at a seismic scale

The topsets regularly imaged by conventional two-dimensional petroleum industry seismic data commonly comprise sediments deposited in fluvial, paralic and outer shelf environments. The interpreter aims to build a stratigraphic interpretation of the seismic data that addresses lithofacies distribution, depositional environments, locations of petroleum reservoirs and reservoir continuity. This section aims to build on previous chapters and show how sequence stratigraphy and other aspects of stratigraphic prediction can help to address these questions. The discussion is in two parts. The first part addresses basin-scale models of paralic systems. The second part considers seismic facies and the seismic expression of key paralic environments and settings.

8.9.1 Seismic-scale models

Deltaic models

As discussed in section 8.2, two types of deltas can be recognized in basins with a distinct shelf–slope break: shelf-edge deltas and shelf deltas (Fig. 8.24a). On conventional two-dimensional seismic the expression of these two types of delta is quite different. As most shelf deltas prograde into relatively shallow water, say 30–70 m deep, their thickness is small and difficult to resolve on conventional seismic data. At shallow burial depths (< 500 m) the delta fronts of shelf deltas may be resolvable as subtle low-angle clinoforms between parallel horizontal topset reflections. At typical reservoir depths (> 1 km) these clinoforms are generally not resolved and shelf deltas are expressed as parallel seismic reflections.

By contrast, the slope systems that form the delta front of shelf-edge deltas may be several hundreds of metres high

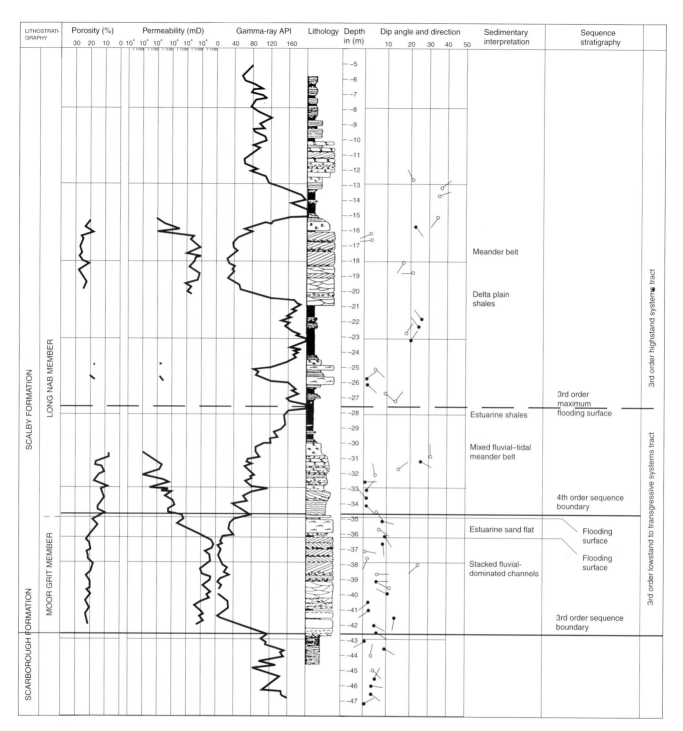

Fig. 8.23 Graphic log, porosity and permeability data showing the close relationship between key sequence stratigraphic surfaces and reservoir quality in the Scalby Formation, Yorkshire. For example, an order of magnitude decrease in permeability occurs across the lowermost flooding surface, and a similar change occurs across the fourth-order sequence boundary (after Eschard *et al.*, 1991)

and clearly resolvable. Shelf-edge deltas also exhibit a range of syn-sedimentary deformational features that can be resolved seismically; growth faults, mud diapirs, sediment slides, canyons and gullies.

Low-angle clastic ramps

Low-angle clastic ramps develop in those basins that lack a pre-existing slope, and are characterized by strong shelfal processes, which redistribute sediment supplied via deltas and prevent the growth of a shelf–slope system (Fig. 8.24b). On regional seismic lines they are expressed as monotonous parallel reflections (e.g. Sangree and Widmier, 1977). Redisplaying conventional seismic lines at reduced horizontal scales (squashing) or increased vertical scales (stretching), commonly by a factor of two to four times, may reveal subtle dips and stratigraphic pinch-outs, or intervals of shelf-delta sedimentation if the resolution is sufficient.

Wide-shelf model

Asquith (1970) suggested that prograding shorelines may feed wide, prograding and aggrading shelf–slope systems (Fig. 8.24c). Although Asquith's original correlations have been questioned (Van Wagoner *et al.*, 1990), other workers have shown similar geometries (e.g. McCave, 1985; Lawrence *et al.*, 1990; S. Sturrock, pers. comm., 1991). The model is important as it indicates that not all prograding siliciclastic slopes are fed directly by deltas — the shoreline may be some distance from the offlap break.

The decision as to which of these three models is appropriate is not straightforward, because the seismic expression of wide-shelf systems is comparable with that of shelf-edge deltas, both in large-scale form (e.g. McMillen and Winn, 1989) and in the potential for generating syn-sedimentary deformational features on the slope. In general though, wide-shelf–slope systems tend to be muddy and to have sigmoidal clinoforms, whereas strongly prograding slopes with toplapping clinoforms are more likely to be fed by shelf-edge deltas.

8.9.2 Seismic resolution

The seismic expression of a single sandstone body isolated in shale is dependent primarily on the relationship between the thickness of the sandstone and the wavelength of the seismic data (see section 3.1.2). For example, at burial depths of 1–2 km. Sandstones with 20% porosity may have a velocity of $2500\,\mathrm{ms}^{-1}$. With a good quality seismic source of 50 Hz, distinct reflections can be seen from the top and the base of sandstones that are thicker than 25 m.

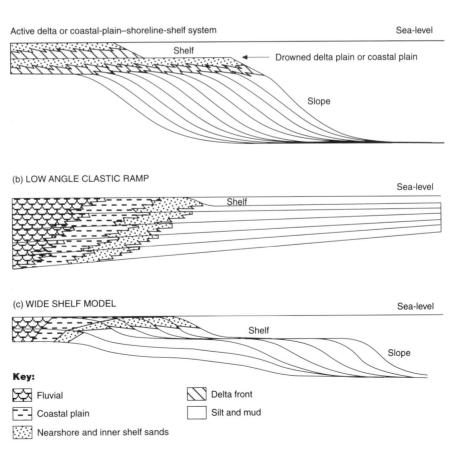

Fig. 8.24 Basin-scale models of paralic systems

However, beds thinner than 6.25 m cannot be distinguished. Beds 25–12.5 m thick will be characterized by constructive interference of reflections from their top and base, and beds 12.5–6.25 m thick will be resolvable only by changes in amplitude (Meckel and Nath, 1977).

Paralic sandstones have mean thicknesses that range from 5 to 30 m (Table 8.2). Most are likely to be less than 15 m in thickness, and many are characterized by thickness changes. As a result, individual sands may be expressed by two reflections, interference, and/or by amplitude changes. In addition, as paralic successions typically comprise a number of closely superposed sandstone bodies, interference of reflections from individual sandstones is likely. In such situations it may be difficult to determine both the number of sandstones and their distribution (Sheriff, 1985). Seismic modelling can be used to help constrain interpretations (Meckel and Nath, 1977).

Horizontal seismic resolution is a function of the wavelength of the seismic data, and the nature of the two-dimensional or three-dimensional data set. For two-dimensional data, seismic migration gives an approximate resolution of 1/4 of the seismic wavelength along the line. Using the parameters stated in the example above this would give a horizontal resolution of 12.5 m. Most paralic sandstone bodies are much wider than this and should be detectable if their thickness is appropriate (Table 8.2). However, unmigrated energy from out of the plane of the line may reduce horizontal resolution. In addition the resolution of two-dimensional seismic data is severely limited by the spacing of the individual seismic lines. By contrast, three-dimensional data migration is performed in three dimensions, giving a resolution of 1/4 of the wavelength in all directions. In addition, three-dimensional data sets are data volumes. There are no line spacing problems. As a result, with good quality data sets, individual sandstones commonly can be tracked on time slices or horizon slices — even thin sands expressed by subtle amplitude changes.

8.9.3 Seismic facies

As a result of the problems of vertical resolution and interference of reflections generated by superposed sand bodies, there are few examples where discrete paralic sandstone bodies have been imaged on conventional two-dimensional seismic data. However, several studies reveal a consistent relationship between paralic depositional environments and seismic facies (Sangree and Widmier, 1977; Bouvier *et al.*, 1989; LeBlanc *et al.*, 1989; Van Vliet and Schwander, 1989). Four key facies are apparent (Fig. 8.25).

1 *High continuity, high amplitude.* Continuous high-amplitude reflections are indicative of extensive uniform shales interbedded with uniform sands, silts, limestones, or coals. Lithological contrasts of this kind are commonly generated at parasequence, parasequence set and sequence boundaries.

2 *Low amplitude.* Low-amplitude reflections indicate either uniform lithologies or thin beds that cannot be resolved. They are typical of offshore marine shales.

3 *Low continuity and variable amplitude facies.* Low continuity and variable amplitude facies are characteristic of fluvial, delta plain and coastal plain successions, which comprise sands of variable size, shape and orientation, scattered in shales.

4 *Low-relief mounded facies.* In addition to being characterized by subtle clinoforms, Sangree and Widmier (1977) argue that shelf deltas may be recognizable by having a broad gently mounded external form.

8.10 Variations in paralic systems within a sea-level cycle

Sequence stratigraphy provides a series of models that allow the temporal and spatial position of broad paralic environments to be predicted (Posamentier and Vail, 1988; Posamentier *et al.*, 1988; Van Wagoner *et al.*, 1990). An understanding of shelf processes, delta position and grain size can be used to refine these models, and to increase their predictive power.

8.10.1 Shelf processes

The interaction between fluvial and shelf processes is critical in determining the spatial distribution of sands within paralic successions. The capacity of shelf processes to redistribute sand is a function of a variety of controls, some of which are unrelated to any sequence stratigraphic framework. For example, for a basin to develop significant tides it must be freely connected to a major ocean: the Mediterranean is essentially tideless because it is too small to generate significant tides, and the straights of Gibraltar are too narrow to allow significant oceanic tidal waves to enter. Nevertheless, several studies have revealed systematic relationships between parasequence stacking pattern and basin process. Where a relationship is apparent, progradational parasequence sets tend to be more fluvially dominated, whereas retrogradational and transgressive parasequence sets tend to be more tide and wave dominated (e.g. Weise 1980; Bhattacharya 1988, 1991; Pulham, 1989; Tyler *et al.*, 1991). The decrease in the effect of fluvial processes in backstepping parasequence sets can be related to a rising base level that drowns and causes the lower courses of rivers to be sluggish. By contrast, shelf width appears to be a major control on marine processes.

Tides

Maximum tidal ranges occur when the shelf is wide enough to allow the shelf sea to resonate with the open ocean tide (Howarth, 1982). This occurs when the shelf has a width of around 300 km, i.e. one-quarter of the wavelength of the ocean tidal wave. Most modern shelves have a width

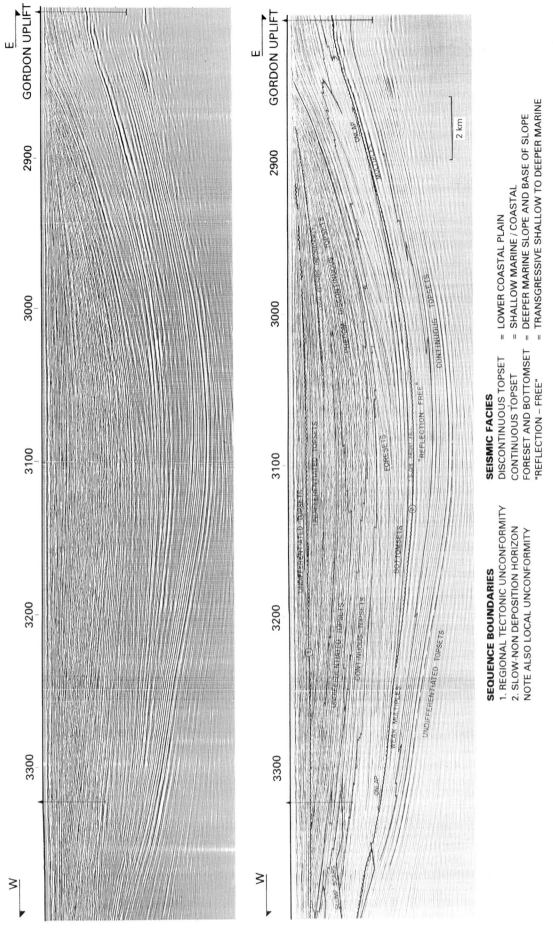

Fig. 8.25 Seismic facies of paralic successions from the Labuan syncline, offshore Sabah, North Borneo (from Van Vliet and Schwander, 1989). Three main topset facies are recognized: a distal 'reflection free' facies, that passes laterally into a 'continuous' facies, which in turn passes up-system into a 'short discontinuous' facies. The facies respectively record: outer shelf shales, interbedded paralic sands and shale, and highly channelized aggrading coastal-plain deposits

less than 300 km and exhibit an approximately linear relationship between shelf width and tidal range, with tidal range increasing as the critical wavelength is approached. Restricted gulfs are often sites of amplified tidal range as they represent areas of increased shelf width. In detail, the basin geometry required for resonance is delicately balanced (Defant, 1961) so that small changes may alter the tidal regime radically (Johnson and Belderson, 1969). This is illustrated by the Bay of Fundy, where changes in basin physiography and sea-level over the last 4000 years have resulted in the tidal regime changing from being microtidal (<2 m tides) to its present macrotidal state. The implication is that temporal changes in the strength of tidal processes are to be expected, particularly when major changes in basin configuration occur, as, for example, across sequence boundaries.

Waves

Coastlines that face the open ocean and the prevailing winds will be more wave dominated than lee-facing coastlines in restricted basins. As shelf width increases, shoaling oceanic waves (i.e. waves that touch the sea bed) are increasingly attenuated and damped by bottom friction. Therefore, wider shelves may be less wave-dominated than equivalent steeper, narrow shelves.

8.10.2 Spatial changes in processes

Individual shelves are not characterized by a single sedimentary process: different processes dominate in different regions. For example, in the modern North Sea the tidal ridges and sand sheets off the English Coast (Stride, 1982) and the Dutch estuaries (Visser, 1980) are well known areas of tide-dominated sedimentation. Yet within the same basin, the Helgoland Bight off Germany is a classic example of a storm-dominated shoreline–shelf (Aigner and Reineck, 1982).

8.10.3 Sequence stratigraphic framework

The relationships between shelf width and sedimentary processes can be combined into a sequence stratigraphic framework, because sequence models suggest a systematic relationship between shelf width and systems tract. The relationships are most clear during the development of a type 1 sequence boundary in a basin with a pronounced shelf–slope break.

In the *highstand systems tract*, the shelf narrows and the system becomes increasingly progradational in response to a decrease in the rate of rise of relative sea-level. The progradational signature favours fluvially dominated deltas that may become increasingly wave-influenced as the shelf width decreases, and the effect of wave damping diminishes.

In the *lowstand systems tract* the rivers initially incise, the shelf is by-passed, and sediment is transported directly to the deep basin. The lowstand prograding wedge, located at the shelf edge and fronted by a distinct slope system, is essentially a shelf-edge delta, located down-dip of the incised valley. The deltaic trunk river forms a point sediment source, fixed at the head of the incised valley, and a strongly prograding shelf-margin delta results. The low rate of generation of shelfal accommodation space favours a sandy shoreline system. Whereas the strong progradation favours a fluvial signature, and the narrow shelf a high degree of wave influence.

An increased rate of relative sea-level rise progressively slows the rate of progradation and may flood the incised valleys so that they take on an estuarine or tide-dominated shelf-delta character. As the interfluves are flooded the rate at which accommodation space is generated increases abruptly, initiating the development of the *transgressive systems tract*. A steady rise in base level causes the rivers to be sluggish in their lower portions, suppressing a fluvial signature. By contrast waves have more time to rework deltaic deposits and the increasing shelf width favours stronger tidal currents, resulting in wave- and tide-dominated deltas. Lateral to deltaic depocentres, the shoreline is likely to be characterized by wave-dominated barrier shorelines, which front large lagoonal areas that can be cut by tidal inlets, depending on the tidal range.

The culmination of the transgression is represented by the maximum flooding surface. In shelfal settings this is represented by a condensed horizon, in the lower delta plain and coastal plain it may be recorded by the maximum landwards penetration of a tidal signature (Shanley and McCabe, 1993).

8.11 Summary

High-resolution sequence stratigraphy is emerging as a powerful tool in the analysis and prediction of paralic successions. It has clear applications in deciphering fundamental stratigraphic controls (namely sediment supply and relative sea-level change) and delineating genetic units both in the subsurface and at outcrop. In the petroleum industry it provides a rigorous framework for subsurface correlation, and can help in the delineation of reservoir flow units. In addition, it can provide a means for classifying analogue data, and it can highlight stratigraphic traps.

Successful high-resolution sequence stratigraphic studies are founded on facies analysis, an approach based on the study of continuous, uninterrupted successions. The advance of sequence stratigraphy has laid increased importance on small-scale facies dislocations, and grain-size changes. It may be time to revisit classic facies models with this in mind. For example, is progradation a steady process, or are small-scale facies breaks that could be attributed erroneously to sea-level fall a common feature?

In contrast with core-based and wireline-log-based

studies, seismic studies are at present relatively crude. This is particularly true for studies based on petroleum industry two-dimensional seismic, where major stratigraphic surfaces, coal prone intervals, and seismic facies may be distinguishable, but key features such parasequences and valley fills cannot be recognized with ease. In the future, the increased resolution of three-dimensional seismic data offers an opportunity for integrating high-resolution sequence stratigraphic concepts with seismic data.

CHAPTER NINE
Deep-marine Clastic Systems

9.1 Introduction

9.2 Deep-marine clastic systems – depositional processes and classification
  9.2.1 Depositional processes
  9.2.2 Classification of deep-marine clastic systems

9.3 Fan development during lowstands
  9.3.1 Lowstand models
  9.3.2 Basin-floor versus slope fan systems
  9.3.3 Lowstand systems tracts and deep-marine clastic systems
  9.3.4 The effect of sediment calibre on the evolution of the lowstand systems tract

9.4 Fan development during highstand and transgression
  9.4.1 Highstand fan systems
  9.4.2 Fan development during transgressions

9.5 Conclusions

9.1 Introduction

Sequence stratigraphy provides a powerful tool for stratigraphic analysis of deep-marine clastic systems when it is combined with an appreciation of the variability in processes and depositional products of deep marine settings. This combined approach based on sequence stratigraphy and sedimentary processes has advanced from early studies of the late 1970s, which emphasized simple model-driven interpretations of deep-marine clastic systems based upon seismic data (e.g. Mitchum *et al.*, 1977a,b; Vail *et al.*, 1977a,b; Mutti, 1985; Posamentier and Vail, 1988; Van Wagoner *et al.*, 1990; Walker, 1992a,b; Posamentier and Weimer, 1993).

This chapter first reviews the range of sediment transport mechanisms in deep marine settings and the controls on coarse clastic deposition within a basinal setting. The principle large-scale deep-marine clastic depositional systems and the controls on their development are discussed. This information is used to develop a suite of sequence stratigraphic models, which show the variety of depositional systems that may develop within a given systems tract.

9.2 Deep-marine clastic systems – depositional processes and classification

9.2.1 Depositional processes

The erosion, transport and deposition of sediment in deep-marine clastic settings are controlled largely by sediment gravity flow processes; flows in which sediment–fluid mixtures move under the influence of gravity (Middleton and Hampton, 1973, 1976; Lowe, 1979, 1982; Middleton and Southard, 1984). Sediment gravity flows form a broad group of genetically related processes, ranging from slumps and slides associated with the downslope translation of cohesive material such as silt and mud, to fully turbulent turbidity currents.

Four basic types of sediment gravity flow are recognized, and include the following (Table 9.1):
1 Turbidity currents, in which the sediment is supported by the upward component of fluid turbulence generated by the density contrast between a sediment–fluid mixture and the surrounding ambient fluid.
2 Fluidized/liquidized flows in which the sediment is supported by the upward movement of the escaping pore fluids.
3 Grain flows in which the particles are supported by the dispersive pressure of colliding grains.
4 Cohesive flows in which particles are supported by matrix density and strength.
Turbidity currents and cohesive flows are considered to be effective agents of sediment transport, and are the principal mechanisms by which turbidite facies* develop within basinal settings (Middleton and Hampton, 1973, 1976; Lowe, 1982; Pickering *et al.*, 1986; Postma, 1986; Pickering *et al.*, 1989). The remaining types of sediment gravity flow are regarded as transient phenomena; occurring between the initiation of sediment movement by slides and slumps, and the final stages of sediment and fluid transport by fully turbulent turbidity currents (Lowe, 1979, 1982; Postma, 1986).

Many authors draw the additional distinction between low- and high-density turbidity current flows (e.g. Middleton

---
* The term turbidite facies is used here in the context of Mutti and Ricci-Lucchi (1972) to refer to all sediments deposited by sediment gravity flows and not just to turbidity currents.

Table 9.1 Characteristic types of sediment gravity flow (modified after Nardin *et al.* (1979) and Cook *et al.* (1982)

Types of mass transport	Internal mechanical behaviour	Transport mechanism and dominant sediment support	Acoustic record characteristics	Sedimentary structures and bed geometry
Rockfall		Freefall and rolling single blocks along steep slopes	Strong hummocky bottom return, hyperbolae and side echoes common. Weak, chaotic internal return: structureless	Grain-supported framework, variable matrix, disorganized. May be elongate parallel to slope and narrow perpendicular to slope
Slide Translational (glide)		Shear failure along discrete shear planes subparallel to underlying beds. Slide may behave elastically at top; plastically at base and thin lateral margins	Internal reflectors continuous and often undeformed; abrupt terminations. Strata of glide blocks may be uncomformable or subparallel to underlying sediment	Bedding may be undeformed and parallel to underlying beds or deformed especially at base and margins where debris flow conglomerate can be generated. Hummocky, slightly convex-up top, base subparallel to underlying beds; 10s to 1000s m wide and long
Rotational (slump)	Elastic	Shear failure along discrete concave-up shear planes accompanied by rotation of slide. May move elastically or elastically and plastically	Internal reflectors continuous and undeformed for short distances with deformation at toe and along base. Concave-up failure plane at head and subparallel to adjacent bedding at toe. Surface usually hummocky	Bedding may be undeformed. Upper and lower contacts often deformed. Internal bedding at angular discordance to enclosing strata. Size variable
Sediment gravity flow Debris flow or mud flow	Plastic	Shear distributed throughout the sediment mass. Clasts supported above base or bed by cohesive strength of mud matrix and clast buoyancy. Can be initiated and move long distances along very low-angle slopes	Sea-floor reflectors may be hyperbolic, irregular, or smooth. Commonly acoustically transparent with few or no internal reflectors. Mounded or lens shaped with blunt termination at head. May be chaotic internally	Clasts matrix-supported; clasts may exhibit random fabric throughout the bed or oriented subparallel, especially at base and top of flow units; inverse grading possible. Clast size and matrix content variable. Occur as sheet to channel-shaped bodies cm^2 to several 10s m thick and 100s to 1000s (?) m long; widths variable
Grain flow		Cohesionless sediment supported by dispersive pressure. Usually requires steep slopes for initiation and sustained downslope movement		Massive; clast A-axis parallel to flow and imbricate upstream, inverse grading may occur near base
Liquefied flow		Cohesionless sediment supported by upward displacement of fluid (dilatance) as loosely packed structure collapses; settles into a tightly packed texture. Requires slopes > 3°	Individual flow deposits very thin; may not be resolvable with present seismic-reflection techniques. Repeated flows may produce a sequence of thin, even, reflectors	
Fluidized flow	Fluid	Cohesionless sediment supported by upward motion of escaping pore fluid. Thin (< 10 cm) and short-lived		Dewatering structures, sandstone dykes, flame and load structures, convolute bedding, homogenized sediment
Turbidity current flow		Clasts supported by fluid turbulence. Can move long distances along low-angle slopes	Thin, even, continuous, acoustically highly reflective units; onlaps slope or raised topography. Discontinuous, migrating and climbing in channel sequences	Bouma sequences. Mm to several 10s of cm thick. 10s to 1000s m in length; widths variable

and Hampton, 1976; Stow and Bowen, 1980; Lowe, 1982; Postma, 1986). The distinction is an important one because the different types of turbidity current flow exert an important control on the final location of sand deposition within a deep marine basin (Reading and Richards, 1994; Richards *et al.*, in press). Where mud-rich provenance areas provide the dominant sediment type to the basin (e.g. mud-rich deltas), high-efficiency turbidity currents and cohesive flows are likely to be more common and any available sand will be transported significant distances away from the base of

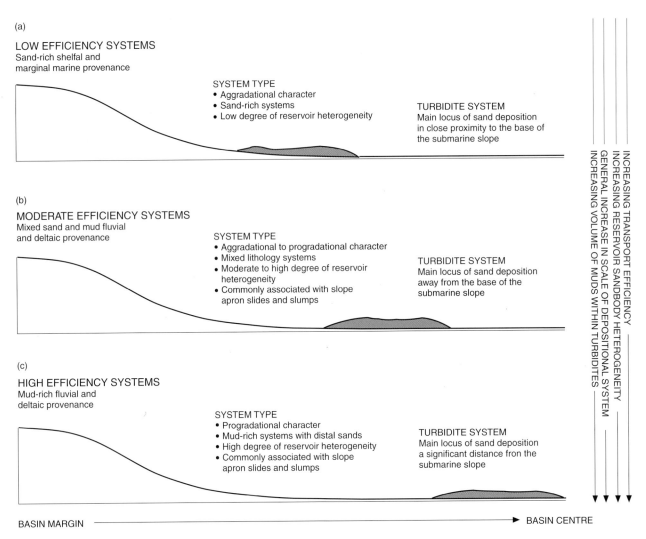

Fig. 9.1 Summary diagram illustrating the concept of transport efficiency and its relationship to the locus of coarse clastic deposition within a basin

the submarine slope to form high-efficiency, mud-rich fans (Fig. 9.1). The depositional records of these systems are likely to comprise thin-bedded and classical turbidites with cohesive debris flows. Conversely, the depositional products of turbidite systems sourced from mixed provenance areas (e.g. fluvial and fluvio-deltaic systems) are likely to show the characteristics of high- and low-efficiency flows. The unstable and highly concentrated nature of high-density flows results in a reduction in transport efficiency and the increasing tendency for sand deposition to occur significantly closer to the basin margin.

Turbidity currents generated from well-sorted sands derived from littoral drift cells or the reworking of relict shelfal and marginal marine sands tend to be deposited close to the base of slope. In these cases the lack of fine grained material results in a reduction in fluid density, fluid buoyancy and turbulence of turbidity current flows such that sands cannot be transported over significant distances (Lowe, 1982). These types of sediment gravity flows equate with the low efficiency or poorly efficient systems of Mutti and Normark (1987).

Finally, bottom currents may provide an additional, but poorly studied mechanism for the transport and redistribution of clastic sediments within deep marine basins and are cited here for completeness. Such currents include contour currents, wind driven surface currents, tides and internal waves (Heezen et al., 1966; Pequegnat, 1972; Bouma and Hollister, 1973). For a detailed discussion on this subject the reader is referred to reviews on this topic presented elsewhere (see Mitchum et al., 1977b; Stow and Holbrock, 1984; Shanmugam et al., 1993).

9.2.2 Classification of deep-marine clastic systems

To be of practical value, sequence stratigraphic concepts and their related predictive models must incorporate some recognition of the variability of turbidite systems† and the range of controls influencing their development. Existing sequence stratigraphic models focus on submarine fans as the sole representative of the deep-marine record of basin fills. However, it is now clear that fans represent one of a wide spectrum of highly variable deep-marine clastic systems including the 'classic' fan model of Walker (1978) through to submarine ramps (Galloway and Brown, 1972; Chann and Dott, 1983; Heller and Dickinson, 1985), channel–levee complexes (Damuth *et al.*, 1983) and slope aprons (Gorsline and Emery, 1959).

The genesis and character of deep-marine clastic systems reflect the complex interplay between a range of autocyclic and allocyclic controls, including sea-level fluctuations, basinal tectonics and the rate, type and nature of sediment supply (Fig. 9.2). These controls are rarely mutually exclusive and are more commonly interdependent (Reading and Richards, 1994). As a result, no single universal model can be used to describe and predict the facies and stratigraphic architecture of deep marine systems. The use of sequence stratigraphy in the prediction and characterization of these systems is best achieved by understanding the variability and controls on the development of turbidite systems and evaluating them in the context of their related depositional systems tract.

† The term turbidite systems is used here to denote the range of clastic systems developed within deep marine settings and include submarine fans, submarine ramps and slope aprons (Reading and Richards, 1994; Richards *et al.*, 1996; Richards and Bowman 1996)

Deep-marine turbidite systems can be categorized into a number of end-member types on the basis of the volume and grain size and the nature of their supplying system (Reading, 1991; Reading and Orton, 1991; Reading and Richards, 1994; Richards and Bowman, in press). These three factors control the sand-body architecture, geometry and internal facies distribution of deep-marine turbidite systems. For example, the method of sediment supply may control the volume and internal facies character of the fan system, whereas the number of entry feeder points to the fan will control its gross geometry and distribution, all other things remaining equal. Finally, the dominant grain size of the deep-marine clastic system directly reflects the underlying control of provenance and source area composition on the depositional processes (e.g. transport efficiency), sedimentation patterns and distribution of coarse and fine-grained facies within the fan.

Four main groups of deep-marine clastic system are identified; gravel-rich, sand-rich, mixed sand–mud and mud-rich systems (Fig. 9.3). The apices of each ternary diagram represent both the nature of sediment supply and the number of entry points feeding the system. The scheme distinguishes between single point-source submarine fans, multiple point-source submarine ramps, and line sourced, slope-apron systems (Reading and Richards, 1994; Richards *et al.*, 1996). The depositional character and subsurface attributes of these systems are summarized in Tables 9.2 and 9.3.

As a general rule, submarine fans and ramps appear to

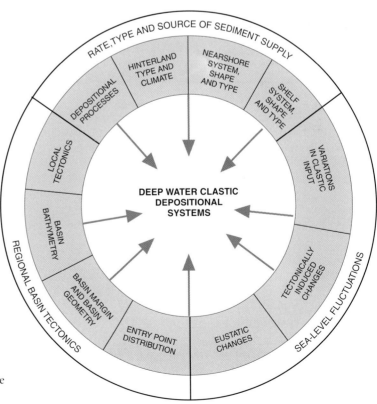

Fig. 9.2 Controls on the development of deep-marine clastic systems (after Richards *et al.*, in press)

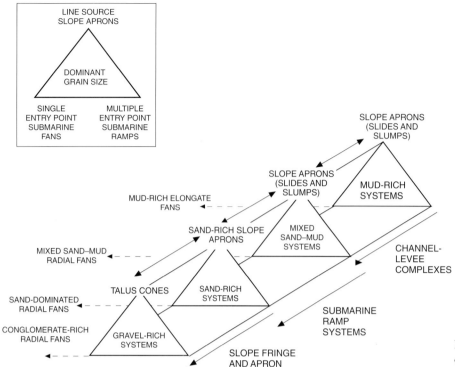

Fig. 9.3 Classification of deep-marine clastic systems by grain size and feeder type (after Richards et al., 1996)

relate to more stable drainage/feeder systems and tend to display more organized and relatively predictable internal architectures and facies distributions. As a result, they are commonly the principal targets for hydrocarbon exploration. Conversely, line-sourced slope aprons are commonly associated with slope failure, local sediment sources and/or ephemeral drainage and feeder systems, resulting in greater facies variability and degrees of internal disorganization. These systems form less attractive targets in sub-surface oil and gas exploration. At the broad scale, each of the systems displays a predictable arrangement of architectural elements, sand-body geometries and depositional facies (Fig. 9.4). These in turn influence the seismic architecture, acoustic expression, and wireline log character of the system, together with sand-body geometry, reservoir architecture and facies recognized in core (Reading and Richards, 1994; Richards and Bowman, in press).

SYSTEM TYPE	WEDGES	CHANNELS	LOBES	SHEETS	CHAOTIC MOUNDS
GRAVEL-RICH SYSTEMS	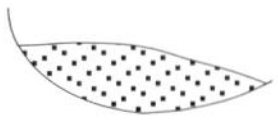 ▨	CHUTES 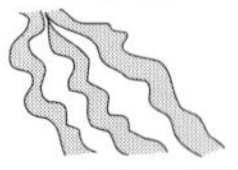			
SAND-RICH SYSTEMS		BRAIDED 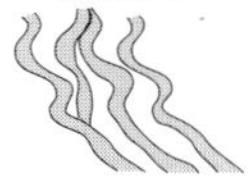	CHANNELIZED LOBES 		
MUD/SAND-RICH SYSTEMS		CHANNEL-LEVEE 	DEPOSITIONAL LOBES 		SLUMPS AND SLIDES 
MUD-RICH SYSTEMS		CHANNEL-LEVEE 	DEPOSITIONAL LOBES 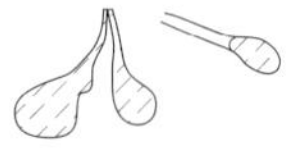 →	 ▬	SLUMPS AND SLIDES 

Fig. 9.4 Principal architectural elements of deep-marine clastic systems (after Reading and Richards, 1994)

Table 9.2 Sedimentological characteristics of deep-marine clastic systems (after Reading and Richards, 1994)

	Feeder system type												
	Single point-source submarine fans categorized by dominant grain size				Multiple-source submarine ramps categorized by dominant grain size				Line-source slope aprons categorized by dominant grain size				
	Mud (MF)*	Mud/sand (MSF)	Sand (SF)	Gravel (GF)	Mud (MR)*	Mud/sand (MSR)	Sand (SR)	Gravel (GR)	Mud (MA)*	Mud/sand (MSA)	Sand (SA)	Gravel (GA)
Size	Large	Large–moderate	Moderate	Small	Moderate	Moderate	Moderate	Small	Large	Moderate	Small	Very small
Slope gradient (m km⁻¹)	Low 0.20–18	Low–Moderate 2.5–18.0	Moderate 2.5–36	High 20–250	Moderate 2.5–25	Moderate 7–35	Moderate >35	High 20–250	Moderate–High 40–150	Moderate–High 40–150	Moderate–High 40–150	Very high 20–500
Shape Radius/length (km)	Elongate 100–3000	Lobate 10–450	Radial/lobate 10–100	Radial 1–50	Lobate 50–200	Lobate 5–75	Linear belt 1–50	Linear belt 1–10	Linear belt 10–100	Linear belt 10–100	Linear belt 1–10	Linear belt 1–5
Source area: size	Large	Moderate	Moderate–small	Small	Moderate	Moderate–small	Small	Very small	Large	Moderate	Small	Very small
gradient	Low	Moderate	Moderate	High	Low	Moderate	Moderate	High	Low	Moderate	Moderate	High
distance	Distant	Moderate	Close	Close	Distant	Moderate	Close	Close	Moderate	Moderate	Close	Close
Feeding systems	Large, mud-rich, river delta	Large mixed-load, river delta and/or down-dip canyon	Shelf failure or shelf canyon	Fan-delta or alluvial cone	Large mud-rich river delta	Mixed-load delta, linear shoreline	Sand-rich clastic shoreline/shelf	Alluvial fan/braid plain/fan-delta	Wide starved shelf	Relict/narrow shelf	Narrow shelf	Braid plain
Supply mechanism	Infrequent slumps and slump-initiated low-density turbidity currents. Contour currents	Mainly high- and low-density turbidity currents	Reworking or direct access to shelf clastics. Low-efficiency turbidity currents	Frequent mass flows, slumps, river generated turbidity currents	Infrequent slumps and slump-initiated low-density turbidity currents. Contour currents	Mainly high- and low-density turbidity currents	Reworking or direct access to shelf clastics. Low-efficiency turbidity currents	Frequent mass flows, slumps, river generated turbidity currents	Major slumps and related low-density turbidity currents, contours currents	Combination low- and high-density turbidity currents	Collapse of shelfal clastics generating low-efficiency turbidity currents	Frequent mass flow slumps, river generated turbidity currents
Size of flows	Very large	Moderate	Moderate–small	Very small	Large	Moderate	Moderate–small	Small	Large	Moderate	Small	Small
Channel system	Large, persistent; meandering to straight with well-developed stable levee system	Moderate scale, meandering to braided systems; laterally migrating with levees	Braided to low-sinuosity impersistent channels and chutes. Rapid lateral migration	Braided, small impersistent chutes	Moderate sizes; channel–levee systems	Multiple, leveed channels with meandering to straight planform	Multiple, laterally migrating braided to low-sinuosity channels	Small, impersistent chutes	None; dominance of failure chutes	Multiple, straight to meandering leveed channels	Multiple chutes and poorly developed impersistent channels	Small impersistent chutes
Distal slope/lower fan sediments	Thin, sheet-like flows forming interbedded sands, silts and muds. Coarse intervals forming thin clastic sheets	Mixed-load turbidity current flows forming lobes of interbedded sands and muds	Sand-rich turbidity current flows forming low-relief lobes and sand sheets	Thin, dilute turbidity current flows forming thin distal turbidites	Thin, sheet-like flows forming interbedded sands, silts and muds	Mixed-load turbidity current flows forming lobes of interbedded sands and muds	Sand-rich turbidity current flows forming low-relief lobes and sand sheets	Thin, dilute turbidity current flows forming thin distal turbidites	Slumps and debris flows, largely mud dominated	Rare sandy turbidity current flows, slumps	Impersistent sand-rich turbidity current flows	Dilute turbidity current flows depositing largely mud and occasional sands
Principal basin plain deposits	Turbidites > hemipelagics	Hemipelagics > turbidites	Hemipelagics	Hemipelagics	Turbidites > hemipelagics	Hemipelagics > turbidites	Hemipelagics	Hemipelagics	High-density mud flows	Hemipelagics	Hemipelagics	Hemipelagics

* MF, mud fan, etc.; MR, mud ramp, etc.; MA, mud apron, etc.

Table 9.3 Subsurface characteristics of deep-marine clastic systems (after Reading and Richards, 1994)

	Feeder system type											
	Point-source submarine fans categorized by dominant grain size				Multiple-source submarine ramps categorized by dominant grain size				Line-source slope aprons categorized by dominant grain size			
	Mud (MF)*	Mud/sand (MSF)	Sand (SF)	Gravel (GF)	Mud (MR)*	Mud/sand (MSR)	Sand (SR)	Gravel (GR)	Mud (MA)*	Mud/sand (MSA)	Sand (SA)	Gravel (GA)
Principal architectural elements: proximal area distal area	Channel–levees Distal sheets	Channel–levees Lobes	Channels Channelized lobes	Wedges Distal sheets	Channel–levees Distal sheets	Channel–levees Lobes	Channels Channelized lobes	Wedges Distal sheets	Slumps, slides Chaotic mounds	Slides Offset lobes	Channels Slumps	Wedges Debris flows
Seismic architecture	Channel–levees, distal parallel reflectors	Channel–levees and mounds	Constructional, low-relief mounds	Wedges	Channel–levees and distal parallel reflectors	Channel–levees and mounds	Constructional, low-relief mounds	Wedges	Chaotic mounds	Chaotic mounds	Mounds	Wedges
Sand percentage	≤30	≥30 to ≤70	≥70	Variable 5–50 (>50% gravel)	≤30	≥30 to ≤70	≥70	Variable 5–50 (>50% gravel)	Highly variable, 0–20	Highly variable	Highly variable	Highly variable
Sand-body geometry	Large, lenticular channels with multiple, variable scale sand, silt and mud fills. High degree of heterogeneity. Distal fan dominated by thin sand, silt and mud sheets	Lenticular channels dominated by sand or mud-fill. Down-dip lobes formed of interbedded and alternating sands, silts and muds	Broad sheet-like to low-relief-lobate sand-body geometries dominated internally by channelized sandstone units	Irregular interconnected gravels. Proximal areas dominated by conglomerates and breccias. Sands dominant within medial to distal parts of system	Moderate size sand bodies within overall large channel form. Sands commonly isolated in both down-dip and up-dip directions	Offset stacked, lenticular channel sand bodies bounded by levee fines, passing down-dip into offset stacked lobate sand bodies formed of sandstones and mudstones	Broad sheet-like to low-relief lobate sand-body geometries dominated internally by channelized sandstone units	Irregular interconnected gravels. Proximal areas dominated by conglomerates and breccias. Sands dominant within medial to distal parts of system	Limited sand development. Commonly confined to slide scars and slump generated lows	Laterally extensive, separated by silts and muds	Lobate sand bodies dominated by interconnected channelized units	Laterally extensive, distally limited
Turbidite facies (after Mutti and Ricci-Lucchi, 1972)	A,B,E,F	C,D	B,C	A,B,F	A,B,E,F	C,D	B,C	A,B,E,F	F	D,F	D,E	A,B,F
Reservoir heterogeneity	High	High–moderate	Low	High	High	High–moderate	Low	High	High	High–moderate	Low	High
Communication: vertical lateral	Poor Poor	Moderate Poor	Good Good	Good Poor	Poor–moderate Moderate	Moderate Moderate	Good Good	Good Moderate	Poor Poor	Moderate Moderate	Very good Very good	Good Very good
Common reservoir trap type	Stratigraphic	Stratigraphic	Structural	Structural	Stratigraphic	Stratigraphic	Structural	Structural	Stratigraphic	Stratigraphic	Structural	Structural
Importance and position on relative sea-level cycle	Important, lowstands	Potentially important, highstand and lowstands	Not important	Not important	Important, lowstand	Potentially important, highstand and lowstand	Not important, highstand and lowstand, major flooding events	Important, lowstand and highstand	Highly important, highstand and rising base level	Low importance, rising/falling base level	Low importance, falling base level	Low importance, falling base level

* MF, mud fan, etc.; MR, mud ramp, etc.; MA, mud apron, etc.

Gravel-rich systems

Gravel-rich submarine fan and ramp systems are generally small in scale and commonly form the down-dip marine equivalents of coarse-grained fan-delta, alluvial cone and braid plain settings (Fig. 9.5, Tables 9.2 and 9.3). Gravel-rich slope aprons are derived from reworking, mass-wasting and catastrophic submarine rock fall avalanches associated with high-angle slopes and scarps (Fig. 9.5a). They commonly occur adjacent to relict or active submarine fault scarps. Sediment is often poorly sorted prior to deposition. The depositional products of these systems reflect a wide variety of mass flow processes and include non-channelized chaotic boulder/cobble beds, and intraformational rotational slumps of fine grained interbedded sandstone and mudstones with exotic clasts. These coarse-grained facies commonly interbed with turbidite sandstones and mudstones (Surlyk, 1978; Ineson, 1989). Ephemeral input from up-dip alluvial fans and fan-delta systems may liberate sand-grade material to the apron, resulting in the formation of more sand-rich facies fringing the distal edges of the system. In seismic data the slope apron shows little coherent acoustic character and forms wedge-shaped aggradational packages which thicken towards relict fault scarps.

Gravel-rich submarine fan and ramp systems typically form broad, wedge-shaped sediment bodies, characterized in their proximal parts by conglomerates and sandstones, derived from rock-fall and debris flows (facies A and F of Mutti and Ricci-Lucchi, 1972; Surlyk, 1978; Kessler and Moorhouse, 1984; MacDonald, 1986; Ineson, 1989; Ferrentinos et al., 1988; Piper et al., 1990) as well as thick-bedded and erosively based gravel- and sand-rich high-density turbidites (Fig. 9.5b,c). Sand-rich facies are best developed within the medial parts of the system, where they occur as stacked, massive or graded sandstones characterized by dewatering and fluid-escape structures typical of high-density turbidites (facies B, Mutti and Ricci-Lucchi, 1972; Lowe, 1982). The distal margins of the systems pass abruptly into thin sandstone turbidite beds and interbedded, hemipelagic shales (Ineson, 1989; facies D, Mutti and Ricci-Lucchi, 1972; Prior and Bornhold, 1989). The fan and ramp systems show wedged-shaped seismic geometries with little or no coherent internal seismic character. The wedge commonly thickens into the hanging-wall areas adjacent to up-dip transfer segments. Rapid lateral and vertical facies variations are the dominant depositional motif of gravel-rich fans and ramps. This facies variability is illustrated by the internal architecture of gravel-rich submarine ramps from the Jurassic syn-rift successions of the Wollaston Foreland, East Greenland (Surlyk, 1978) and the South and North Brae fields of the North Sea Basin (Harms et al., 1981; Stow et al., 1982).

Sand-rich systems

The term sand-rich denotes those submarine fan, slope-apron and ramp systems that exhibit a sand content exceeding 70%, when measured throughout the system (Fig. 9.6). This sand content defines the lower limit for a major change in the seismic character and facies architecture of deep marine turbidite systems (Reading and Richards, 1994; Richards et al., 1996). Sand-rich systems form relatively small-scale features (generally 1–50 km radius) sourced from the incision or failure of relict sand-rich shelves or by direct canyon access to littoral drift cells (Reading and Richards, 1994).

Sand-rich slope aprons are derived from the reworking and mass wasting of relict shelf–slope systems, and form isolated wedge-shaped sand bodies paralleling the basin margins (Fig. 9.6a). The slope apron is limited in basinward extent, reflecting the local nature of sediment source areas to the turbidite system. Coarse-grained high-density turbidites dominate the proximal facies of the system, whereas more distal areas show an increasing basinward interbedding of hemipelagic mudstones (Kumar and Slatt, 1984). The systems form lobate to aggradational wedged-shaped seismic packages that bank up against basin-margin slopes. The apron is internally seismically opaque with rare development of clinoform geometries.

Sand-rich fan and ramp systems are dominated by up-dip channel sand bodies, which pass down-dip into channelized fan lobes (Nelson and Nilsen, 1984; Busby-Spera, 1985; Kleverlaan, 1989; Reading and Richards, 1994). Studies of outcrop and subsurface examples demonstrate similar high sand : shale ratios throughout the extent of the fan and ramp system (Fig. 9.6b,c). The mid-fan and medial ramp comprise elongate, channelized lobe systems, which coalesce to form a broad sand-sheet (Link and Nilsen, 1980; Link and Welton, 1982; Chann and Dott, 1983; Busby Spera, 1985; Heller and Dickinson, 1985; Guardado et al., 1989; Kleverlaan, 1989). The systems are largely dominated by massive to internally graded high-density turbidites, with only subordinate thin-bedded turbidites forming fan-abandonment and fringe deposits. Sand-body continuity and connectivity are typically very good, although more widespread shales may occur, reflecting lobe abandonment intervals (e.g. McGovney and Radovitch, 1985). Mid-fan deposits are dominated by abrupt changes from shale into thick, amalgamated sandstones. Cleaning upward, coarsening upward trends are noticeably absent, whereas the transition from fan/ramp to basin plain is generally abrupt.

On seismic data the fan and ramp systems occur as a single seismic package, with poor to well-developed mounding. The seismic envelope defines the full extent of the turbidite system. Internal reflectors may be subhorizontal to low-angle inclined with rare clinoform geometries. Examples of sand-rich ramps and fans include the Tortonian sandy fans of the Tabernas Basin (Kleverlaan, 1989), the Eocene Matilija and Rocks Sandstone (Link and Nilsen, 1980; Link and Welton, 1982), and the Jurassic Magnus and Miller systems of the early post-rift fill of the North

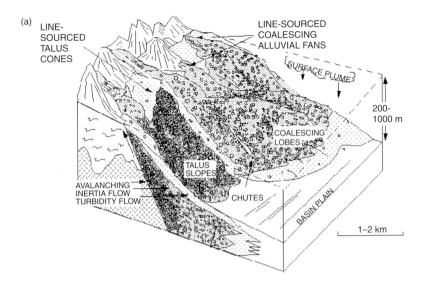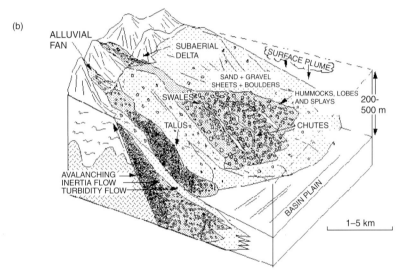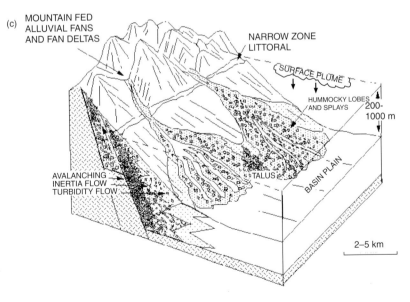

Fig. 9.5 Summary block diagrams of gravel-rich, deep-marine clastic systems (after Reading and Richards, 1994)

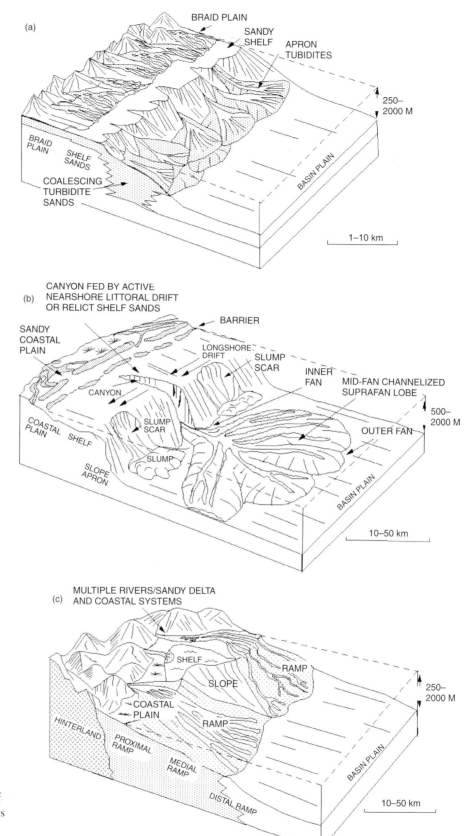

Fig. 9.6 Summary block diagrams of sand-rich, deep-marine clastic systems (after Reading and Richards, 1994)

Sea and Palaeogene Frigg, Cod, Balder, Andrew and Forth Fans, also of the North Sea Basin (Heretier et al., 1979; Kessler et al., 1980; De'Ath and Schuyleman, 1981; Sarg and Skjold, 1982; McGovney and Radovitch, 1985).

Mixed sand—mud systems

The term mixed sand-mud is used here to denote fan and related systems that exhibit sandstone contents between 30% and 70% throughout the system (Fig. 9.7). These systems commonly are derived from large mixed-load delta, shoreline and coastal plain provenances. Mixed sand—mud slope aprons, fans and ramps form moderate-scale features (10–350 km radius) and account for a major portion of the ancient deep-marine clastic record (Reading and Richards, 1994).

Mixed sand—mud slope aprons are characterized by a wide variety of mass flow processes, resulting in complex, irregular and often disorganized lithofacies distributions (Fig. 9.7a). The apron system is dominated largely by slump packages composed of deformed hemipelagic shales and contorted thin-bedded turbidites, slide blocks and chutes infilled with slope-related mudstones or thick- and thin-bedded, discontinuous turbidites and debris flows (Hill, 1984; Nelson and Maldanado, 1988; Alonso and Maldanado, 1990). Gullies and constructional channel systems may traverse the disrupted slope-apron surface and pass laterally and basinward into more stable areas, where laminated mudstones and sandy mudstones predominate. Local development of turbidites within these gullied and channelled areas may lead to isolated channel-fills of heterogeneous turbidites and sandstones with interbedded mudstones. The systems commonly form well-developed mounded seismic packages dominated internally by hummocky to chaotic reflectors (Nardin et al., 1979). The mounds are located in close proximity to the base of the submarine slope owing to sediment input to the apron from mass wasting and failure. Detailed mapping of the internal seismic facies commonly reveals a chaotic seismic character.

Mixed sand—mud submarine fans and ramps are dominated by two main architectural elements; a channel—levee system and down-dip depositional lobes (Fig. 9.7b,c; Normark, 1978; Walker, 1978; Normark et al., 1979; Droz and Bellaiche, 1985). The dominance of one or either of these architectural elements in seismic data is controlled by the grain size of the system. In general, increases in mud content within the fan or ramp result in a reduction in the relief and acoustic expression of lobes and the increasing dominance of channel—levee systems.

Channel—levee systems form the conduits through which sediment is distributed to the main area of the fan. These may be mud-filled, reflecting rapid fan abandonment, or contain a central core of coarse-grained, highly heterogeneous channel-fill deposits, flanked by levee siltstones and mudstones (Walker, 1978; Winn and Dott, 1979; 1985; Tyler et al., 1984; Mutti et al., 1985a; Weuller and James, 1989; Schuppers, 1992). Channel fill facies vary from sandy conglomerates and pebbly sandstones with thick bedded, high-density turbidites to fine-grained thin-bedded turbidites and hemipelagic mudstones. Individual sandstone beds are commonly lenticular and erosively based. In high-resolution and multichannel seismic data, the channel system may display a poorly to well-developed wedge-shaped or 'gull wing' geometry dominated by diffuse or well-defined low-angle reflectors, which downlap away from the apex of the channel axis.

Constructional lobes form the down-dip equivalents of the channel—levee system (Fig. 9.7b,c). They comprise overlapping, layered sand bodies characterized by complex and variable sand-body geometries and architecture. Sand : shale ratios are highest within the core or apex of the lobe, but decrease markedly towards the lobe margins. The lobe may be dominated by massive, thick-bedded high-density turbidites (Kleverlaan, 1989; Kulpecz and Van Geuns, 1990) or display more classic turbidites and interbedded hemipelagic shales (Stevens Fan, MacPherson, 1978; Webb, 1981; Marnosa-Arenacea Fan, Ricci-Lucchi and Valmori, 1980) with poorly developed cleaning upward log cycles or coarsening upward grain-size trends. Such fan systems display a high degree of variability at the small to medium scale. On seismic data, mid-fan depositional lobes and associated channel systems show poor to well-developed hummocky acoustic character. These features may occur as isolated mounds developed a significant distance from the submarine slope. The depositional character and facies of mixed sand—mud fans and ramps are illustrated in studies by Mutti (1972, 1985, 1992), Walker (1978, 1985), Kleverlaan (1989) and Pickering (1981). Subsurface examples include the Forties, Nelson and Gannet fan and ramp systems, North Sea Palaeogene (Armstrong et al., 1987; Kulpecz and Van Geuns, 1990), the Miocene Yowlumne field, San Joaquin Basin, California (Berg and Royo, 1990) and the Permian Spraeberry Trend of West Texas (Tyler and Gholston, 1988).

Mud-rich systems

Mud-rich turbidite systems contain less than 30% sand (Fig. 9.8). They form a common feature of basins with mature drainage patterns, large source areas and river and deltaic systems dominated by fine-grained suspended load (Reading and Richards, 1994). Mud-rich systems are volumetrically the most important deep-water clastic systems occurring in the world's oceans today (Damuth and Kumar, 1975; Kolla et al., 1980; McHargue and Webb, 1986; Kolla and Coumes, 1987; Weimer, 1990), and form large-scale slope aprons, fans and ramps 50–3000 km in radius (Reading and Richards, 1994) off many continental margins.

Mud-dominated slope aprons are broadly similar in character to their sandier, mixed sand—mud counterparts

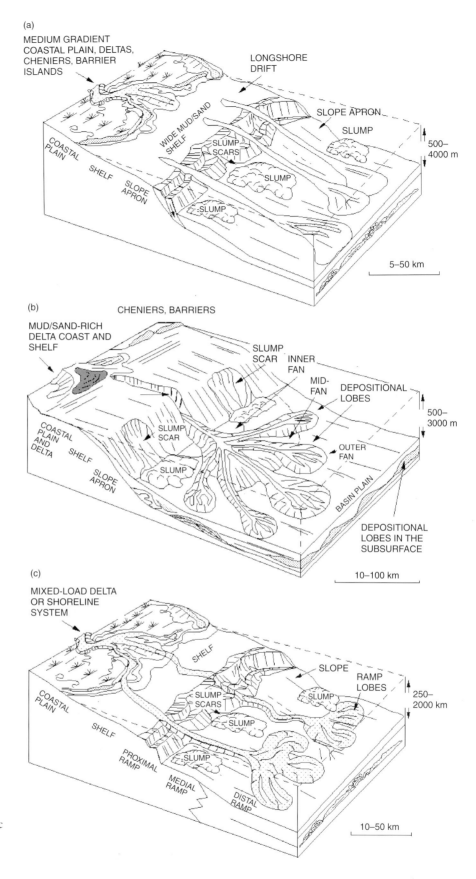

Fig. 9.7 Summary block diagrams of mixed sand–mud, deep-marine clastic systems (after Reading and Richards, 1994)

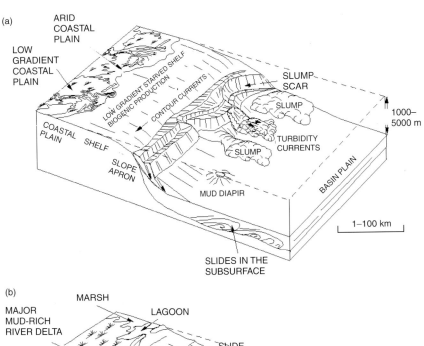
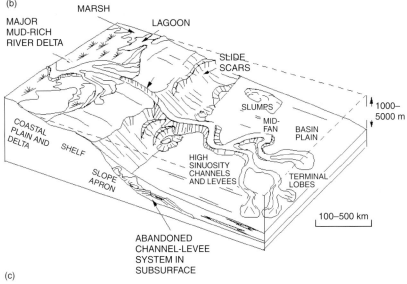
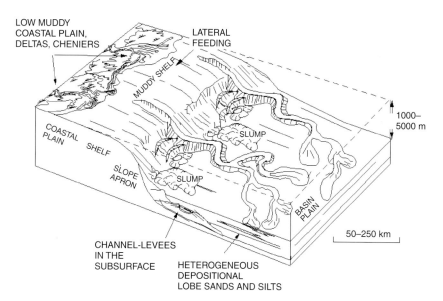

Fig. 9.8 Summary block diagrams of mud-rich, deep-marine clastic systems (after Reading and Richards, 1994)

(Fig. 9.8a). The slope margin is characterized by erosional gullies and rotational slumps generated by sediment loading, and foundering together with curvilinear extensional faults associated with the downslope translation of submarine slides (Gorsline and Emery, 1959; Nardin et al., 1979). The base of the mud-rich apron may be characterized by silt- and mud-dominated lobes, debris flows and uneven topography generated by the compressional toes of slides. The nature of depositional processes within the system leads to a complex and irregular distribution of fine-grained lithofacies. Irregularly distributed sandstone turbidite packages may develop where retrogressive slumping exhumes relict coarser grained shelf and upper slope deposits. Slide gullies located along the margins of the slope apron may form conduits for basinward sediment transport in shelf-edge delta systems. Slope aprons show a dominance of mounded seismic packages characterized by hummocky to chaotic reflectors, which may pass basinward into discontinuous parallel reflectors.

Mud-rich fan and ramp systems are typically large systems (up to thousands of kilometres) dominated by well-developed channel–levee systems (Kolla et al., 1984; Manley and Flood, 1988; Weimer, 1990; Fig. 9.8). Sands are poorly developed within the system and are confined to areally restricted, highly heterogeneous channel-fill sandstones within the axis of the channel–levee system (Imperato and Nilsen, 1990; Weimer, 1990). The outer fan area is dominated by thin, distal sheet-sands (Pickering, 1983; Nelson et al., 1992; Twitchell et al., 1992). Mud-rich fans and ramps are generally readily distinguished on seismic data because of their large scale and the characteristic development of extensive channel–levee systems (Bellaiche et al., 1981; Kolla et al., 1984; McHargue and Webb, 1986; Kolla and Coumes, 1987; Damuth et al., 1988). The upper or inner fan area is dominated by chaotic and mounded seismic reflector packages, reflecting slope failure and slumping, separated by more organized channel–levee systems. These latter features show a characteristic 'gull wing' seismic geometry with an orientation perpendicular to the channel axis. The mid-fan area displays smaller scale, gull wing geometries, reflecting channel–levees with limited seismic relief. The outer or lower fan system is manifest seismically as more continuous, parallel to subparallel reflector sets (Damuth, 1980; Droz, 1983; O'Connell, 1986).

9.3 Fan development during lowstands

9.3.1 Lowstand models

Over the past two decades, sequence stratigraphic models have been used widely to predict and locate submarine fan reservoirs in frontier and mature basin areas. These models assume that the growth and deposition of a turbidite system are intimately tied to a cycle of eustatic or relative sea-level change. Early sequence stratigraphic models envisaged fan development to be related to eustatic sea-level fall, where base level fell below the offlap break, causing entrenchment and erosion of fluvial systems into a subaerially exposed shelf, and the bypass of coarse-grained clastic sediment into the basin (Fig. 9.9, Mitchum et al., 1977a,b; Vail and Todd, 1981; Posamentier and Vail, 1988; Posamentier et al., 1988; and see Chapter 2). The implications of this model were clear; sequence boundaries generated by eustatic sea-level fall should be associated with submarine fan complexes in the basin. The model went further in suggesting that the timing of fan deposition could be predicted from the eustatic sea-level curve (Vail et al., 1977a,b,c; Shanmugam and Moiola, 1985). Early lowstand fan models largely ignored the role played by tectonics, sediment type and sedimentation rate in influencing the growth, architecture and evolution of these systems through time (Shanmugam et al., 1985; Kolla and Macurda, 1988; Posamentier and Vail, 1988; Posamentier et al., 1988; Galloway, 1989; Reading and Richards, 1994).

For the Quaternary stratigraphy record there is some evidence to support a eustatic control on fan development. The Bengal and Indus fans, two of the largest modern fans, both show an abrupt decrease in terrigenous clastic input following the Holocene sea-level rise. A similar picture is seen in recent deep-sea drilling evidence from the Mississippi and Amazon fans, which show comparable changes in sedimentation patterns from Holocene pelagic deposition ($3-30$ cm $10\,000$ years^{-1}) to rapid turbidity current deposition (1200 cm $10\,000$ years^{-1}) during the glacial lowstand (Fig. 9.10). This pattern is also repeated in the Astoria fan system (Shanmugam et al., 1985).

In the geological record, most of the hydrocarbon-bearing fans are interpreted to relate to periods of relative sea-level fall, if not global eustatic lowstands (Fig. 9.11). However, in many of these cases proving correlation of fan development with eustatic lowstand events is difficult. It is clear that base level fall must have a profound effect on the potential for erosion of continental, marginal marine and shallow marine depositional systems. Whilst fan development may occur at any time, the volumes of clastic sediments delivered to a basin for fan development during highstand or transgression are unlikely to match the potential volumes of clastic input liberated during periods of relative sea-level fall.

9.3.2 Basin-floor versus slope fan systems

Models for deep-water, lowstand depositional systems recognize two major types of fan. These comprise a 'basin floor' or lowstand fan and a 'slope fan' forming part of the lowstand wedge (Mitchum et al., 1977a,b; Posamentier and Vail, 1988). These terms require some clarification in the light of the classification scheme above.

The lowstand or basin-floor fan is considered to form a

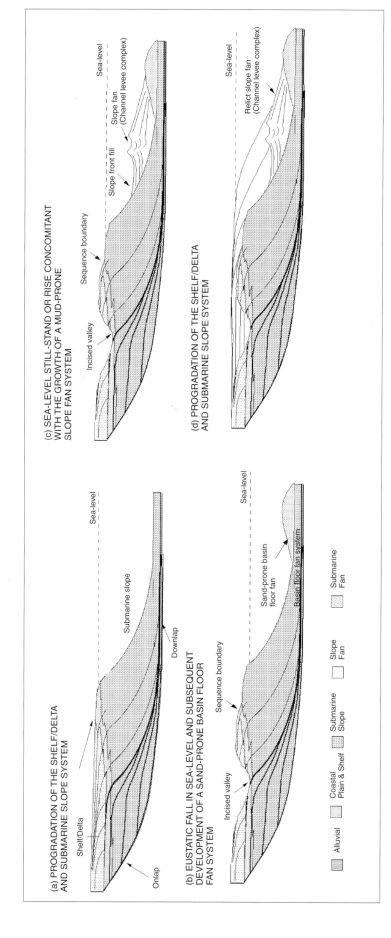

Fig. 9.9 Early models for fan development during eustatic falls in sea-level (after Vail and Todd, 1981). See text for detailed explanation

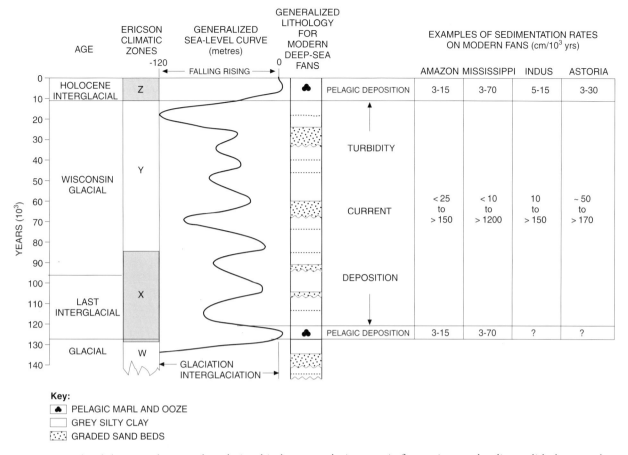

Fig. 9.10 Generalized diagram showing the relationship between glacio-eustatic fluctuations and sediment lithology on the Amazon and other deep-sea fans and adjacent areas during the last 130 000 years (from Shanmugam et al., 1985). Age boundaries, Ericson climatic zones, sea-level curve and lithologic log are modified from Damuth (1977). Data from other areas based on Damuth and Kumar (1975), Damuth (1977), DSDP Leg 96 Scientific Party (1994) and Ewing et al. (1958)

more sand-prone system in response to the increased stream capacity and enhanced alluvial gradients associated with the early phases of relative sea-level fall (Van Wagoner et al., 1990). During this period, the shelf system is largely bypassed and the major locus of coarse sand deposition is shifted to the basin (Posamentier and Vail, 1988; Van Wagoner et al., 1990). This increase in sediment calibre and load may be enhanced further by the reworking and entrenchment of incised valleys on the shelf. This seaward shift in deposition associated with the relative sea-level fall may be recognized as a downshift in coastal onlap in seismic data, a change in shelfal and shoreline parasequence stacking patterns and the abrupt change from hemipelagic mudstones to clean sandstones in logs penetrating the basinal succession (Fig. 9.12; Van Wagoner et al., 1990; Vail and Wornardt, 1990; Posamentier and Erskine, 1991).

The slope fan is developed as relative sea-level stabilizes and later starts to rise. During this period, progressive reductions in alluvial gradients associated with the increasing shelfal accommodation result in a decrease in the volume and calibre of sediment supplied to the basin. This change in the character of basin depositional systems is accompanied by coastal onlap and the progressive landward shift in deposition associated with increasing shelfal accommodation and the progressive infilling of up-dip incised valleys. The relative sea-level rise may be recognized by a change in shelfal and shoreline parasequence stacking patterns and the transition on logs from clean sandstones of the lowstand fan into fine-grained, interbedded sandstones and mudstones reflecting levee channel and overbank deposits of the slope fan (Fig. 9.13; Posamentier and Vail, 1988; Vail and Wornardt, 1990; Van Wagoner et al., 1990; Posamentier and Erskine, 1991).

Seismic expression

Basin-floor fan and slope fan are terms used widely in the literature. Studies suggest that these two fan types can be recognized in both seismic and wireline-log data based on the simple criteria outlined above (Posamentier and Vail, 1988; Vail and Wornardt, 1990; Posamentier and Erskine, 1991; Sangree et al., 1991). Furthermore, each fan type is

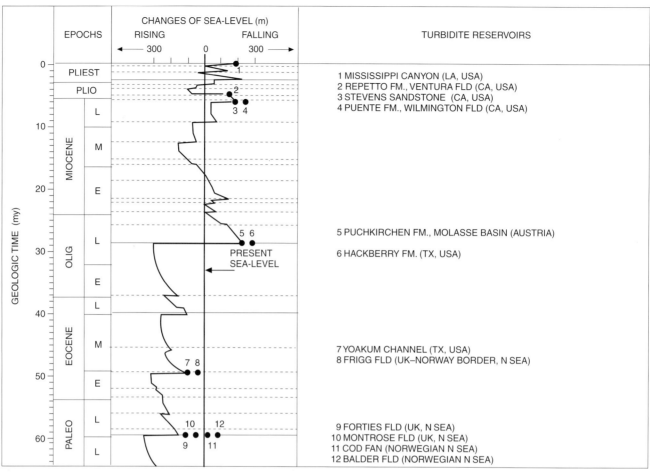

Fig. 9.11 Apparent correlation of hydrocarbon-bearing submarine canyon and fan deposits with periods of low sea-level (after Shanmugam and Moiola, 1982)

considered to exhibit a different sand body architecture, sand : shale ratio and facies.

Basin-floor and slope fans are considered to occur at different times during the relative sea-level cycle and are regarded therefore as separate, mutually exclusive entities. Mitchum (1985) has argued that good seismic evidence exists in a number of examples to show a distinct difference in the timing of basin-floor and slope fan development. This interpretation has been questioned in numerous papers in the sedimentological literature, which suggest that basin-floor and slope fans may form the components of the same fan system, rather than being true fans in their own right (Walker, 1978; Mutti and Normark, 1987; Walker, 1992b; Kolla, 1993). To date, the evidence in favour of contemporaneity between basin-floor and slope fan systems has been equivocal because of the lack of stratigraphic resolution in outcrop data needed to resolve the timing of fan development.

The terms basin-floor fan and slope fan are misleading and inappropriate for describing fans for two main reasons. Firstly, they refer to the seismic expression of a specific architectural element of a fan system, rather than representing fans in their own right. Secondly, the terms assume that the spectrum of deep-marine clastic systems can be represented by two single depositional models (Kolla, 1993; Reading and Richards, 1994).

The seismic expression of a fan system is largely controlled by the acoustic properties of the dominant grain size, and lithology of the fan and its contrast with hemipelagic basinal sediment. As discussed earlier, the lithological character of the fan is controlled by sediment source area composition (Reading, 1991; Reading and Richards, 1994; Richards and Bowman, in press). Furthermore, coarse-grained fan systems are composed of a different set of architectural elements to their finer grained counterparts (Fig. 9.4). Sand-rich fan systems are expressed as a single seismic envelope (mound) developed within a basin-floor location, where the boundaries of the envelope define the whole fan system and its component architectural elements (Figs 9.6 and 9.14). This type of fan system would be referred to as a basin-floor fan.

In contrast, mixed sand-mud fans contain architectural

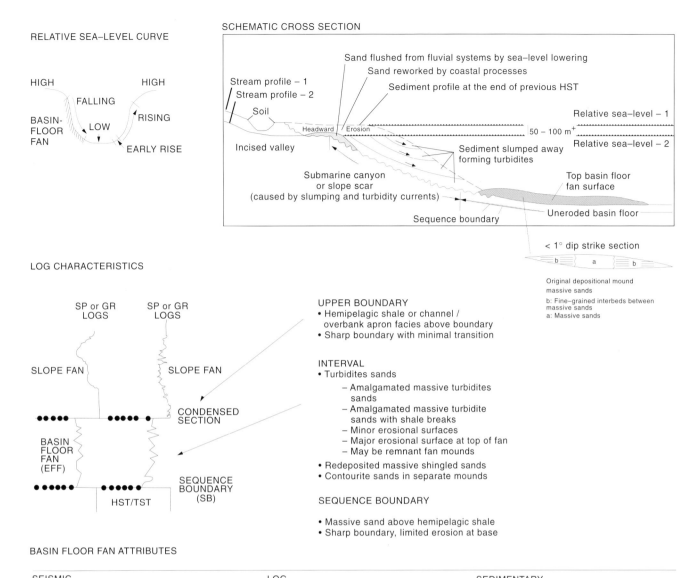

Fig. 9.12 Key characteristics and log responses of a basin-floor fan system (after Vail and Wornardt, 1990; Kolla, 1993). SP, spontaneous potential; GR, γ-ray; HST, highstand systems tract; TST, transgressive systems tract

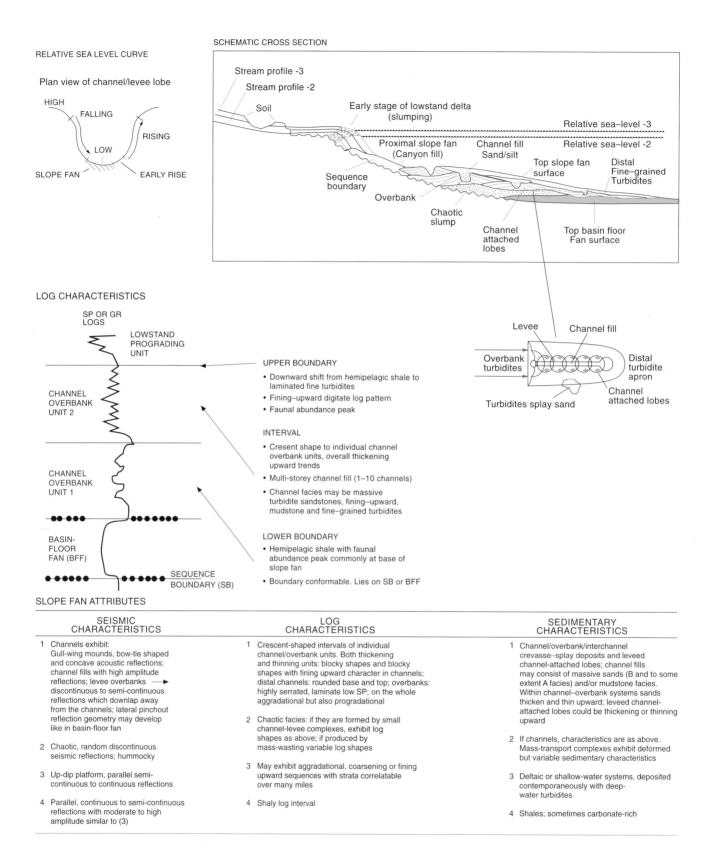

Fig. 9.13 Key characteristics and log responses of a slope-fan system (after Vail and Wornardt, 1990; Kolla, 1993) SB, sequence boundary; BFF, basin-floor fan

elements that mimic the seismic expression of both basin-floor and slope fans (Figs 9.7 and 9.15; Normark, 1978; Droz and Bellaiche, 1985). The levee and channel-fill elements may not be imaged seismically where their lithological character is similar to the background hemipelagic sediments. Where sufficient lithological and acoustic differences exist, the channel–levee system is imaged as a short, stubby 'gull-wing' seismic geometry dominated internally by poor to well-developed downlapping reflector sets, which pass laterally into a channel-fill characterized by either hummocky clinoforms (sand-filled) or parallel reflector sets (mud-filled channel). Down-dip, the constructional lobes of the fan are commonly expressed as a series of overlapping or mutually exclusive constructional mounds characterized internally by clinoforms or parallel reflector sets. In this case the depositional lobes of the fan form constructional mounds located within the basin (basin-floor mounds) that can be traced up-dip into the channel–levee system. In conventional terminology the channel–levee complex would be referred to as a 'slope fan', whereas the basinal, time equivalent facies would be represented by 'basin-floor' lobes.

For more mud-rich fan systems, the dominant architectural elements of the fan are large-scale (> 10 km) channel–levee systems, within an overall gull-wing geometry (Figs 9.8 and 9.16). The features show characteristic wedged-shaped seismic geometries, dominated internally by well-developed downlapping reflectors oriented at right angles to the main axis of the channel systems (McHargue and Webb, 1986; Kolla and Coumes, 1987; Weimer, 1990). The acoustic response of mud-rich systems therefore is characteristic of a slope fan.

Basin-floor fans and slope fans on wireline logs

The application of wireline log analysis to basinal environments, and the main criteria for recognizing major stratigraphic surfaces in the basin are outlined in section 4.4.5. Criteria also have been developed to recognize 'basin-floor' and 'slope-fans' on the basis of core-calibrated wireline log response (Pacht *et al.*, 1990; Vail and Wornardt, 1990; Vail *et al.*, 1991; Figs 9.12 and 9.13). Sequence boundaries are considered to occur at the base of log patterns indicating an abrupt increase in silt and/or sand overlying hemipelagic shales. Sequence boundaries at the base of basin-floor fans are easy to define by the abrupt juxtaposition of a blocky 'box car' sand-log pattern over 'rail road' log patterns typical of hemipelagic shales (Vail and Wornardt, 1990; Vail *et al.*, 1991). Where the slope fan rests directly on older sequences and no basin-floor fan exists, the sequence boundary is more difficult to identify and is placed at the first significant increase in silt content above the hemipelagic shales. The sequence boundary separates the railroad track pattern of the underlying hemipelagic shale section from the crescent-shaped log pattern of the coarsening/fining upward packages of the slope fan (Vail and Wornardt, 1990).

These models should be used with some caution, and they require calibration to core and seismic data before final interpretations are made (Kolla, 1993). Without calibration, log-based models for fan prediction are imprecise. More recent studies illustrate the relationship between log character and the depositional elements of the fan system based upon the integration of core, wireline log and seismic data. These studies show that sand-rich fan systems show log responses characteristic of 'basin-floor fans', whereas log patterns for mud-rich fans compare favourably with 'slope fans'. This apparently simple relationship is invalidated in mixed sand–mud fans, where the system may show both 'basin-floor fan' and 'slope fan' log signatures (Fig. 9.17).

9.3.3 Lowstand systems tracts and deep-marine clastic systems

Three principal models have been documented in the literature for lowstand systems tracts developed within shelf–slope, ramp and extensional-basin-margin settings (Mitchum *et al.*, 1990; Vail and Wornadt, 1990).

In basins with a pronounced shelf–slope break, the lowstand systems tract deposited in a basin is commonly separated into a lowstand fan unit and a lowstand prograding wedge (see section 2.4.3 and Fig. 9.18a). The high sand : mud ratio lowstand fan unit is followed by a finer grained lowstand wedge dominated by slope deposits that have been described variously as slope fan, slope-front-fill, wedge and submarine fans. These slope-fan systems are characterized by active leveed channel deposits at the apex of the canyon mouth.

In ramp margins, the lowstand systems tract is considered to form a two-part wedge (see section 2.4.7 and Fig. 9.18b). The lowstand wedge therefore comprises up-dip incised valley deposits, which may be traced down-dip to one or more progradational parasequence sets of the lowstand prograding complex. The top of the lowstand wedge is the transgressive surface; the base of the lowstand wedge is the lower sequence boundary.

Van Wagoner *et al.* (1990) have argued that basin-floor and slope fans are absent on ramp margins due to the lack of long submarine slopes required to promote turbidity current development. This assumption is questionable given that turbidity current flows may generate on extremely low-angle slopes of less than 1°. Indeed, certain Plio-Pleistocene sections of the Gulf of Mexico margin show the characteristic features of a ramp margin, yet contain prolific hydrocarbon accumulations within turbidite reservoirs of the slope system. Although these reservoir systems do not represent classic basin-floor fans, the presence of turbidite facies within the intraslope basins of the Gulf of Mexico demonstrates the importance and activity of turbidity current processes in ramp settings.

Within structured mature passive-margin settings, overall depositional patterns are strongly affected by extensional growth faults and folds related to sediment loading, salt

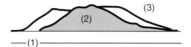

INNER FAN MOUND STRATIGRAPHY

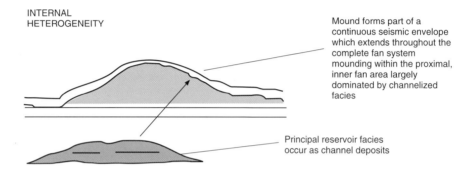

(1) Condensed section or fine-grained facies reflecting distal toesets of preceding systems

(12 Seismic envelope defining submarine fan channel and channel margin. Mounding generated by differential compaction of channel sandstones and channel-margin siltstones and mudstones

(3) Shale/mudstone drape reflecting fan abandonment

INTERNAL HETEROGENEITY

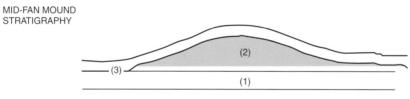

Mound forms part of a continuous seismic envelope which extends throughout the complete fan system mounding within the proximal, inner fan area largely dominated by channelized facies

Principal reservoir facies occur as channel deposits

SEISMIC CHARACTER OF MID-FAN LOBES

MID-FAN MOUND STRATIGRAPHY

(1) Condensed section or fine-grained clastics reflecting distal toesets of preceding systems

(2) Seismic envelope of mid-fan system

(3) Shale/mudstone drape reflecting fan abandonment

INTERNAL HETEROGENEITY

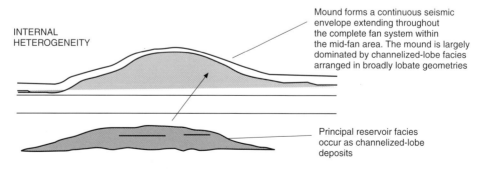

Mound forms a continuous seismic envelope extending throughout the complete fan system within the mid-fan area. The mound is largely dominated by channelized-lobe facies arranged in broadly lobate geometries

Principal reservoir facies occur as channelized-lobe deposits

Fig. 9.14 (*continued opposite*) Seismic expression of sand-rich fans (after Richards and Bowman, in press)

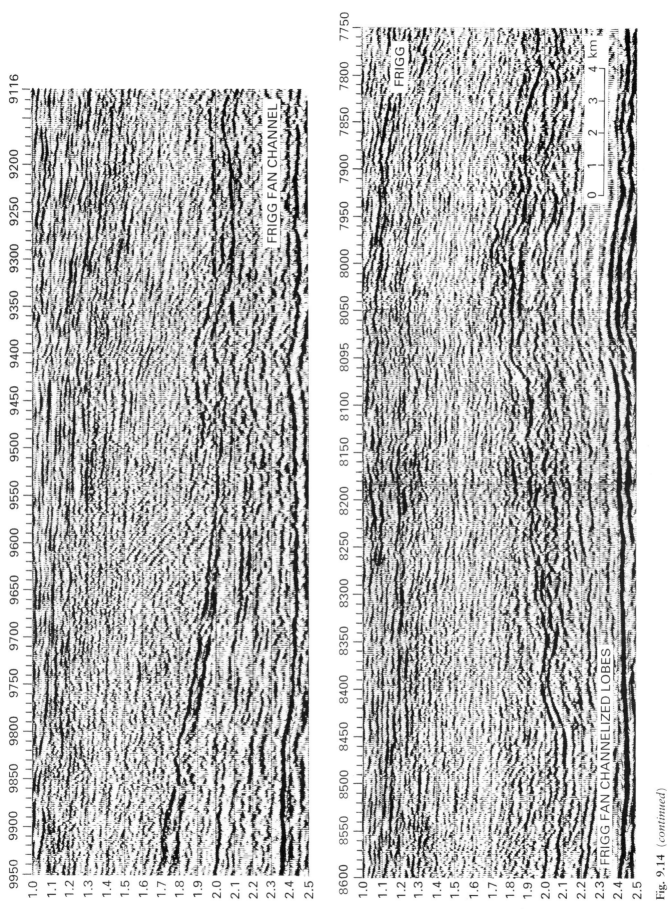

Fig. 9.14 (continued)

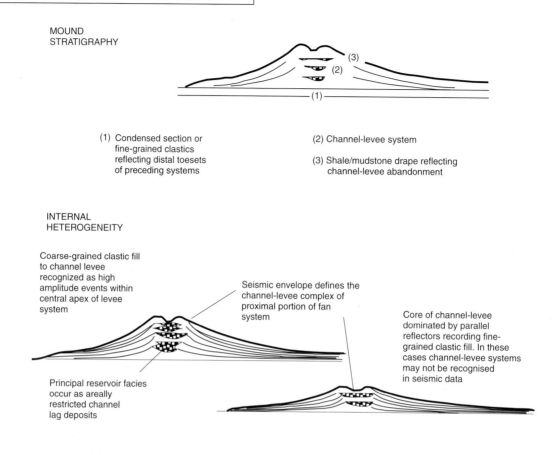
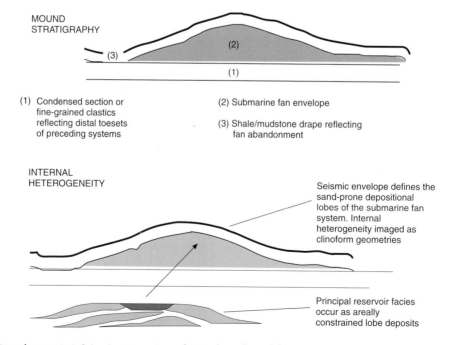

Fig. 9.15 (*continued opposite*) Seismic expression of mixed sand-mud fans

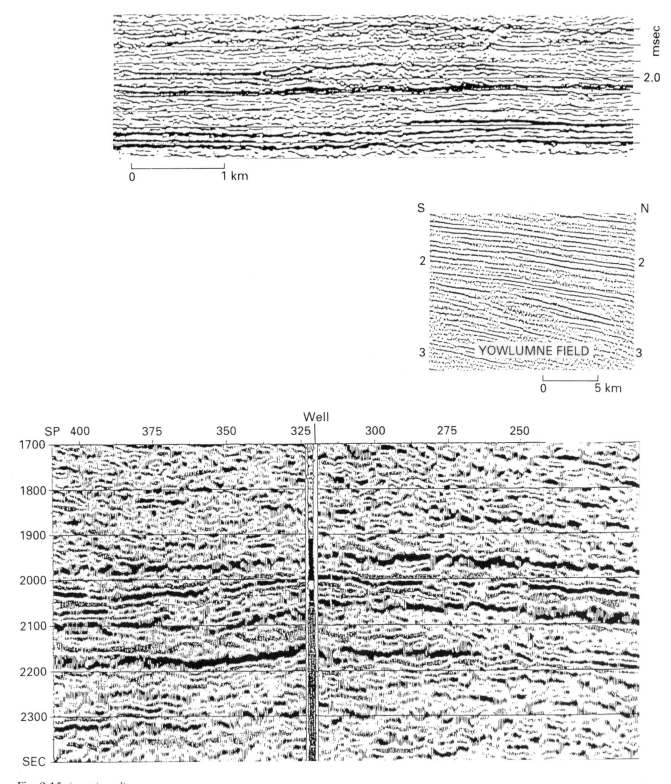

Fig. 9.15 (continued)

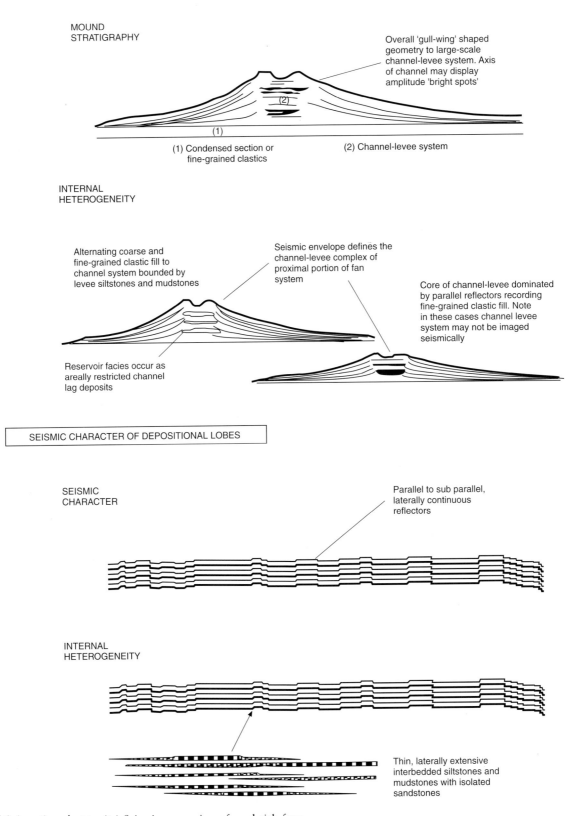

Fig. 9.16 (*continued opposite*) Seismic expression of mud-rich fans

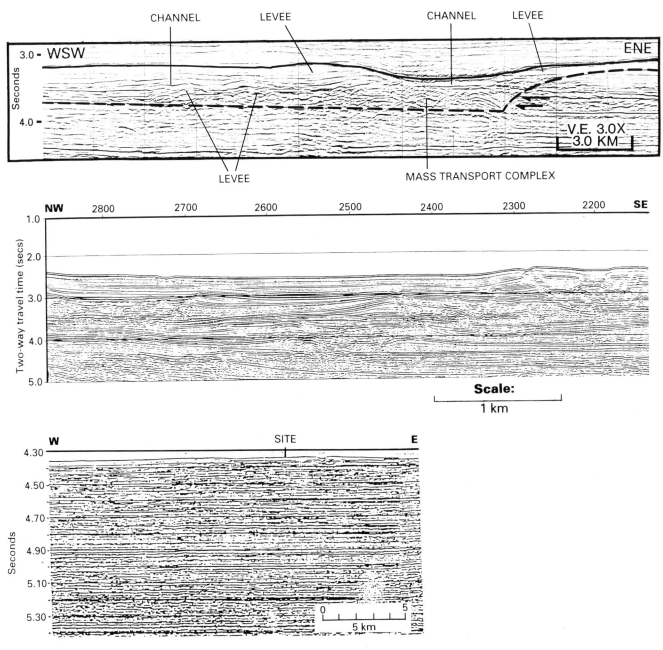

Fig. 9.16 (*continued*)

diapirism and sediment creep of mobile clay (Fig. 9.18c). The interaction of these syn-sedimentary structural features with varying rates of sediment supply and relative sea-level change introduces further complexity into sequence stratigraphic models. Mitchum *et al.* (1990) suggest that conventional sequence stratigraphic models can be modified to include a threefold history in growth-fault development related to the relative sea-level cycle in the Cenozoic successions of the Gulf of Mexico.

In the case of shelf-margin growth faults, fault initiation is related to active sedimentation and loading associated with highstand progradation. For slope-related growth faults, the triggering mechanisms may relate to periods of relative sea-level fall, when the shelf was exposed and sediment was fed directly on to the slope, causing sediment loading and instability and growth fault development (Mitchum *et al.*, 1990). In the Gulf of Mexico, the history of individual growth faults commonly spans a number of eustatic sea-level cycles. A phase of hanging-wall displacement results in significant topographic relief between hanging-wall and footwall areas. Earliest sequences develop the greatest thickness of lowstand units with the deposition of basin-floor and slope fans. A second intermediate phase of fault movement is characterized by slow hanging-wall collapse or creep. At this stage, reductions in fault movement are considered to reflect the development of a contrac-

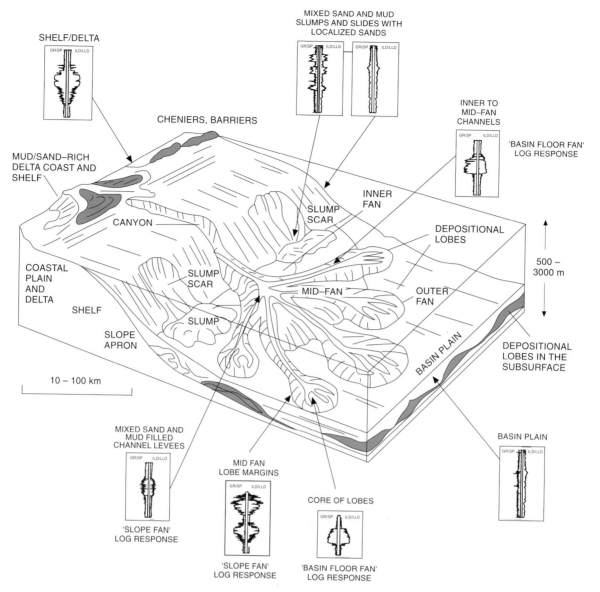

Fig. 9.17 Log signatures from selected depositional lobe and channel–levee elements of a single point-source, mixed sand–mud fan system. Note variation in log character with respect to the position within the fan (after Richards and Bowman, in press)

tional toe in the hanging wall and reduction in fault-plane dip angle, resulting in increased resistance to fault movement. This phase is characterized by the development of lowstand prograding complexes within which parasequences may display an overall trend from progradation to aggradation. Finally, a phase with no differential growth is characterized by filling of the remaining topography between the hanging-wall and footwall areas. The lack of relief between footwall and hanging-wall areas results in continuity of systems tracts across the fault.

This type of model assumes an a priori link between the relative sea-level cycle and growth fault development and that the rate and extent of fault movement varies through the relative sea-level cycle. Studies of these systems demonstrate the complex nature of the sedimentary record of the hanging-wall fill and footwall areas of growth-fault systems, e.g. slope aprons, failures, slumps and slides (Kolla et al., 1990; Reading and Richards, 1994). Thus, although the broad observations made from the Gulf of Mexico may be applied to similar structured passive-margin settings, the model is not universal in its application. The genetic relationship between growth-fault development and sediment loading in response to active deposition and progradation of depocentres on to unstable mud- and clay-rich slope deposits is generally accepted (Mandl and Crans, 1981), but additional work is required to define the relationship between the initiation of thin-skinned extensional faulting to specific periods of the relative sea-level cycle.

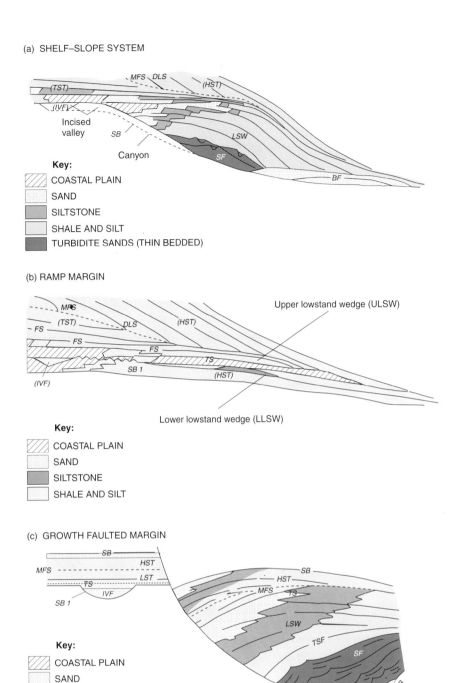

Fig. 9.18 Sequence stratigraphic model showing the idealized development of systems tracts within a (a) shelf–slope system, (b) ramp margin and (c) growth-faulted margin (after Vail and Wornardt, 1990). HST, highstand systems tract; TST, transgressive systems tract; LST, lowstand systems tract; IVF, incised valley-fill; SB, sequence boundary; MFS, maximum flooding surface; FS, (marine) flooding surface; DLS, downlap surface; SF, slope fan; BF, basin-floor fan; LSW, lowstand wedge; TS, transgressive surface; TSF, top of slope fan

9.3.4 The effect of sediment calibre on the evolution of the lowstand systems tract

Currently published models for lowstand systems tracts assume a relatively simple evolution of fan systems in response to a cycle of relative sea-level fall, stillstand and subsequent relative sea-level rise. However, these models ignore the impact of sediment calibre on deep-marine clastic systems within a basin fill (Kolla and Macurda, 1988; Reading and Richards, 1994; Richards et al., 1996). Furthermore, studies of modern systems suggest that fan deposition can occur at all stages of the relative sea-level cycle (Bouma et al., 1989; Galloway, 1989; Kolla, 1993; Kolla and Perlmutter, 1993; Richards and Reading, 1994). The classification scheme or 'ternary diagram' in Fig. 9.4 can be used as a framework for modifying conventional lowstand models by linking changes in sediment provenance to the evolutionary trend of deep marine clastic systems during a period of relative sea level fall and lowstand. Conventional and more ordered systems are represented at the bases of the ternary diagram. More chaotic systems are represented at the ternary diagram's apex. The approach is not restricted to basin margins with pronounced shelf–slope breaks and can be applied equally to more structured basin settings associated with dip-slip extension or growth faulting.

Sand-rich provenance areas

Relative sea-level fall and access to sand-rich source areas may generate an initial sand-prone phase of deposition in the form of a basin-floor fan or ramp. This situation may occur where the relative sea-level fall results in the erosion of coarse-grained shelf and marginal clastic sediments. In this case the starting point on the ternary diagram lies within the sand-rich end of the spectrum (Fig. 9.19). Subsequent stillstand and relative sea-level rise may result in an overall transition from sand-rich, to mixed sand–mud, to mud-rich systems (Fig. 9.19). The diagram additionally attempts to limit the potential range of fan systems occurring at any point in time, within a specific phase of the relative sea-level cycle.

Mixed sand–mud provenance areas

This case may be analogous to basins characterized by large-scale prograding shelf–slope and delta-slope systems where a relative sea-level fall results in the erosion of topsets or foresets. Relative sea-level fall and access to mixed sand–mud source areas may generate an initial sand-prone phase of clastic input to the basin. In this case, the starting point on the ternary diagram lies within the mixed sand–mud part of the spectrum (Fig. 9.20). A mixed sand–mud fan or ramp system is characterized at the base of slope by a channel–levee system ('slope fan'), which may be traced down-dip into sand-prone depositional lobes ('basin-floor fan'). Subsequent stillstand and relative sea-level rise may result in an overall reduction in drainage basin gradients, decrease in coarse clastic input and an overall transition from mixed sand–mud to mud-rich systems (Fig. 9.20). In this case, care must be used in interpreting the origin and genesis of basinal constructional mounds associated with mixed sand–mud basin-margin systems. It is possible that mounded packages developed at, or in close proximity to, the base of the slope record slump and slide systems associated with failure and slope collapse (namely slope aprons).

Mud-rich provenance areas

In turbidite systems with a fine-grained provenance the conventional lowstand models require significant modification. The initial phases of deposition associated with a relative sea-level fall will be represented by a channel–levee complex typical of a mud-rich slope fan (Fig. 9.21). In this case the slope fan would form the lowstand fan, and the record of the stillstand and relative sea-level rise would be manifest in the increasing scale and aspect ratio of channel–levee complexes observed on seismic data. The lowstand systems tract would not therefore contain a sand-rich 'basin-floor' fan.

9.4 Fan development during highstand and transgression

9.4.1 Highstand fan systems

There are a number of notable exceptions to the relationship of active fan sedimentation and lowstand events. The Congo and Magdelana fans currently receive large volumes of clastic material via turbidity current and mass flow transport processes because the continental shelves associated with these fans are relatively narrow, and the submarine canyons feeding the fans extend across their respective shelf systems into their adjacent river mouths (Heezen *et al.*, 1964). The Monterey, Avon and Calabar fans provide further cases in which canyons feed directly into the littoral drift cells of a shoreface or deltaic system (Burke, 1972; Normark, 1978; Dingus and Galloway, 1990). Similarly, significant clastic input to the Crati fan in the Coriglioni Basin, Italian Mediterranean, is achieved because the channel–levee systems of the fan feed directly into a coarse-grained delta system on the shelf (Ricci-Lucchi *et al.*, 1984). In this latter case, active fan sedimentation reflects the lack of shelfal accommodation, resulting in bypass to the basin.

A number of studies demonstrate the development of active fan sedimentation during phases of highstand where clastic input overwhelms any potential eustatic signature and results in direct and substantial feed of coarse-grained clastics into ancient deep marine basins. (Coleman *et al.*, 1983; Busby-Spera, 1985; Heller and Dickinson, 1985; Mitchum, 1985).

9.4.2 Fan development during transgressions

Galloway (1989) provides a further variant on the theme of fan development by suggesting that turbidite systems may also develop during periods of regional transgression of the basin margin (Fig. 2.23). Transgressions, following periods of active outbuilding of the continental margin, are considered to be optimum periods in which extensive mass wasting and retrogradation of the upper slope and continental margin may occur. The initial phase of fan development would occur at a time diametrically opposed to the lowstand event on the relative sea-level cycle. Because such models involve slope reworking and regrading mechanisms as the source of sediment supply, the type of fan, ramp or slope-apron system developed is controlled largely by the composition of the adjacent depositional slope and shelf

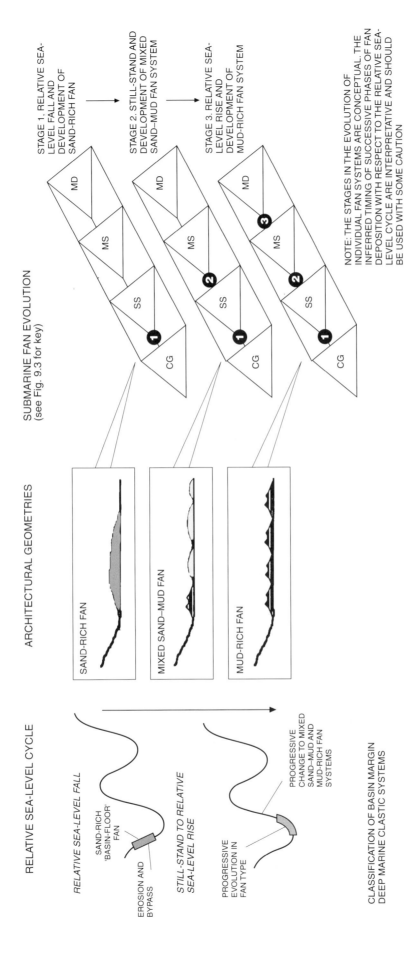

Fig. 9.19 Sequence stratigraphic model for the evolution of deep-marine clastic systems during a period of relative sea-level fall, stillstand and initial rise assuming an initial sand-rich provenance area. In this example deep-marine clastic sedimentation shows a general evolution from sand-rich to mixed sand–mud and mud-rich systems. This change should be recognized by variations in the architectural elements and seismic character of successive depositional systems. The time-related change in style of turbidite sedimentation may include an evolution from single-point-source fans to multiple- or line-source ramps and slope aprons. CG, gravel-rich systems; SS, sand-rich systems; MS, mixed sand–mud systems; MD, mud-rich systems

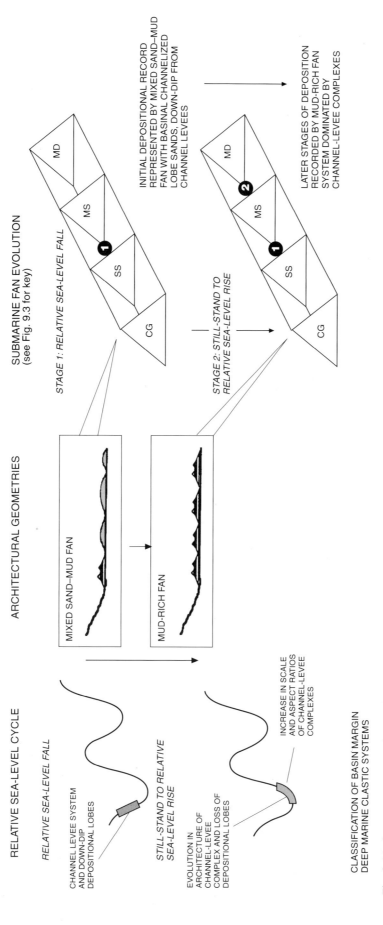

Fig. 9.20 Sequence stratigraphic model for the evolution of deep-marine clastic systems during a period of relative sea-level fall, stillstand and initial rise assuming an initial mixed-sediment provenance area (e.g. mixed-load fluvial-deltaic system). In this example deep-marine clastic sedimentation shows a general evolution from mixed sand–mud-rich to mud-rich systems. The development of fine-grained clastic sediments within the basin may lead to slope failure and slumping associated with slope-apron development. This change should be recognized by variations in the architectural elements and seismic character of successive depositional systems. The time-related change in style of turbidite sedimentation may include an evolution from single-point-source fans to multiple- or line-source ramps and slope aprons. CG, gravel-rich systems; SS, sand-rich systems; MS, mixed sand-rich systems; MD, mud-rich systems

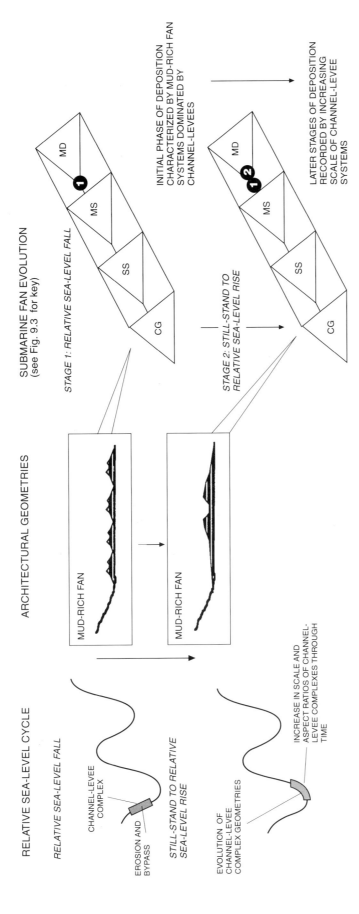

Fig. 9.21 Sequence stratigraphic model for the evolution of deep marine clastic systems during a period of relative sea-level fall, still-stand and initial rise assuming an initial mud-rich sediment provenance area (e.g. suspended-load dominated fluvio-deltaic or coastal plain system). In this example deep marine clastic sedimentation shows a general evolution from mixed sand—mud-rich to mud-rich systems. The development of fine-grained clastic sediments within the basin may lead to slope failure and slumping associated with slope apron development. This change should be recognized by variations in the architectural elements and seismic character of successive depositional systems. The time-related change in style of turbidite sedimentation may include an evolution from single-point-source fans to multiple- or line-source ramps and slope aprons. CG, gravel-rich systems; SS, sand-rich systems; MS, mixed sand-rich systems; MD, mud-rich systems

system. In this case no potential exists for shelf bypass and the access of sediment from hinterland source areas. Subsequent deep marine clastic deposition during highstand, however, may occur preferentially down-dip of major sites of slope failure (Dixon *et al.*, 1993). Studies of modern and ancient fan systems show that slope failure may result in geographically concentrated slide and failure scars on the slope. During subsequent active deltaic and shelfal progradation, sediment may be funneled preferentially through slope-failure scars such that they become the conduits to the basin.

More recent studies of the Mississippi fan suggest that transport of sandy turbidites and fan development may continue into the late stages of rising sea-level. Such cases occur where the canyon system extends into mid-shelf water depths and accesses sediment directly from up-dip fluvial valley systems during rising base level or from the interception of longshore drift cells as the canyon erodes landward (Kolla and Perlmutter, 1993). In addition, continuation of sandy turbidite sedimentation during the late sea-level rise has been invoked for the pre-Holocene depositional record of the Amazon fan (Flood *et al.*, 1991).

9.5 Conclusions

Deep-marine clastic systems are not developed exclusively during periods of sea-level fall. As a result, significant caution should be used in interpreting all turbidite systems as lowstand deposits. The value of sequence stratigraphic models for predicting turbidite systems lies in understanding whether their development is linked genetically to a cycle of relative sea-level fall. Even with this potential link established, the interpreter must always remember the potential pitfalls and consider the options offered by alternative models.

Nevertheless, sequence stratigraphy provides an extremely powerful tool for developing a coherent time-stratigraphic framework for subdividing deep-marine clastic successions. A three-stage process for stratigraphic prediction in deep-marine clastic systems is recommended. Firstly, develop a stratigraphic framework that links systems tracts from basin margin to basin centre. Second, delineate the tubidite system using careful outcrop or seismic mapping to define its key architectural elements. Finally, characterize the turbidite system based on an understanding of depositional processes and sand-body architecture as observed or inferred from outcrop, core, wireline logs and seismic data.

CHAPTER TEN
Carbonate Systems

10.1 Introduction

10.2 Controls on carbonate sedimentation
    10.2.1 Organic and inorganic carbonate production
    10.2.2 'Highstand' shedding
    10.2.3 Carbonate platform drowning
    10.2.4 Carbonate platform exposure

10.3 Carbonate slopes, platform classification and facies belts
    10.3.1 Slopes and platform classification
    10.3.2 Facies belts on carbonate platforms

10.4 Sequence stratigraphic models for carbonate platforms
    10.4.1 Introduction
    10.4.2 Ramps
    10.4.3 Rimmed shelves
    10.4.4 Escarpment margins
    10.4.5 Isolated platforms

10.5 Cyclicity and parasequences on carbonate platforms

10.6 Conclusions

10.1 Introduction

'Carbonate sediments are born, not made'. Noel James' apt quotation neatly sums up why carbonates, as living systems, require a rather different sequence stratigraphic approach to their inert siliciclastic counterparts. Fortunately, this has not resulted in a proliferation of new terms and concepts, as many of the existing sequence stratigraphic principles discussed in the first four chapters are valid. However, the range of environmental factors and of carbonate platform geometries add two new dimensions that extend the range of interpretation of any carbonate succession.

The objectives of this chapter are to outline the broad controls on carbonate sediment production, the role of sea-level and the major differences compared with siliciclastic systems. This introductory section is followed by a brief discussion of carbonate platform types, and of the geometries that further distinguish carbonates from siliciclastics. The final two sections outline sequence stratigraphic models for different carbonate platform types, with subsurface and outcrop examples, followed by a discussion of cyclicity in carbonate platform settings and the concept of the parasequence in carbonates.

Several terms used in this chapter require defining. The term 'carbonate platform' is used in a general sense to denote any large body of shallow-water carbonate production, such as the Bahama Banks, the Maldive Atolls and the Great Barrier Reef. The term 'reef' is generally used here in a narrower sedimentological sense, as a biologically influenced carbonate accumulation that was large enough during formation to possess some topographic relief. This contrasts with the geophysical 'seismic' reef, which usually is defined by mounded or chaotic seismic reflector geometry, but may or may not have any biological reefal component. The term 'margin' is used in a broad sense to denote the change from shallow-water high-energy deposits to slope or deeper shelf sediments. The margin will coincide with the shelf break on carbonate platforms that exhibit a break in slope (rimmed shelves or escarpment margins; see section 10.3), but on low-angle, homoclinal carbonate ramps the margin is simply a facies change. Finally, some familiarity with carbonate depositional environments is assumed before reading this chapter. For an introduction, the reader is referred to the following excellent texts on general carbonate sedimentology, modern and ancient; Bathurst (1975), Scholle *et al.* (1983), Reading (1986), Scoffin (1987), Tucker and Wright (1990) and Walker and James (1992).

10.2 Controls on carbonate sedimentation

10.2.1 Organic and inorganic carbonate production

Carbonate sediments are produced today in several marine and some terrestrial settings, but their locus of maximum production, the 'carbonate factory', is in shallow tropical seas with low siliciclastic input. Carbonate sediment types can be split into a skeletal component, dominated at present by corals and algae, and a non-skeletal component comprising coated grains (including ooids), peloids, aggregates and clasts (Folk, 1959). It is important to distinguish between organic and inorganic carbonate production, as their relative productivities can be very different, with organic carbonate production generally able to outpace inorganic production in shallow water environments (Schlager, 1981). Because of this productivity difference, inorganically dominated carbonate systems tend to evolve sequence stratigraphic geometries more akin to siliciclastic systems than their organic counterparts.

The major contributors to modern shallow-water organic carbonates are corals and algae. The obvious manifestation of vigorous coral growth is a reef, but reefs are very sensitive to environmental conditions, notably light and temperature, and only thrive between the tropics. Bosscher and Schlager (Schlager, 1992) have shown that the distribution of modern reefs is effectively restricted by the minimum winter temperature, and major reef systems are rare north and south of the 20°C isotherm. Light intensity is as significant as temperature in controlling reef productivity, owing to the symbiotic relationship between photosynthetic algae and shallow-marine reef growth (Tucker and Wright, 1990). Because light penetration into water decays exponentially with depth, there is a shallow zone of light penetration, the photic zone, rarely greater than 10–20 m below the sea-surface, below which coralgal growth rates diminish. In short, the most productive loci of carbonate production are the upper few tens of metres of shallow tropical seas (Fig. 10.1).

At higher latitudes, major reef frameworks become less common, and are replaced by smaller reef complexes and bioclastic sand shoals. Molluscs and bryozoa, although present in tropical reef communities, increase in importance as latitude increases and temperature falls. This change in carbonate sediment producers is shown schematically in Fig. 10.2 (Schlanger, 1981).

Modern organic carbonate factories are not faunal and floral representatives of ancient systems. Figure 10.3 illustrates schematically the major framework builders through time (see also Wilkinson, 1979). Organisms that are extinct today, stromatoporoids, provided the major Devonian framework builders. The margins and slope sediments of the Triassic carbonate platforms of the Italian Dolomites include a significant component of calcisponge debris (Bosellini, 1984), and the extinct rudists, aberrant cup-shaped bivalves, were the main framework building organisms in the Cretaceous.

Irrespective of stratigraphic age, the organic margin of a carbonate platform does not dictate the growth potential of the entire platform. Platform interior facies will, in general, have a lower production rate than the margin (Bosence, 1989; Schlager, 1992). For example, rapidly rising sea-level may be matched by the growth of a margin, but not of a platform interior lagoon, leaving the lagoon relatively sediment-starved and deep. Present-day examples of this differential platform growth are the Belize lagoon (Ginsburg and James, 1974; Purdy, 1974), with water depths of up to 60 m, and the Maldive Atolls, with water depths between the coral islands or 'Faros' of several tens of metres (Purdy and Bertram, 1993).

The discussion thus far has centred on organic carbonate production. As mentioned above, inorganic carbonate production, notably that of ooids, tends to be lower than reefal production, and may be up to two orders of magni-

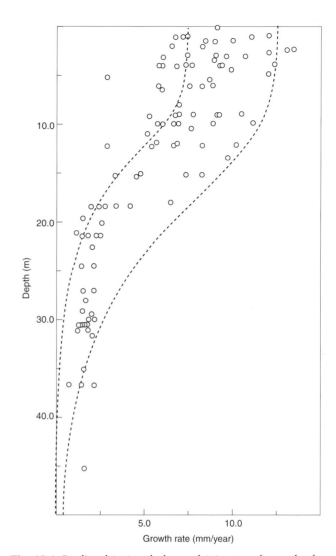

Fig. 10.1 Predicted (---) and observed (○) rates of growth of the Caribbean reef coral *Montastrea annularis*. The predicted rates are derived from a light–growth equation (Schlager, 1992), and compare well with the observed values, indicating that light exerts a strong control on coral carbonate production (from Schlager, 1992)

tude slower than the healthiest Recent reef corals (Fig. 10.4). The corollary of this is that oolitic margins are less able to keep pace with sea-level rise than modern reefal margins. Combined with a lack of a rigid framework, this will tend to cause oolitic margins to develop less aggradational geometries than their organic counterparts. But can the highly productive organic margins always keep pace with rising sea-level? Figure 10.4 suggests that *most* healthy coral systems can keep pace with any rate of sea-level rise recorded in the Holocene, and that *most* carbonate-producing systems can outpace *most* rates of subsidence-induced sea-level change. It is tempting to generalize and to conclude that carbonate growth can keep pace with sea-level rise, but Schlager's (1981) 'paradox of drowned reefs and carbonate platforms' shows that this

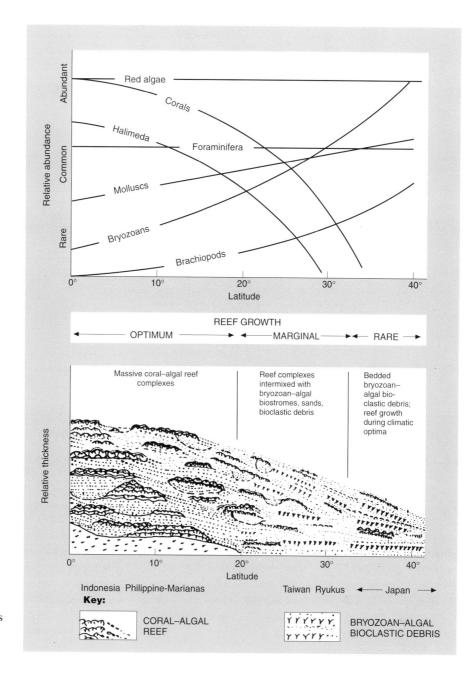

Fig. 10.2 Latitudinal change from tropical to temperate carbonate facies in the northern Pacific (from Schlanger, 1981)

clearly is not the whole story. In fact, the most rapid, but short-lived, sea-level rises (corresponding to fifth-order, or even higher, sea-level fluctuations discussed in Chapter 2) can deliver the knock-out blow to a carbonate system, particularly if this rapid sea-level rise occurs just as the system is beginning to develop, or 'start-up' (Neumann and Macintyre, 1985). As these authors point out, there is a sigmoidal growth pattern to organic carbonate systems, where the creation of new living space (or 'accommodation', Chapter 2) is at first more rapid than the ability of the new population to fill the space. Following this start up or 'lag' phase, the population growth may exceed the rate of accommodation creation in the 'catch up' phase, and finally the ability of the population to grow will be limited by the creation of accommodation in the 'keep up' phase (Fig. 10.5). Figure 10.6 illustrates the start-up, catch-up and keep-up phases for an idealized isolated carbonate platform experiencing transgression.

Once the carbonate-producing system has survived the start-up phase, it will grow vigorously if environmental conditions remain favourable, keeping pace with rising sea-level, which can lead to the development of strongly aggradational margins. This contrasts with siliciclastic systems, which will tend to backstep rather than accrete vertically during transgression. If, however, the rate of sea-level rise in the early start-up phase is so rapid that the photic zone pulls away from the earliest start-up population prior to the catch-up phase, the carbonate system will drown, or

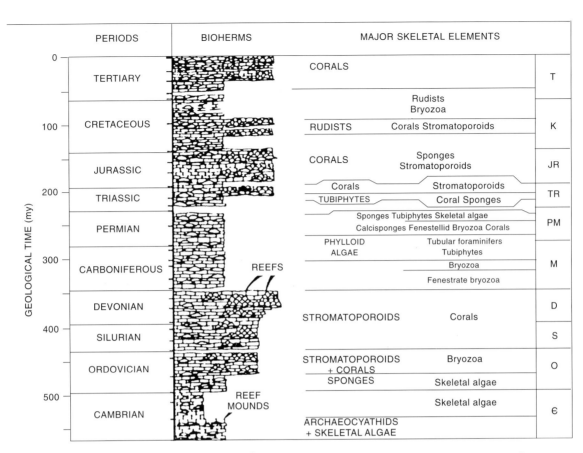

Fig. 10.3 Simplified stratigraphic column indicating the major framework-building organisms through time (from James, 1983)

step back to a shallower water position. As Schlager (1992) comments, losing this race for the photic zone can extinguish carbonate platforms in the short term, but as Fig. 10.4 illustrates, longer term third- and second-order processes are unlikely to be able to drown healthy platforms. Some other mechanism must be responsible for massive platform carbonate extinctions.

10.2.2 'Highstand' shedding

The 'keep up' phase of Neumann and Macintyre (1985) mentioned above is a period of platform development during which the production of carbonate exceeds the space available to accommodate it (Fig. 10.6). Under such circumstances, produced carbonate may be redeposited elsewhere, in deeper water, on tidal flats and as aeolianites. The Bahamas carbonate platform at present is such an overproducing system. Neumann and Land (1975) were able to demonstrate that the production of carbonate from algal breakdown in a Bahamian lagoon far exceeded the mass deposited *in situ*. The excess production was probably redeposited on neighbouring tidal flats, and in deeper water. Redeposition into deep water of huge quantities of lagoonally produced carbonate also has been demonstrated by two deep cores taken through a prograding carbonate margin off the Bahamas. The seismically defined prograding clinoforms comprise in excess of 90% pellets, which were produced at water depths of less than 5–10 m on the flat-topped Great Bahama Bank. The importance of the shallow-water carbonate factory as a supplier for deep-water sedimentation also has been demonstrated unambiguously by other workers in the Bahamas (Droxler *et al.*, 1983; Reijmer *et al.*, 1988), the Great Barrier Reef (Davies *et al.*, 1989) and the Indian Ocean. Furthermore, ancient calciturbidite successions demonstrably deposited in deep water contain a very high proportion of grains that can have originated only in shallow water (Reijmer *et al.*, 1991). These observations suggest that deeper water carbonate accumulation can proceed to any great degree only when a carbonate system is healthy and actively overproducing sediment, and this situation will apply only when carbonate platform tops are flooded to a few tens of metres, such as the Bahamas today. Redeposition of carbonate from platform top to the slope and the basin is referred to as 'highstand shedding'. The term is a little unfortunate in a strict sequence stratigraphic sense because a carbonate platform will also shed sediment during transgression (Driscoll *et al.*, 1991) and sea-level fall, but all other factors being equal, a platform

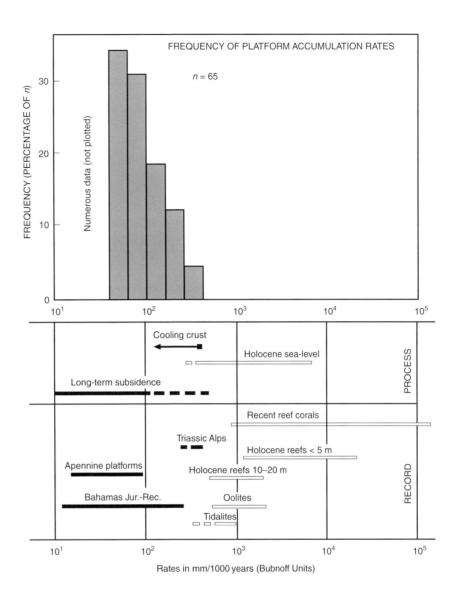

Fig. 10.4 Growth rates of carbonate systems compared with relative sea-level rise. Solid bars represent the geological record, open bars represent Holocene production rates and sea-level rise (from Schlager, 1981, 1992)

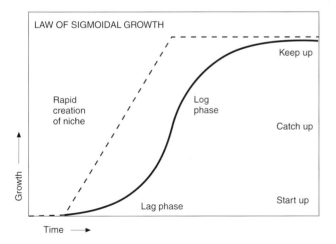

Fig. 10.5 Population growth versus time, indicating 'start-up', 'catch-up' and 'keep-up' phases of Neumann and Macintyre (1985)

will tend to shed most sediment during highstand as the rate of creation of accommodation declines (Chapter 2), and the platform top becomes partially by-passed by over-produced carbonate. In general, carbonate systems are likely to be less productive during lowstand, because the area of shallow-water carbonate production will be reduced. The areal reduction of production may be enormous, particularly on steep-sided platforms such as the Bahamas (Fig. 10.7). Highstand shedding in carbonates provides a major contrast with classic siliciclastic sequence stratigraphy, which predicts that most sediment is shed into the basin during lowstand (Chapter 2). This contrast is illustrated on Fig. 10.8, which shows the highest and most variable basinal sedimentation rates of siliciclastic systems occurring during glaciations when sea-levels were low, and highest and most variable basinal sedimentation rates for carbonate systems occurring during interglacials when sea-levels were high. In both cases the variability of sedimentation rate is

215

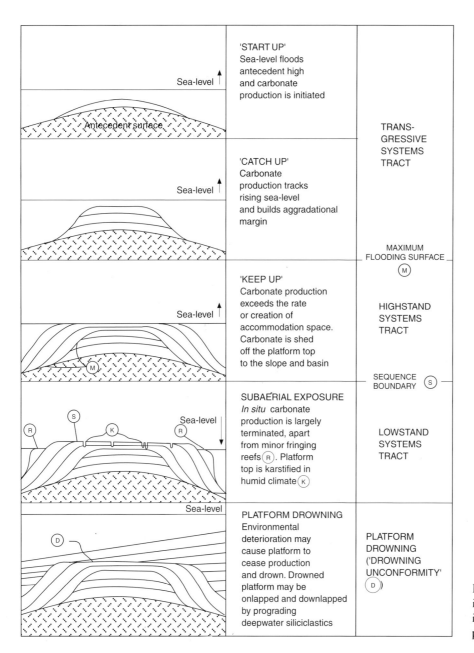

Fig. 10.6 Schematic model for an isolated carbonate platform, showing idealized systems tract geometries and platform drowning

probably due to turbidite-controlled sedimentation, rather than a uniform, continuous rain of pelagic sediment (see the carbonate ooze sedimentation rates for comparison). Another important point is that during low-stands, carbonate systems will not experience the same type of physical erosion as siliciclastic provenances. During subaerial exposure, carbonates will tend to be eroded chemically, resulting in dissolution of grains and cements, and reprecipitation of dissolved material (see further discussion in section 10.2.4). Relatively small quantities of eroded carbonate will be generated during lowstand and the most likely source of lowstand carbonate is sediment production from growing carbonate, such as fringing reefs (Fig. 10.7), or areally smaller platforms that have down-stepped during sea-level fall.

Although highstand shedding is the main way of contributing carbonate sediment to the basin from steep-sided platforms, on lower angle ramps and rimmed shelves, lowstand basinal carbonate deposition can be significant. Possibly the best-constrained example has been demonstrated by Jacquin et al. (1991) in the Cretaceous of the Vercors, France, where lowstand carbonate fans shed from downstepping rimmed shelves are developed. The lowstand fans have a different grain composition from the highstand shedded material.

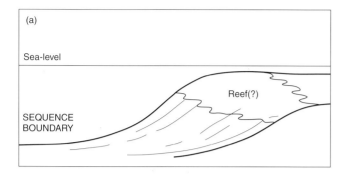

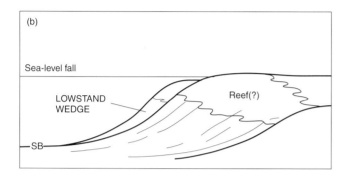

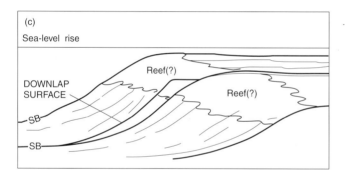

Fig. 10.7 Cartoon of highstand progradation compared with volumetrically smaller lowstand wedge, Bahama platform, Tertiary (after Eberli and Ginsburg, 1987)

10.2.3 Carbonate platform drowning

As well as starting, catching and keeping up, carbonate systems will also give up. Rapid, but short-lived sea-level rises may outpace carbonate platform growth (section 10.2.1), but once the rate of sea-level rise slows down, all other things being equal, the carbonate platform may be restored to productive health. However, there are several environmental factors independent of sea-level change that may seriously impede carbonate development. As well as the appropriate light and temperature requirements, there are at least five additional controls that need to be considered; nutrient supply, clastic input, oxygenation, salinity and predation. Table 10.1 lists the major causes of carbonate platform drowning, with examples from modern carbonate systems and the geological record. For more detail

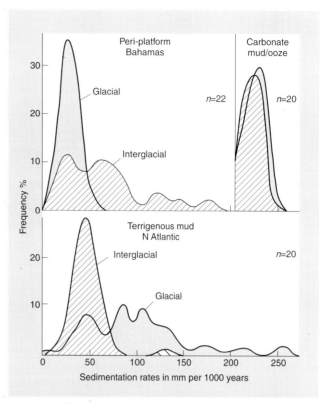

Fig. 10.8 Highstand and lowstand basinal sedimentation rates for carbonate and siliciclastic systems (from Droxler and Schlager, 1985)

on mechanisms of platform drowning the reader is referred to Schlager (1992).

All the mechanisms described in Table 10.1, individually, or in combination, or in concert with rapid sea-level rise, can cause a carbonate platform to drown. In the fossil record, the term 'drowning unconformity' has been coined (Schlager, 1989), and is shown schematically in Fig. 10.6. This 'unconformity' can have developed only when the carbonate platform top was flooded, i.e. within the transgressive or highstand systems tract. Drowning unconformities have several distinctive properties, which, in detail, differentiate them from sequence boundaries. Firstly, there is usually a very rapid change from shallow-water carbonate to deep shelf, slope or basinal deposits. Observations by the author of cores taken from the top of a Miocene carbonate platform, offshore central Vietnam, show a change from shallow-water platform carbonate sedimentation, through a drowning 'sequence' of deeper water Foraminifera, with evidence of marine condensation, including glauconite and phosphates, and finally into basinal mudstones. This transition takes place over several to a few tens of metres, but represents several million years worth of sedimentation. A similar succession has been described from the top of the Liuhua carbonate platform, Pearl River Mouth Basin, offshore China (Erlich et al., 1990). Here, the transition is more gradual, with the drowning unit

Table 10.1 Causes of carbonate platform drowning

Nutrient supply
Low-nutrient environments are most favourable for organic carbonate growth, particularly reefs. In high-nutrient settings it is apparent that carbonate-secreting, framework-building corals are replaced by fleshy algae, sponges or soft corals. Modern examples include the East Java Sea, Indonesia, where corals are scarce below 15 m and are replaced by green algae. The lack of reefs is interpreted to be the result of upwelling nutrient-rich waters, which stimulate the growth of coral competitors (Roberts and Phipps, 1988)

Siliciclastic input
Clay particles in suspension in the water column will inhibit light penetration, significantly reducing or preventing reef growth. In addition, coral polyps are, and many predecessor framework building forms were, unable to cope with clay particles during feeding. Circumstantial evidence of carbonate demise by clastic input in the geological record is demonstrated by seaward stepping Miocene carbonate systems from south to north across the Luconia Province, offshore Sarawak, Malaysia, away from northward prograding delta systems (Epting, 1989; Figure 10.9). Note that changes in water salinity or nutrient content associated with a siliciclastic depositional system such as a delta, may be the prime cause of carbonate production demise

Salinity variations
Changes of salinity may exert a dramatic effect on carbonate productivity, particularly that of reefs. In the Holocene, broad, shallow lagoons on flat-topped carbonate platforms acquired highly variable salinities. The seaward ebb flow of this 'inimical bank water' killed the reefs, which also became overwhelmed by seaward transport of carbonate fines from the lagoons (Neumann and Macintyre, 1985)

Oxygenation
When oxygen is depleted, reefs will die. However, the best evidence for massive carbonate demise resulting from oxygen deficiency is circumstantial. Schlager and Philip (1990) have used evidence for Cretaceous oceanic anoxia and compared this with the frequency of Cretaceous carbonate platform demise to indicate a possible causal link

Predation
The importance of predation on carbonate communities is well-documented from modern day examples. The 'Crown of Thorns' starfish blight, which threatened to wipe out large tracts of the Great Barrier Reef is well-known. However, it is difficult, if not impossible, to demonstrate clearly the effect of predation in the fossil record, and is currently unknown on the scale of an entire carbonate platform

heralded by the appearance of a red algal boundstone, which passes laterally into rhodolith beds of low faunal diversity. This red algal interval was probably deposited in several tens of metres of water depth, too deep for prolific coral production. The final expiry of carbonate sedimentation is shown by a deepwater condensed section, which is in turn succeeded by basinal muds. Drowning unconformities are also recognizable on seismic data. Figure 10.10 shows a seismic line from the Atlantic margin of the USA, where a flat-topped Early Cretaceous carbonate platform is downlapped by Late Cretaceous to Tertiary siliciclastics. The latter interval must have been deposited on top of the carbonate when sufficient accommodation had been generated on top of the drowned platform to produce the siliciclastic slope.

An intermediate version of complete platform drowning is backstepping or retrogradation. This may take place on the scale of a single platform, where the platform margin retreats away from incoming siliciclastics, such as in the Devonian build-ups of western Canada (Stoakes and Wendte, 1988; Scaturo et al., 1989), or on the scale of a carbonate 'province' such as the Luconia shoals, described in Table 10.1. Backstepping of carbonate margins may precede complete platform drowning, such as in the case of Liuhua, where the margin effectively shrank to the red algal boundstone described above, before its final demise (Fig. 10.11).

In short, the most important point about platform drowning is that it can be controlled entirely by environmental factors; sudden rapid rises in sea-level may outpace carbonate production temporarily, but a deterioration in the environment of the carbonate factory through any of the mechanisms discussed above is a more likely cause of large-scale platform demise. Drowning will take place only when the platform top is flooded, within the transgressive or highstand systems tracts.

10.2.4 Carbonate platform exposure

Subaerial exposure is an obvious alternative method of

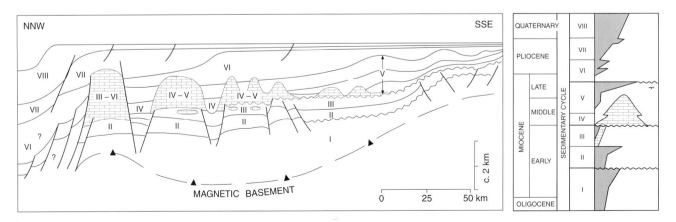

Fig. 10.9 Tertiary carbonates of the Luconia Province, offshore Sarawak. The carbonates young towards the north, as they backstep away from northward directed siliciclastic input (from Epting, 1989)

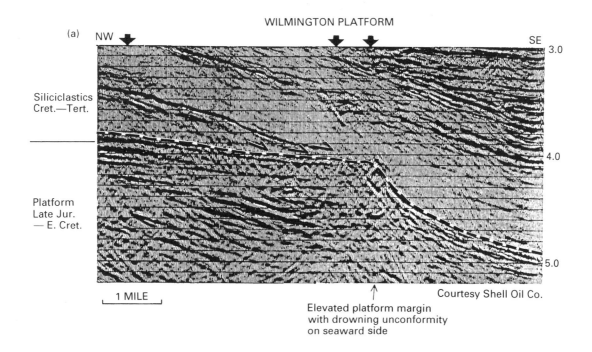

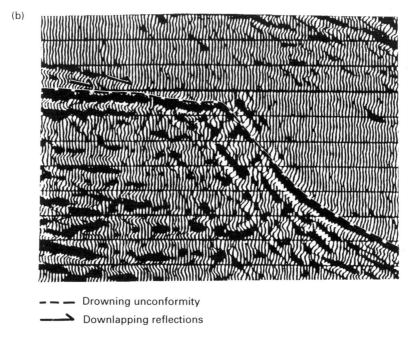

Fig. 10.10 Drowning unconformity on the Wilmington Platform, offshore USA (seismic profile courtesy of Shell Oil Co.; from Schlager (1992))

killing a carbonate platform. Because shallow-marine carbonates build at, or close to, sea-level, they are highly susceptible to frequent, and in some cases, long-lived subaerial exposure. The effect of exposure on carbonates is firstly to turn-off large areas of the carbonate factory and then to erode chemically the exposed carbonate. Depositional carbonate mineralogies are highly unstable, and will experience alteration in the presence of meteoric water, which is acidified by atmospheric and soil carbon dioxide, as well as humic and other biologically derived acids present in soils. Modern marine carbonate is composed chiefly of aragonite, a preferred precipitate of many marine organisms, including corals and molluscs, and high-magnesium calcite, precipitated by echinoderms, benthic Foraminifera and red algae. Dissolution of these grains, and reprecipitation as more stable low-magnesium calcite will occur during meteoric diagenesis, leading to a redistribution of porosity and the formation of zones of considerably enhanced porosity and permeability. Where meteoric diagenesis is very intensive, extensive cavern systems, fissures

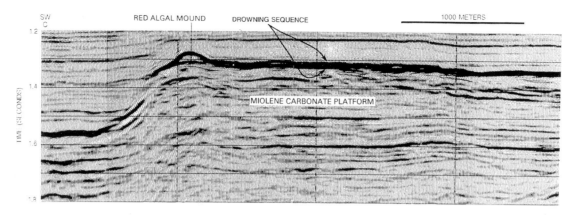

Fig. 10.11 Red algal mound backstepping and preceding final drowning on top of the Liuhua carbonate platform, Pearl River Mouth Basin, South China Sea (from Erlich *et al.*, 1990)

and subterranean streams may form, giving rise to karst (Trudgill, 1985). Karst may provide major secondary porosity and permeability in carbonate reservoirs, and its recognition and prediction is an important practical consequence of identifying carbonate platform exposure (Warson, 1982; Loucks and Anderson, 1985; Craig, 1988).

Not all subaerial exposure will result in meteoric dissolution or karst formation. In more arid climates, where rainfall is limited, exposed carbonates will suffer limited dissolution; instead, carbonate precipitation may give rise to the formation of calcrete crusts, needle fibre calcite and rhizocretions, the last formed by preferential cementation around plant roots. For a more detailed discussion of subaerial-exposure-related diagenesis, the reader is referred to Longman (1980), James and Choquette (1988) and Tucker and Wright (1990).

The most important point from the discussion of carbonate exposure is that most of the processes acting on carbonates during lowstand are chemical. In general, the physical products of carbonate exposure are volumetrically small, although collapse around cave systems and along major fissures can result in carbonate debris cones, and can even control the shape of carbonate margins (Back *et al.*, 1984). If a carbonate platform system is attached to a siliciclastic source area, lowstand may result in siliciclastic transport across the exposed carbonate edifice and into the basin. Examples of this include the Bone Spring Formation of the Permian of New Mexico (Saller *et al.*, 1989), where sandstones onlap the carbonate slope, and the Carboniferous of Anglesey, where fluvial channels incise into shelf carbonates (Walkden and Davies, 1983). In climates that are too arid to sustain major fluvial systems, evaporites may be precipitated close to the surface of the exposed carbonate, and where arid conditions are combined with a restricted basin, subaqueous evaporites may form, onlapping the base of the carbonate slope (Tucker, 1991). On isolated carbonate platforms, remote from a source of siliciclastics, the products of lowstand are likely to be meteoric diagenesis, with associated porosity redistribution, and *in situ* carbonate deposition as fringing reefs, shown schematically on Fig. 10.6. Fringing reefs also may develop on attached platforms, but may be affected adversely by siliciclastic input or evaporative conditions. Note that the volume of *in situ* carbonate produced during lowstand will be highly dependent on the slope of the carbonate margin. This is discussed more fully in the following section.

10.3 Carbonate slopes, platform classification and facies belts

10.3.1 Slopes and platform classification

The geometry of a carbonate platform will exert a major control on its response to sea-level change. For example, on steep sided, isolated platforms the area available for *in situ* lowstand carbonate sediment production will be very small, restricted, at least initially, to a fringing reef around the platform (Fig. 10.12). By contrast, gently sloping carbonate ramps will maintain large areas for carbonate production during sea-level fall. Between these two extremes lie a range of carbonate platform types defined on two main criteria, (i) the slope angle and presence or absence of a slope break, and (ii) whether land-attached or isolated. The classification scheme to be used here follows that of Tucker and Wright (1990), with minor modifications. Carbonate ramps are defined as low-angle (< 1°) homoclinal dipping surfaces, without a significant break in slope (Ahr, 1973), although Read (1985) has described many variants on this theme. Rimmed shelves show a break in slope, and have slope angles between 1° and vertical. Escarpment margins are a subset of rimmed shelves, with the margin having angles of 35° to vertical. These platforms types may be land-attached or detached (isolated). Detached platforms or 'atolls' generally maintain steeper slopes, although such platforms occasionally may develop ramp margins.

Although the above platform categories are convenient for classification purposes, in reality there is a continuum between the different margin types, which has a direct

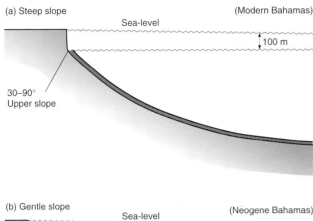

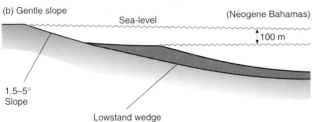

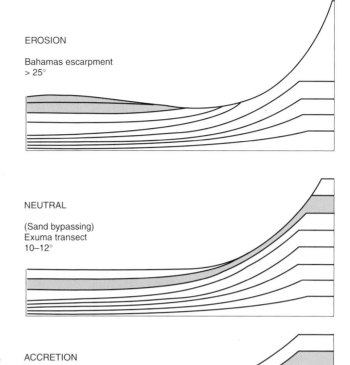

Fig. 10.12 The importance of platform slope angle and geometry on the volume of *in situ* lowstand carbonate production. Steep-sided platforms will permit only a relatively small area of carbonate production during sea-level fall, whereas broader ramps can maintain much larger areas

Fig. 10.13 Erosion, bypass and accretionary carbonate margins (from Schlager and Camber, 1986). ODP, Ocean Drilling Program

bearing on unravelling their sequence stratigraphy (Schlager, 1992). The platform margins of Fig. 10.13 show net accumulation on the Blake transect, bypass of the Exuma transect, and net erosion of the Bahamas escarpment. All of these geometries can be produced during platform growth, in an identical sea-level regime, yet the bypass margin has given rise to a sediment package onlapping the base of slope, and the erosional margin shows truncation of the margin and accumulation of a mounded package on the basin floor (Schlager and Camber, 1986). It would be an unwise seismic stratigrapher who attempted to relate such geometries purely to sea-level change! Net erosion and a mounded basinal package does not necessarily correspond to lowstand. Purdy (pers. comm., 1992) has suggested that erosion of the margin and development of the steepest slopes are more likely during transgression and early highstand, when a carbonate system is aggrading and the steepening flanks, subjected to an overlying load of accumulating carbonate, collapse into the basin.

The angle of a depositional carbonate margin will also be controlled by the type of sediment supplied to the slope and the sedimentary process both on and off the platform. Kenter (1990) has shown that coarse-grained sediments can maintain much higher angles of repose, and hence steeper slopes, than finer grained deposits. He was also able to show that cohesionless coarse-grained sediments could build slopes up to 35°, whereas muddy cohesive sediments tended to maintain low slope angles characterized by repeated massive slumping (Fig. 10.14). The type of sediment supplied to a slope will be dependent on the type of carbonate factory, and on the orientation of the slope with respect to current regimes. For example, the carbonate platforms of the Italian Dolomites show slope angles up to 30°, some of the steepest examples on Fig. 10.14, because of gravel-sized particles of calcisponge supplied from the carbonate factory. This contrasts with the low slope angles evident from seismic sections across the muddy northern flanks of the modern Little Bahama Bank. The effect of current activity on carbonate platforms will also influence sediment supply to the slope and basin. On windward margins, wave activity will tend to push most of the fine sediment on to the platform, leaving the coarser fraction to accumulate on the windward slope. Leeward margins will tend to accumulate finer sediment and, in general, build shallower slopes (Hine *et al.*, 1981; Tucker, 1985).

10.3.2 Facies belts on carbonate platforms

Using the general platform classification described above, idealized facies belts for ramps, rimmed shelves and escarpment margins are outlined below. Although carbonate-producing organisms have evolved through geological time,

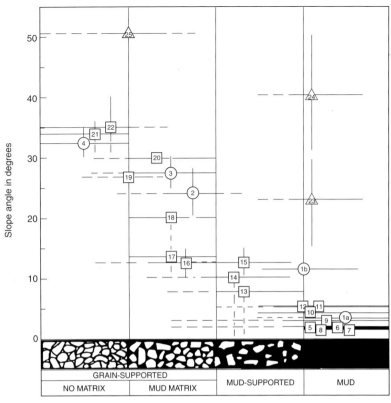

Fig. 10.14 Sediment composition versus slope angle on carbonate margins (from Kenter, 1990)

shallow-water carbonates have tended to develop broadly similar facies belts through time.

Carbonate ramps

Ramps are low-angle seaward dipping surfaces, with no continuous elevated rim or clear break in slope. The transition between low energy, lagoonal facies through high-energy grainstone or reefal facies to deeper water facies is thus very diffuse, and ramp sedimentation may be dominated by basinal processes (storms and tidal currents) over wide areas. The generally uniform dip on a ramp surface is maintained largely by sediment redistribution by hydraulic processes across its broad but shallow expanse. Deeper ramp facies commonly contain sediment of shallow-water origin transported offshore during storm surges (Aigner, 1984; Wright, 1986). This 'regrading' process is important in allowing the ramp to maintain its inherently unstable geometry. Ramps are often seen to develop into rimmed shelves as the differential build-up potential of the more productive margin outpaces that of adjacent deeper and shallower water facies.

The idealized facies belts of a land-attached ramp are illustrated on Fig. 10.15. A broad area of peritidal carbonates, controlled largely by climatic and hydraulic setting, comprising evaporites and tidal flat deposits is developed adjacent to the ramp topography. Evaporites and evaporitic structures, such as tepees, laminated crusts and enterolithic folds (Shinn, 1983; Scoffin, 1987; Tucker and Wright, 1990) are characteristic of tidal flat deposition in arid areas, but in more humid climates evaporites may be absent, replaced by soil and palaeokarsts developed upon subaerially exposed beach and lagoon deposits, such as in

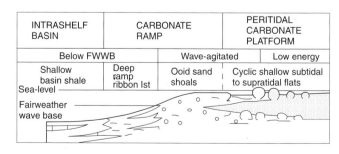

Fig. 10.15 Carbonate ramp facies model (after Read, 1982). FWWB, fair-weather wave base

the Carboniferous of South Wales (Wright, 1986). Seaward of the tidal flat environment, lagoonal sediments pass into the wave-dominated ramp margin, a diffuse belt of bioclastic and/or oolitic shoals, up to several kilometres in width. Reefs also may develop in this area, although most reefs in the modern Arabian Gulf ramp system tend to occupy submarine swells and areas between active ooid sand sedimentation (Purser, 1973). No slope environment as such exists on a ramp, and the deeper ramp environment is characterized by muddy facies, often with well-developed storm beds of originally shallow platform sediment developed above storm wave base.

Rimmed shelves

Carbonate rimmed shelves have a distinct slope, with angles of up to 35°, separated from a broad, shallow platform top. As with carbonate ramps, the landward portion of modern rimmed shelves may comprise evaporitic or mangrove/microbial peritidal facies, depending on climate. The major differences between rimmed shelves and ramps are (i) the thinner margin of a rimmed shelf, often only several hundred metres or less in width, and (ii) the presence of a well-defined slope on which slope sedimentation processes act. Idealized facies belts for a land-attached rimmed shelf would thus be intertidal and lagoonal deposits passing seaward into a well-defined margin, comprising reefal build-ups and/or oolitic and bioclastic sand shoals (Fig. 10.16). The lagoon itself may be subject to wave and tidal agitation where gaps in the margin exist. Alternatively, only occasional agitation may affect the lagoon and it may become hyper- or hyposaline depending on the climate and runoff. On rimmed shelves, the margin is a very turbulent zone, where storm and oceanic waves and tidal currents impinge on the sea-bed. This zone is much better defined than on a ramp, where shallowing towards the margin is more gradual and wave energy is dissipated more gently over a wider area. Seaward of the rimmed shelf-margin, proximal slope facies may develop, comprising large reef talus blocks. These pass downslope into debris flow deposits and turbidites, which fine distally. The deeper rimmed shelf facies may consist of very distal turbidites, interbedded with basinal mudstones of pelagic or hemipelagic origin.

Escarpment margins

Escarpment margins can be thought of as rimmed shelves without a significant accretionary slope facies (the 'escarpment by-pass' type of Read, 1982), with proximal lagoonal, inter- and supratidal facies belts, as described for rimmed shelves in the preceding section. On such margins, at slope angles in excess of 35–40°, most sediment will tend to bypass the upper slope and accumulate on the lower slope or in the basin (Fig. 10.13). A steep shelf-edge cliff is developed through rapid aggradation, which may be partly eroded by gullies down which shallow-water carbonate debris passes. Alternatively, the cliff may be highly eroded, cutting back into older carbonates. A good example of the latter are erosional shelf-margins off Belize, where Pleistocene limestones are being exposed along the erosional scarp (James and Ginsburg, 1979). The sediment accumulating at the base of the escarpment margin is usually coarse-grained and disorganized, and may form a continuous apron of peri-platform talus, particularly where

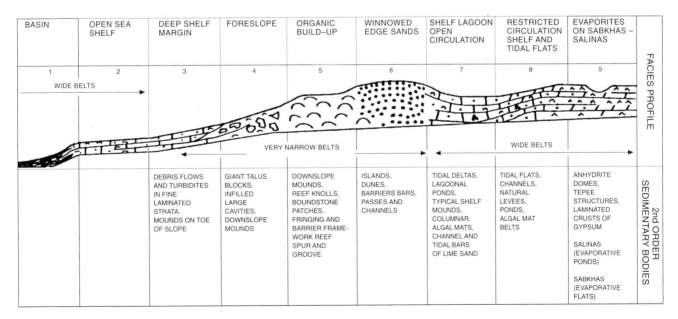

Fig. 10.16 Standard facies belts on a rimmed shelf (after Wilson, 1975)

the shelf margin is highly erosional. Deeper into the basin, more organized proximal carbonate turbidite facies may develop, which pass basinward into distal turbidites interbedded with pelagic and argillaceous carbonate (Fig. 10.17).

Isolated carbonate platforms

Isolated platforms are areas of shallow-water carbonate accumulation surrounded by deep water. Modern isolated platforms range from Pacific atolls a few kilometres across to massive areas of carbonate production such as the Bahama Banks, 700 km from north to south. Isolated platforms may develop any of the three categories of margin described above, although steeper rimmed shelf or escarpment margins tend to be more common. The major difference between large and small isolated platforms is the area of overall carbonate productivity. In both cases, the margin usually will have roughly the same dimensions, but on larger platforms the extensive carbonate production in the lagoon will provide a massive additional source of sediment to shed into deeper water.

10.4 Sequence stratigraphic models for carbonate platforms

10.4.1 Introduction

The foregoing sections hopefully have convinced the reader that carbonate systems may behave quite differently from siliciclastics, and that a modification of the standard siliciclastic sequence stratigraphic model is necessary to encompass carbonate idiosyncrasies. To this end, a series of general sequence stratigraphic models, with examples from outcrop and the subsurface, are outlined in the following sections. This section also draws on previously published carbonate sequence stratigraphic models; Sarg (1988), Calvet *et al.* (1990), Tucker and Wright (1990), Hunt and Tucker (1991) and Tucker (1991).

Each carbonate platform type will be presented in an idealized sea-level cycle of transgression, highstand and lowstand or drowning, using the facies models described in section 10.3. Climate also exerts an influence on the sedimentology of systems tracts, particularly lowstand, and the important differences between arid and humid lowstand systems tracts will be presented. It is important also to bear in mind that for all these models, only the gross systems-tract geometries will be discussed, i.e. those visible on exploration seismic data (Chapter 3) or at mountainside outcrop scale (Bosellini, 1984). The finer scale cyclicities observed on a several to a few tens of metre scale will be the subject of section 10.5.

10.4.2 Ramps

Once a ramp system (Fig. 10.18) is established, rising sea-level in the transgressive leg of an idealized sea-level cycle will commonly cause the ramp to step landward because of its low carbonate productivity. Both the coastal onlap and downlap points will migrate landward, with a progressive superimposition of deeper-water on shallow-water facies. At this time, the deeper ramp may become sediment starved, and as water depth increases over the deep ramp, the potential for significant accumulation of organic matter increases. An example of petroleum source-rock accumulation during ramp transgression may be the distal facies of the Jurassic Smackover ramp system in the Gulf of Mexico (Moore, 1984; Sassen, 1988), which forms a major source-rock interval. More proximal facies of a ramp transgressive systems tract are shown by the Jurassic Lincolnshire Limestone Formation in eastern England (Ashton, 1980; Emery

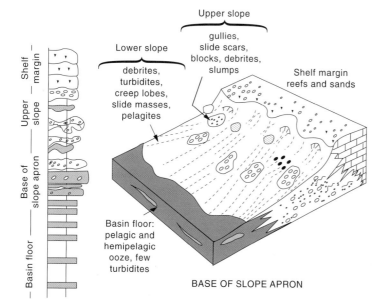

Fig. 10.17 Simplified facies model for a carbonate escarpment margin, showing bypass of the upper slope and sediment accumulation at the toe of slope and in the basin (after Mullins, 1983)

and Dickson, 1991). Here, the carbonate ramp was initiated on estuarine muds and silts as pre-ramp topography was infilled during transgression and clastic facies belts were pushed landward. The carbonate succession, only up to 40 m thick in total, comprises oolitic and bioclastic shoal facies overlying back-barrier and lagoonal facies, indicating overall transgression (Fig. 10.19). Other examples of landward stepping ramp systems during transgression include the Lower and Middle Anisian ramps of the Lower Muschelkalk, Catalan Basin, Spain (Calvet et al., 1990), and the Upper Muschelkalk ramps of the intracratonic German Basin (Aigner, 1984).

As the rate of sea-level rise decreases, carbonate ramp systems will tend to prograde basinwards. In early highstand, when accommodation is still being created landward of the margin, supra-, intertidal and lagoonal deposits may accumulate as topsets, but during late highstand, very little or no topset deposition will occur, and a 'strandplain' ramp may result (Tucker and Wright, 1990). The seismic expression of the highstand systems tract of carbonate ramps is typically as a series of shingled clinoforms, with rarely resolvable topsets, such as the Smackover Formation (Moore, 1984), Lower Permian ramps of the Midland Basin, Texas (Mazzullo and Reid, 1989) and the Upper Devonian Nisku Formation, Alberta, Canada. In the latter example, deep ramp build-ups sitting on low-angle clinoforms are the main exploration targets. In outcrop, progradational oolitic facies are recognized from the Castell Coch Limestone of the Carboniferous of South Wales (Burchette, 1987), the Muschelkalk of Europe and the Middle Jurassic of France, Spain (Aurell, 1991) and southern England.

During falling sea-level the landward portion of the ramp will be exposed. Where clastic sediment supply is low and the ramp system was prograding into an open shelf setting, the ramp may offlap, stepping downward such that its onlap point falls below the earlier ramp margin. Alternatively, if the siliciclastic sediment supply is significant, it may overwhelm carbonate production and a lowstand siliciclastic system may be established across the ramp

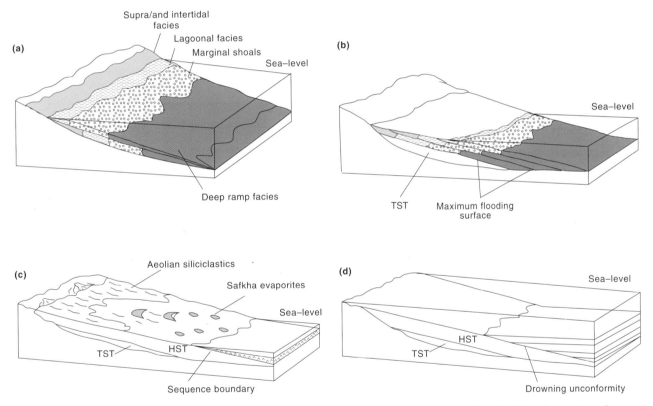

Fig. 10.18 Sequence stratigraphic models for ramp systems. (a) Transgressive systems tract shows landward stepping of ramp facies and sediment starvation in the basin, with the potential for the development of organic-rich mudstones. (b) Highstand systems tract shows seaward progradation of the margin and progressive thinning of topsets. In late highstand, topsets may not be developed and a 'strandplain' system (cf. Calvet et al., 1990) may result. (c) Lowstand systems tract in an arid, restricted basin. Here, the basinal facies is developed as a subaqueous evaporite wedge onlapping deeper ramp facies of the previous highstand systems tract above a sequence boundary. The exposed ramp may be the site of extensive sabkha evaporite development and may be encroached upon by aeolian siliciclastics. In a humid setting, the exposed ramp may be incised by fluvial channels and karstified, or if the siliciclastic input is relatively low, a new lowstand ramp may nucleate below the margin of the previous ramp system. (d) Drowning of a carbonate ramp followed by siliciclastic progradation over the drowned ramp. The ramp and overlying siliciclastic wedge will be separated by a drowning unconformity

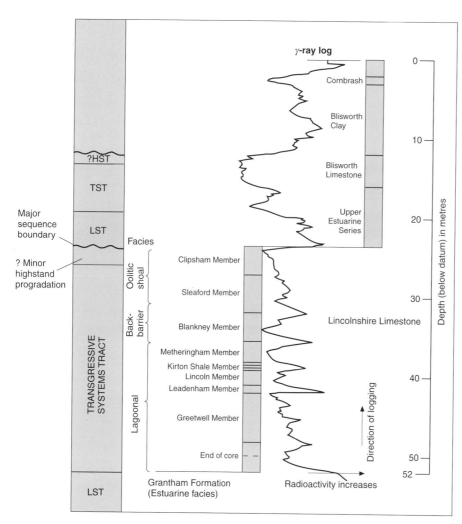

Fig. 10.19 Summary log and facies model for the Lincolnshire Limestone Formation, eastern England. The lowstand system tract (LST) is developed in a clastic system, the estuarine Grantham Formation, which is succeeded by a landward stepping transgressive systems tract (TST; the Lincolnshire Limestone), as shown by the superimposition of more distal shoal facies over lagoonal and back-barrier sediments (after Emery and Dickson, 1991). HST, highstand systems tract

surface. Note that, in general, non-skeletal carbonate margins are better able to persist during phases of siliciclastic input than reef-dominated margins. An example of lowstand siliciclastic deposition replacing carbonate ramp deposition during lowstand is from the Carboniferous of Anglesey (Walkden and Davies, 1983). In more humid climates, proximal ramp deposits that become exposed during lowstand will develop soils and karst surfaces. Ramsay (1987) has described palaeokarsts developed on Carboniferous ramp oolites of South Wales deposited during slight sea-level falls. By contrast, more arid environments may develop calcretes or extensive sabkha evaporites on the exposed ramp, with encroachment of aeolian facies if the climate is too dry to develop fluvial systems. Mixed siliciclastic and evaporitic lowstand systems tracts on exposed carbonate ramps have been described by Calvet *et al.* (1990) from the Triassic of Spain. In an arid climate, basin isolation may occur during lowstand, resulting in evaporation. A lowstand evaporite wedge may onlap the distal ramp facies with cessation of carbonate production because of hypersaline conditions.

The idealized sea-level cycle described above does not account for drowning on a carbonate ramp, which may take place during transgression or highstand due to environmental deterioration. Unequivocal examples of drowning on ramps are not numerous. Perhaps the best-documented example is again from the Triassic of the Catalan Basin in Spain (Calvet and Tucker, 1988). Here, deeper ramp mounds show evidence of exposure during an abrupt sea-level fall. Subsequent transgression led to the development of a stratified sea, with oxygen-depleted waters. Instead of continued mud-mound growth, organic-rich carbonate accumulated. Strictly speaking these Triassic mud mounds were not really drowned, but rather failed to re-establish themselves in the transgression that followed sea-level fall and karstification.

10.4.3 Rimmed shelves

Carbonate rimmed shelves (Fig. 10.20) may be initiated in several ways. They may start up during transgression over a non-carbonate substrate, particularly over a pre-existing shelf break or structural high. In so doing, a barrier may become established whereby wave energy can be dissipated and lagoonal and inter- and supratidal carbonate deposition can proceed in the protected lee of the barrier. Rimmed shelves also may be initiated over ramps. As transgression continues, the productive margin may be the only element of the platform able to keep up with sea-level, and a well-defined slope may become established. At this time, lagoonal carbonate production also may be outpaced by rising sea-level, leaving a deepening hole in the lee of the rim, such as in the present-day Belize lagoon. More commonly observed, however, is 'keep up' of the entire rimmed shelf up to and including the margin, to give a flat-topped carbonate rimmed shelf overlying a low-angle ramp. Excellent examples of this transition include the Lower Permian of the Midland Basin, Texas, where Wolfcampian ramps are overlain by rimmed shelves of the Wichita and Lower Clear Fork Formations (Mazzullo and Reid, 1989; Fig. 10.21). In outcrop, the ramp-to-rimmed-shelf transition observed in the Triassic of the Italian Dolomites is quite spectacular (Fig. 4.5). One of the key differentiating factors between siliciclastic and carbonate transgressive systems tracts is the ability of the carbonate to aggrade, rather than backstep during rapid sea-level rise. Aggradation, and even progradation, during transgression is common in rimmed shelves (Pomar, 1991). However, if transgression was to continue uninterrupted, with no change in carbonate productivity, the rimmed shelf would be unable to keep building vertically and maintain its slope angle, and would evolve into an escarpment margin.

During highstand the rimmed shelf can shed overproduced carbonate on to the slope and into the basin. In early highstand, topsets may continue to accumulate as lagoonal and inter- and supratidal facies. However, during late highstand, topsets may be poorly resolved, particularly on exploration seismic data. It also may be impossible to identify a maximum flooding surface in the platform interior facies because of the lack of backstepping in the underlying transgressive systems tract. However, seaward of the aggraded rim of the transgressive systems tract it may be possible to identify the major downlap surface and hence maximum flooding surface of the rimmed shelf interval, but this is often equivocal. Examples of prograding rimmed shelves are common, and are particularly well shown on seismic data. Figure 10.22 shows examples from the Tertiary of the Dampier Basin, Northwest Shelf, Australia. Progradation of rimmed shelves in outcrop is shown superbly by the Triassic of the Italian Dolomites (Fig. 4.5).

Sea-level fall on a rimmed shelf, as on a ramp, may result in a range of different lithologies, facies and geometries. Saller *et al.* (1989) have shown major clastic input during lowstand in the Permian of New Mexico, where a series of lowstand sandstones interdigitate with carbonate slope deposits in the Mescalero Escarpe Field. All the sands appear to sit above type 1 sequence boundaries. The importance of lowstand evaporite precipitation both in the topsets of rimmed shelves and as wedges onlapping carbonate slope facies has been described by Tucker (1991) for the Permian of northeastern England. Here, aridity restricted significant fluvial input during lowstand and hypersalinity prevented further carbonate growth. Karst formation on rimmed shelf topsets in more humid systems is also highly significant as a porosity modifying process. Examples include karstic carbonate reservoirs of the Gargano Peninsula, eastern Italy, where spectacular cavern and fissure systems provide the porosity, and, most significantly, the permeability for subsurface oil reservoirs such as the Rospo Mare Field (Zezza, 1975).

Type 2 sequence boundaries and associated shelf-margin wedge systems tracts also may be developed on attached rimmed shelf systems, where sea-level falls over the platform top, but sediment accommodation remains at the platform edge (Sarg, 1988, and Chapter 2). Shelf-margin wedge systems tracts are likely to be developed only if there has been differential subsidence of the top of the flat-topped carbonate platform to create a dipping surface on to which the shelf-margin wedge can onlap back on to (Fig. 10.20). Note that the shelf-margin wedge systems tract will not be developed on ramps, because ramps lack a shelf margin (Calvet *et al.*, 1990).

Drowning of rimmed shelves is better documented than for ramps. The example of the Wilmington Platform is illustrated on Fig. 10.10, where a siliciclastic system overlies a drowned, flat-topped rimmed shelf. Another example is the Jurassic of Jebel Bou Dahar of the Moroccan Atlas Mountains, where outcrop exposure shows a carbonate margin and slope onlapped by basinal shales, with evidence for progressive environmental deterioration on the platform top provided by a decreasing diversity and productivity of carbonate (Campbell, 1992).

10.4.4 Escarpment margins

As discussed in the foregoing section, aggradation of a rimmed shelf can give rise to an escarpment margin (Fig. 10.23). The margin initially may be of the bypass type, but oversteepening can lead to erosion and apparent backstepping of the margin as transgression proceeds (Fig. 10.13). The products of platform erosion usually will be large blocks of talus derived from the margin, which will accumulate at the base of the carbonate slope, passing distally into more organized calciturbidites derived from the platform interior. On the carbonate platform itself, it usually will be very difficult, if not impossible, to differentiate the aggradational transgressive systems tract from the overlying highstand systems tract. Because no accretionary slope is developed, the platform will be unable to prograde during highstand, and the platform top facies will continue to build vertically, albeit with thinner topsets. Figure 10.24 shows the steep-walled margin of the Cretaceous West Florida platform, which accumulated during transgression and highstand and has remained exposed since the Cretaceous. Erosion of the 'margin' has been demonstrated by dredge hauls from the wall of the platform, which reveal platform interior deposits (Corso *et al.*, 1988).

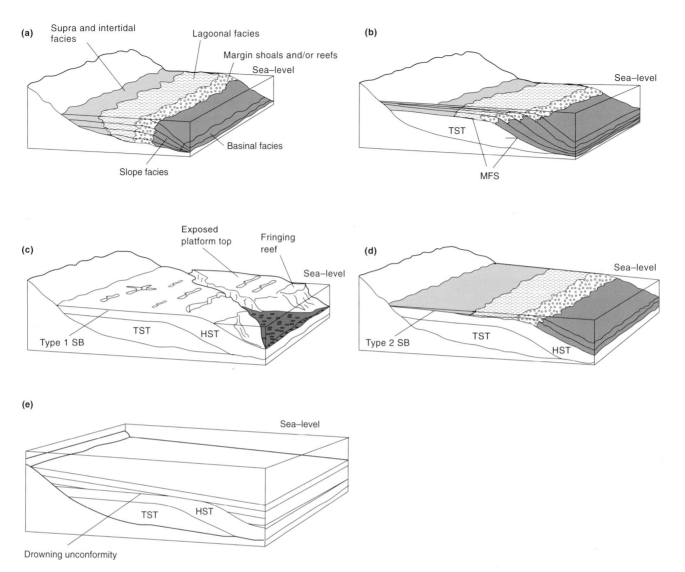

Fig. 10.20 Sequence stratigraphic models for rimmed shelves. (a) Transgressive systems tract (TST) of a rimmed shelf system showing aggradation of the margin. (b) Highstand systems tract (HST) of a rimmed shelf showing progradation and progressive thinning of topsets. (c) Type 1 sequence boundary and development of the lowstand systems tract on a humid rimmed shelf. Here, sea-level has fallen significantly below the margin of the rimmed shelf, and the exposed platform top is karstified and may become incised by fluvial channels. Siliciclastic sediment may be deposited in the basin, onlapping the carbonate slope. *In situ* carbonate production may take place as fringing reefs if the siliciclastic input is sufficiently low and directed away from favourable carbonate production sites. (d) Type 2 sequence boundary and development of a shelf-margin-wedge systems tract on a rimmed shelf. Sea-level has not fallen significantly below the previous margin, and the platform top is not extensively exposed to meteoric diagenesis. The shelf-margin wedge expands landward and seaward with rising sea-level, but note that this can occur only if there is greater seaward subsidence of the originally flat platform top to provide a dipping surface that can be onlapped. (e) Drowning of a rimmed shelf as carbonate production is terminated by environmental deterioration while still submerged. The top of the carbonate is characterized by a drowning unconformity, which separates underlying platform carbonate from onlapping and downlapping deep-water siliciclastics. BF, basinal facies; SF, slope facies; SB, sequence boundary; MFS, maximum flooding surface

On the basin floor, highstand and lowstand deposits can be differentiated better. Highstand shedding from the platform will supply platform interior material to the basin, which may accumulate as calciturbidites or settle out of suspension. These deposits may onlap and cover the underlying talus cones developed during the previous transgression. Glaser and Droxler (1991) documented thick Holocene highstand wedges onlapping the base of the erosional margin of Pedro Bank, Nicaragua Rise. Note that margin failure also may occur during highstand, as well as during transgression, but highstand talus cones are more likely to be interbedded intimately with platform interior facies. The angular contrast arising from the contact between talus cones and finer grained sediment can give rise

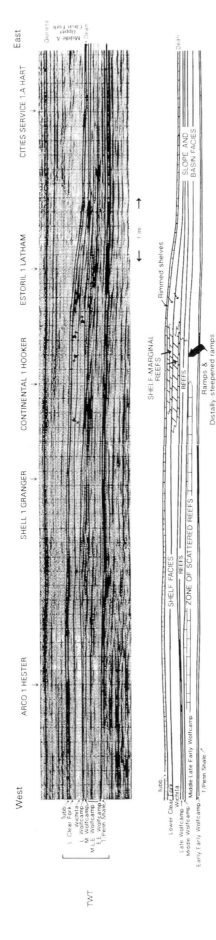

Fig. 10.21 Lower Permian ramps of the Midland Basin, West Texas. The ramps show a generally progradational geometry, with a diffuse margin, whereas the overlying rimmed shelves show a stationary aggradational margin, steepening through time (from Mazzullo and Reid, 1989).

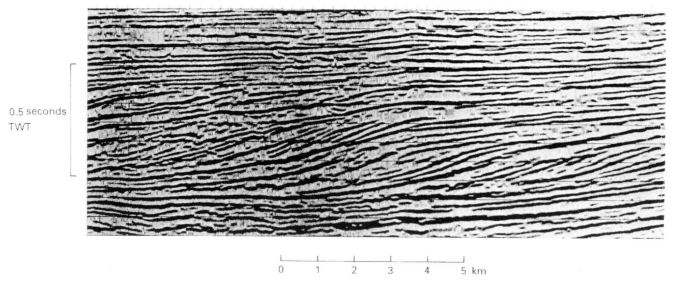

Fig. 10.22 Seismic section showing progradation of Tertiary carbonates of a rimmed shelf system, Dampier Basin, northwest Australia

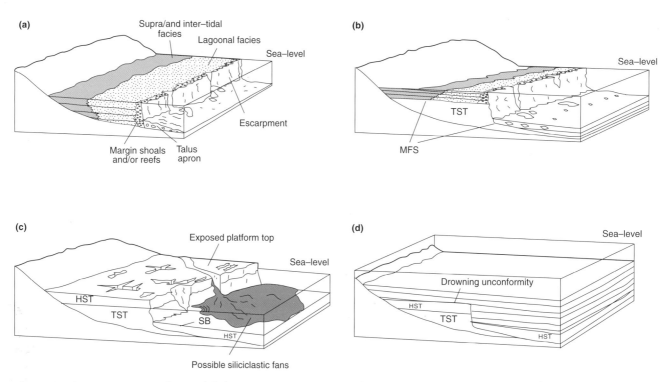

Fig. 10.23 Sequence stratigraphic models for escarpment margins. (a) Transgressive systems tract (TST) of an escarpment margin showing aggradation and failure of the margin. The products of margin failure onlap the steep slope as a talus apron. (b) Highstand systems tract (HST) of an escarpment margin showing continued aggradation, but a progressive thinning of topsets. In the basin and at the base of slope, carbonate shed from the platform top during highstand onlaps the talus apron of the transgressive systems tract. Some talus also may be shed off the carbonate margin during highstand. (c) Lowstand systems tract (LST) on a humid escarpment-margin system. Here, sea-level has fallen significantly below the margin, and the exposed platform top is karstified and may become incised by fluvial channels. Siliciclastic sediment may be deposited in the basin, onlapping the carbonate slope, and failure of the margin also may result in the deposition of talus cones at the base of slope. *In situ* carbonate production is likely to be very minor on very steep margins. (d) Drowning of an escarpment margin, as carbonate production is terminated by environmental deterioration while still submerged. The top of the carbonate is characterized by a drowning unconformity, which separates underlying platform carbonate from onlapping and downlapping deep water siliciclastics

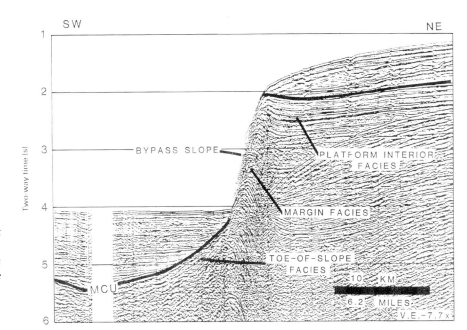

Fig. 10.24 Seismic profile of the West Florida Cretaceous platform margin. Although the margin shows a 'seismic reef', labelled as margin facies, dredge hauls show that the margin facies actually comprise platform interior deposits (from Corso *et al.*, 1988)

to geometries that resemble onlap on to sequence boundaries. Grammer *et al.* (1990) and Grammer (1991) have described such geometries from submersible dives over slopes of Bahamian platforms, and Bosscher and Southam (1992) have generated computer simulations of grain-size-related 'sequence boundaries', which are clearly not sequence boundaries *sensu stricto* (Chapter 2).

Lowstand on the platform top of escarpment margins will be similar to that described for rimmed shelves. Karstification and fluvial incision may occur in humid climates, evaporite precipitation may occur on the platform top or onlap the platform margin in arid, restricted settings (Tucker, 1991). On very steep erosional margins, very little *in situ* carbonate deposition is possible, but on lower angle bypass slopes fringing reefs may become established. The low rates of sedimentation during lowstand on an escarpment margin provide an opportunity for older slope deposits to become lithified by marine cementation. Grammer and Ginsburg (1991) have dated thick marine cements from lithified onlapping wedges off the Great Bahama Bank. Carbon-14 ages from these cements range from 11 000 to 13 000 years BP, corresponding to the last sea-level lowstand. Ginsburg *et al.* (1991) also have been able to link sequence boundaries to intervals of intense submarine cementation off the Great Bahama Bank.

Drowning of escarpment margins proceeds much as described for rimmed shelves. Excellent examples of drowned steep-sided platforms can be seen in the Cretaceous of southern Europe, where Schlager (1992) has suggested that their demise may be linked to global oceanic anoxic events.

10.4.5 Isolated platforms

Isolated carbonate platforms may show one or more of the different types of margin described above, depending on the tectonic setting, windward–leeward orientation of the platform and the platform size. To avoid repeating the systems tract geometries described above, this section will be devoted to short descriptions of published examples of sequence stratigraphy of isolated platforms. A simplified model for isolated platform sequence stratigraphy has been outlined in Fig. 10.6.

Symmetrical platforms

Epting (1980) has described two symmetrical isolated build-up types. The 'pinnacle' type build-up is essentially an aggraded mound which shows continuous gradual backstepping (Fig. 10.25). It is impossible to identify different systems tracts from seismic data alone in this example. Pinnacle build-ups are particularly common in the Tertiary of southeast Asia (Hatley, 1980; Grainge and Davies, 1985). Rudolph and Lehmann (1989) describe a giant gas accumulation from the Miocene of offshore Indonesia, the Natuna D-Alpha Field, which differentiated from a broad underlying carbonate platform into several discrete pinnacles. Epting's second category was the 'platform type' build-up, which shows aggradation in the underlying transgressive systems tract, followed by progradation during highstand, and a well-developed maximum flooding surface separating the two systems tracts (Fig. 10.26).

Asymmetrical platforms

The Liuhua Oilfield in the Miocene of the Pearl River Mouth Basin, South China Sea, shows a strong asymmetry. The steep southwest-facing escarpment margin was fault controlled (Fig. 10.11), and built up vertically as an escarpment margin through time prior to eventual drowning. The

Fig. 10.25 A 'pinnacle' type carbonate platform from the Miocene of offshore Philippines. It is impossible to break out different systems tracts from seismic data alone in this example, which shows continuous gradual backstepping and confused internal seismic character. The top of the platform is very rugose. This may represent the effects of karstification, or the differentiation of the platform top into several small pinnacles prior to drowning

northeast-facing margin, by contrast, probably evolved from a ramp to a rimmed shelf through time, and shows evidence of aggradational and progradational geometries (Erlich et al., 1990).

Eberli and Ginsburg (1987) contrasted the development of different platform margin types across the Bahamas. Figure 10.27 shows a windward margin to the west and a leeward margin to the east. The windward margin was reefal and aggradational and built a steep escarpment margin during transgression and highstand. The leeward margin, with an accretionary slope consisting chiefly of highstand shedded material and some shingled lowstand reflectors, prograded westwards, ultimately infilling the embayment between two previously isolated platforms. Thus the extensive flat-topped Great Bahama Bank seen today is a result of the coalescence of separate carbonate platforms.

10.5 Cyclicity and parasequences on carbonate platforms

The discussion thus far has centred chiefly on seismic-scale expressions of carbonate sequence and systems-tract development. Carbonates also exhibit very well-developed cyclicity on scales of decimetres to several metres. These cycles form the building blocks of the carbonate systems tracts and sequences. The stacking patterns of these carbonate rhythms for different systems tracts on different carbonate-margin types are illustrated schematically on Figs 10.18, 10.20 and 10.23, and although the stacking patterns of these cycles can provide some clues as to the overall behaviour of sea-level and/or carbonate productivity, individual cycles may be quite different geologically, and may originate in several different ways.

Carbonate cycles and parasequences — examples

Several types of carbonate cycle can be recognized. The most common is the shallowing upward cycle, which comprises a lower energy, deeper water base that may have been deposited below wave base, shallowing up into a shoal or reefal facies. The shoal facies may be overlain abruptly by a deeper water facies and a further shallowing upwards facies succession. This is a carbonate parasequence. Examples of stacked carbonate parasequences such as this have been described from the Cambrian of the USA (Markello and Read, 1981; Montanez and Osleger, 1993), from the Lower Carboniferous of Wales by Gray (1981), and the Jurassic of southern and eastern England (Ashton, 1980). These cycles are a few to several metres thick and are inferred to have been deposited over tens of thousands of years, at least in the latter two examples, by seaward progradation of shoal facies interrupted by rapid deepening events.

Shallowing upward cycles deposited in more proximal positions are also abundantly described in the literature. However, these cycles are not parasequences under the definition used in Chapter 2, although they may link-up distally with parasequences. The Middle and Upper Triassic of the Dolomites in northern Italy is particularly notable for its metre-scale cycles (Goldhammer et al., 1990). In the Latemar Limestone of the Ladinian, the cycles are usually less than 1 m thick and comprise limestone–dolomite pairs, with shallow subtidal grainstones and packstones capped by dolomite containing pisoids, tepee structures and vadose cements, indicating subaerial exposure (Fig. 10.28). Up to 500 such cycles have been described, but these thinner cycles may be disturbed by thicker horizons of disrupting tepees, reflecting longer periods of subaerial exposure. The

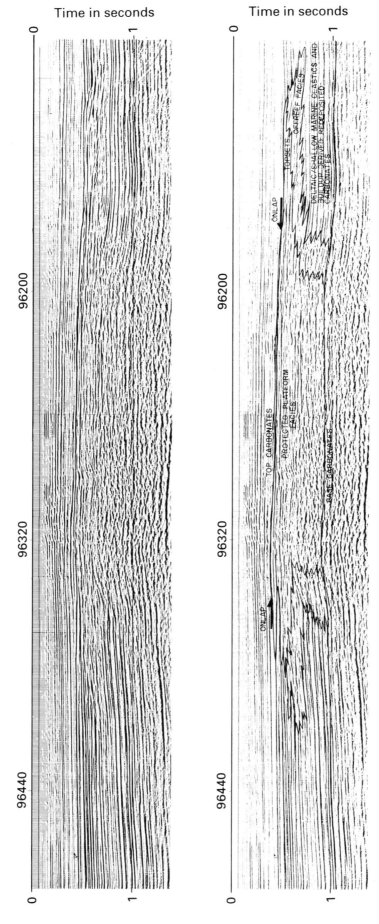

Fig. 10.26 A flat-topped carbonate platform from the Miocene of the Luconia Province, offshore Sarawak. The earliest phase of platform development is essentially aggradational, with coeval siliciclastic sedimentation in the basin adjacent to the platform. Following aggradation of the transgressive systems tract, the platform progrades over basin facies during highstand, and the position of the maximum flooding surface can be recognized off the platform flanks, but less easily in the platform interior (from Epting, 1980)

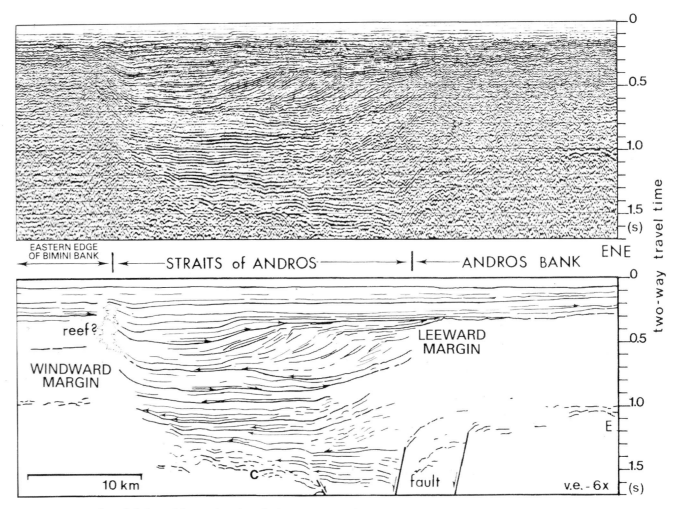

Fig. 10.27 Windward (left) and leeward (right) platform margins of the Bahamas (from Eberli and Ginsburg, 1987). See text for details

absence of any tidal-flat facies in these carbonates rules out the possibility of tidal-flat progradation as a mechanism for producing the cyclicity, suggesting that an external control such as sea-level fluctuation may have caused the rhythms. Other proximal shallowing upward cycles include grainstone-based cycles shallowing upwards into tufa, and tufa-based cycles shallowing upwards into tepees and breccias from the Early Proterozoic of the Rocknest Formation, Canada (Grotzinger, 1986).

In arid settings carbonate cycles may be associated with evaporite precipitation, usually where gypsum or anhydrite occurs at the top of the cycle, overlying dolostones or limestones with intertidal sedimentary structures. This type of cycle has been interpreted as the development of supratidal sabkhas with attendant evaporite precipitation over intertidal carbonates, and again represents a shallowing upward rhythm. Such units are usually a few metres thick, and have been observed in the Jurassic Arab Formation of Abu Dhabi by the author, and in the same Formation in Saudi Arabia (Wood and Woolfe, 1969). In both areas the evaporites provide major intraformational sealing units overlying reservoir bioclastic and oolitic shoal facies. Sabkha sulphate seals and carbonate–evaporite cycles are also well described from the Permian carbonates of the Permian Basin, Texas (Handford, 1982; Craig, 1990).

In more humid settings, carbonate cycles on attached platforms may include a significant siliciclastic component. The Yoredale Series of the Upper Dinantian of northern England comprise limestones deposited below wave base during overall transgression when siliciclastic input was more muted, overlain by mudstones and sandstones deposited by prograding delta systems. Abandonment of the prograded delta lobes led to the development of swamp environments and the subsequent formation of coal (Leeder and Strudwick, 1987). In the Upper San Andres Formation of the Permian of New Mexico, mixed carbonate and siliciclastic cycles that contain all the elements of a sequence on the scale of a few metres have been described by Sonnenfeld (1991). Here, siliciclastics represent lowstand deposition on the slope and in the basin, and carbonates

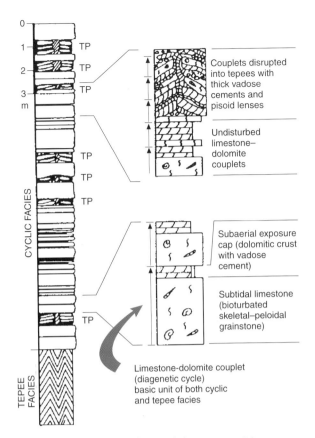

Fig. 10.28 Cyclic carbonates of the Latemar Limestone, Italian Dolomites (after Hardie *et al.*, 1986). See text for details

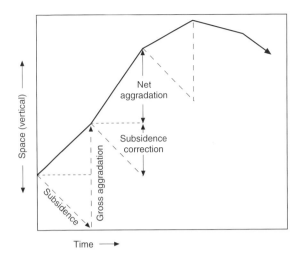

Fig. 10.29 Fischer plot of four hypothetical high-frequency carbonate platform cycles (from Sadler *et al.*, 1993)

represent transgressive and highstand deposits. These cycles have been termed fourth- and fifth-order sequences, because they have a greater degree of symmetry than a typical parasequence, and show all the systems tracts of a typical sequence. Goldhammer *et al.* (1991) have recognized 'fourth order' sequences from the Carboniferous of the Paradox Basin, comprising carbonates, siliciclastics and evaporites, with a mean thickness of 35 m and periodicity of about 260 000 years.

Carbonate cycle stacking patterns and Fischer plots

In recent years, there has been a revival of interest in analysing the stacking patterns of high-frequency platform-top carbonate cycles. Much of this analysis has been achieved by the use of accommodation or 'Fischer plots' (Fischer, 1964; Sadler *et al.*, 1993). A Fischer plot is a series of sawtooth waves, with the vertical line of the sawtooth representing the vertical thickness of a carbonate cycle, linked by a diagonal line drawn from the top of the cycle down to the right of the plot. The bottom right of the diagonal links to the base of the next vertical line representing the next carbonate cycle (Fig. 10.29). The only factual information contained in the Fischer plot is the cycle thickness. The diagonal line subjectively represents the passage of time and subsidence prior to the deposition of the next cycle, and is based on two simplifying assumptions of a fixed cycle period and linear subsidence, and, as such, Fischer plots are 'purely descriptive representations of stratigraphic sections that emphasize stacking patterns' (Sadler *et al.*, 1993). Despite the simplifying assumptions, Fischer plots combined with detailed facies analysis have been used to show how platform-top carbonate-cycle stacking can help to distinguish different systems tracts and the location of sequence boundaries within otherwise monotonous platform-top carbonate systems. Goldhammer *et al.* (1993) were able to use Fischer plots from Ordovician carbonates of the El Paso Group, West Texas, to differentiate systems tracts. Lowstand systems tracts comprised thin quartz arenites, transgressive systems tracts comprised thickening upward subtidal cycles, and highstand systems tracts comprised dolomitic peritidal cycles containing some quartz sand. Fischer plots of the stacking pattern of the West Texas cycles and other coeval Ordovician cycles in the USA are illustrated on Fig. 10.30. Other studies that have used Fischer plots to define sequences and systems tracts include Read and Goldhammer (1988) and Montanez and Osleger (1993). Sadler *et al.* (1993) provide a good review of Fisher plots, with recommendations for their construction and interpretation.

Carbonate cycles — origin

The origin of carbonate and mixed cycles is the subject of considerable and continuing debate (Chapter 2). It is a vast subject in its own right, and will be considered only briefly here. Three mechanisms are usually invoked; eustasy, and two types of autocyclic mechanism. The eustatic control involves changes in sea-level that are governed by orbital perturbations (Milankovitch cyclicity, Chapter 2).

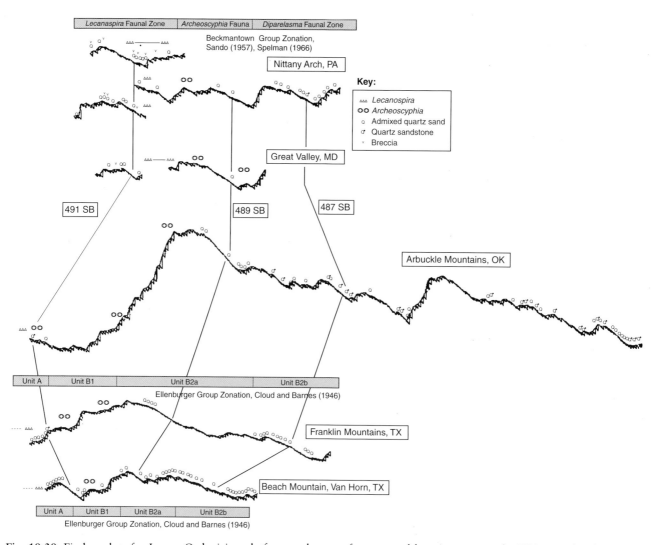

Fig. 10.30 Fischer plots for Lower Ordovician platform carbonates from several locations across the USA, correlated biostratigraphically. Third-order sequence boundaries (491, 489 and 487 Ma) were correlated using high-frequency cycle stacking patterns and the distribution of quartz sand (from Goldhammer et al., 1993). SB, sequence boundary

Goldhammer et al. (1987) have observed that, in detail, the cycles observed in the Triassic of the Dolomites comprise asymmetric packages of five cycles that thin upwards. This 'pentacycle' has been interpreted to reflect the 20 000 year precession periodicity superimposed on the 100 000 year cycle. Read et al. (1992) and Read and Horbury (1993) have attempted to relate carbonate cyclicity and diagenesis to Milankovitch forcing, using many examples drawn from the Phanerozoic. However, it is evident that throughout geological time the volume and extent of polar ice-caps has been highly variable. Invoking rapid and high-amplitude sea-level change for cyclic carbonate development in 'ice-house' time is reasonable, but it becomes more difficult in 'greenhouse' times. In addition, any cycles developed by an orbital forcing mechanism ought of be of great lateral extent, but cycles of limited lateral extent have been described in the literature (e.g. Pratt and James, 1986).

A popular autocyclic mechanism to explain carbonate rhythms, attributable to R.N. Ginsburg, is the tidal-flat progradational model (James, 1984). Here, the area of carbonate production is reduced gradually and overwhelmed eventually by tidal-flat progradation. Because the carbonate factory can no longer supply material to the tidal flats, progradation ceases, and the reduced-productivity carbonate platform is unable to keep pace with continuing relative sea-level rise, subsides, and is flooded during the ensuing transgression. During transgression, the carbonate factory is re-established and the cycle recommences. This mechanism has been used to explain intervals that cannot be interpreted easily in terms of orbital forcing, and where tidal-flat progradation is evident. Lankester (1993) has suggested that a similar mechanism may be applicable to the platform-top facies of the Miocene of Majorca for the highest frequency cycles, which show no evidence for sea-level change, and Goldhammer *et al.* (1993) have speculated that such an autocyclic mechanism may be responsible for high-frequency carbonate cycles of the Ordovician of Texas described above. An additional autocyclic mechanism proposed by Pratt and James (1986) is the tidal island model. Here, a carbonate platform is never exposed or flooded entirely, but is covered with tidal-flat islands. These islands may prograde or build vertically, and may coalesce with one another. On doing so, the subtidal area for carbonate production may decrease, and the coalesced island may subside. Hydrographic forces also may shift the site of sedimentation to a new area, and a new island may develop as the old one subsides. This model was invoked originally to explain the lack of lateral continuity of carbonate cycles in the Ordovician St George Group, Newfoundland (Pratt and James, 1986).

10.6 Conclusions

Carbonate sequence stratigraphy has elements in common with siliciclastic systems. However, there are many features that are specific to carbonates that need to be considered carefully before a sequence stratigraphic interpretation of a carbonate system is undertaken. The following check-list summarizes the most important points.

1 Carbonate is generally produced in platform-top environments. Carbonate production usually exceeds the rate of creation of accommodation, especially during highstand, and this carbonate will bypass the platform top and be discharged into adjacent basins, a process termed 'highstand shedding'. Most carbonate is shed into deeper water during highstand in contrast to siliciclastic systems.

2 Carbonate platforms accumulate at or close to sea-level, and provide excellent depth indicators for interpreting relative sea-level changes. Carbonate platforms also are susceptible to repeated subaerial exposure because of the water depths in which they accumulate, and abundant small-scale (a few to several metres) carbonate cycles commonly are developed in carbonate platform-top environments. Analysis of the stacking patterns of carbonate cycles can enable a sequence stratigraphic interpretation to be made. Fischer plots may be of some use in stacking pattern analysis.

3 Organic carbonate production usually can keep pace with sea-level rise, except where the rise is very rapid or where the environment of carbonate production deteriorates. Environmental deterioration may be caused by changes in light penetration, salinity, oxygenation, nutrient levels, siliciclastic input and predation. It is usually very difficult to identify the cause of environmental deterioration in the rock record.

4 Organic carbonate systems may build much more aggradational margins than siliciclastic systems, especially during transgression. Backstepping is less common in organic carbonate systems, except where environmental deterioration occurs.

5 'Drowning unconformities' developed over carbonate systems may share certain geometrical characteristics with sequence boundaries. From seismic data alone it may be difficult to distinguish between these two surfaces.

6 Apparent sequence boundaries can be produced by changes in the grade of carbonate sediment accumulating on depositional slopes. These apparent boundaries may be difficult to distinguish from genuine sequence boundaries, especially on seismic data.

7 Relative sea-level fall usually will result in exposure of much of the carbonate platform area, depending on the geometry of the carbonate system. Subaerial exposure will tend not to yield much carbonate debris to the basin, because carbonates will tend to be eroded chemically, rather than physically. If the carbonate platform is attached, siliciclastic material may be transported into the basin at this time.

8 The effects of subaerial exposure on the carbonate platform will be climate dependent. Humid climates will result in major dissolution and reprecipitation of carbonate, and may form karst terrains. Arid climates will result in less intense carbonate diagenesis, but potentially more evaporite precipitation on the exposed carbonate shelf and in the basin.

CHAPTER ELEVEN

Organic-rich Facies and Hydrocarbon Source Rocks

11.1 **Introduction**
    11.1.1 Controls on organic richness and source potential
    11.1.2 Source rocks, tectonics and sea-level change

11.2 **Delta/coastal plain organic-rich facies and source rocks**
    11.2.1 Sequence stratigraphic significance of coals

    11.2.2 Geochemistry of delta plain organic-rich facies

11.3 **Organic-rich facies and systems tracts in clastic systems**
    11.3.1 Lowstand systems tract
    11.3.2 Transgressive systems tract
    11.3.3 Highstand systems tract

11.4 **Marine Carbonate Source Rocks**
    11.4.1 Genetic classification scheme
    11.4.2 Intercarbonate build-up
    11.4.3 Intraplatform depression
    11.4.4 Unrestricted basin margin
    11.4.5 Deep ocean basin

11.5 **Conclusions**

11.1 Introduction

Sequence stratigraphy provides a useful geological framework for considering the distribution of organic-rich facies (source-rocks*). It can allow a seismic, well or outcrop section to be subdivided into systems tracts in which the factors controlling source-rock deposition can be considered. Understanding the sequence stratigraphic context of a source-rock is necessary in order to predict its lateral extent and variability. However, it should be emphasized that source-rocks cannot be predicted from stratal geometries alone because there are too many variables involved (Fig. 11.1). This chapter first explores the controls on deposition of organic-rich facies and to what extent these can be predicted from sequence stratigraphic analysis. It then considers in detail the sequence stratigraphic context of coals and transgressive marine black shales and concludes by describing stratigraphic models for organic-rich carbonate facies.

11.1.1 Controls on organic richness and source potential

Enhanced organic matter preservation is a function of many factors, the most important being the physiogeography of the basin, climate, terrestrial organic productivity, marine aquatic organic productivity, oceanic circulation, sedimentation rate and water depth (Fig. 11.1). A number of these factors clearly are not predictable from systems tract analysis alone, e.g. climate and oceanic circulation.

---

* An oil-prone source-rock comprises sediments that are rich in organic carbon and contain organic material sufficiently hydrogen-rich to convert mainly to oil on maturation (Tissot et al., 1974).

Terrestrial organic productivity

Terrestrial organic productivity is a primary influence on the development of coals and coaly sediments deposited in coastal/delta-plain environments. McCabe (1984) considers that the potential for modern peat accumulation is a complex function of climate, which controls the balance between the rates of plant production and decay. Hot, humid climates favour plant production and cool climates favour plant preservation. The nature of the plant ecosystem has a strong influence on the type of organic matter preserved and hence the potential for oil- or gas-prone source-rock development (section 11.2.2).

Terrestrial organic matter supply

The rate of terrestrial organic matter supply to marine sediments is controlled principally by the nature of the floral ecosystem in the hinterland, the grain size of the sediment and the distance from the shoreline (Schlesinger and Mellack, 1981). Terrestrial organic productivity was negligible in pre-Devonian times and therefore little is preserved in marine sediments. Terrestrial organic matter will, because of its low density and grain size, tend to be concentrated in the fine-grained mud and silt facies of the systems tract. Terrestrial organic matter supply rates will be highest in fine-grained sediments in close proximity to well-vegetated deltas. Swamp and marsh areas are particularly important sources of terrestrial organic matter. Supply decreases exponentially with increasing distance from the shoreline and increasing water depth. However, shelf-edge deltas potentially may deliver high terrestrial organic matter fluxes to the upper slope, and submarine canyons may tap

238

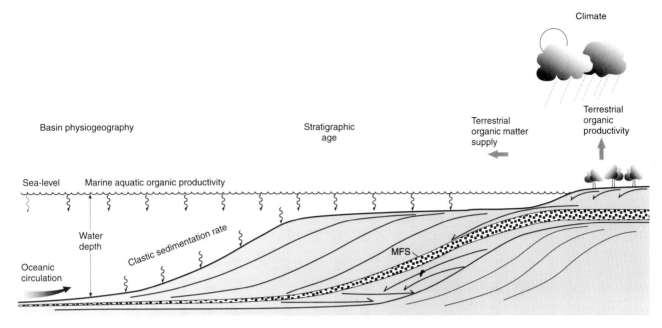

Fig. 11.1 The factors that influence the organic richness and source potential of sediments in clastic depositional systems. Note that most marine petroleum rocks were deposited under anoxic bottom water conditions. The development of anoxic conditions is controlled by the factors shown. Stippled zone highlights the transgressive systems tract. MFS, maximum flooding surface

sources of terrestrial organic matter in the coastal plain and transport it directly into deep water settings.

Primary productivity

Oil-prone kerogen in marine source-rocks originates predominantly from marine phytoplankton. The flux of algal organic matter to marine sediments is a function of primary organic productivity in the photic zone and water depth (Calvert, 1987; Schwartzkopf, 1993). Zones of high organic productivity usually occur in the vicinity of continents. Zones of particularly high productivity are located in areas of coastal upwelling, where shore-parallel winds result in upwelling of nutrient-rich deep water. Modern upwelling areas are concentrated on the western margin of modern continents (Pelet, 1987). Note, however, that palaeoproductivity prediction in the geological record is difficult. Palaeoclimatic models can allow prediction of prevailing wind patterns and hence upwelling areas in the geological record. The accuracy of these predictions of upwelling areas is very variable (R. Miller, pers. comm.).

Water depth

The flux of carbon from surface productivity to bottom sediments is strongly influenced by water depth. Degradation and recycling of organic matter in the water column (whether oxic or anoxic) sharply reduces the carbon flux at shallow water depths. At 1000 m water depth, the carbon flux is <10% of the value at 100 m water depth (Suess, 1980; Betzer *et al.*, 1984). As a general rule-of-thumb, the supply of marine algal organic matter to bottom sediments is likely to decrease with increasing water depth and distance from the shoreline, owing to lower surface productivity and remineralization in the water column (Schwartzkopf, 1993).

Organic matter preservation

Aquatic organic matter derived from phytoplankton is the main precursor of oil-prone kerogen. Preservation of oil-prone organic matter in sediments is greatly enhanced under anoxic† bottom-water conditions (Demaison and Moore, 1980) which develop where oxygen demand from decaying organic matter exceeds supply. Anoxic conditions result from restricted vertical circulation of sea-water and/or high biological productivity.

Physiogeographic restriction helps limit the water column circulation, and hence the resupply of oxygen to bottom waters. It is a favourable element for predicting anoxic environments. Physiogeographic restriction may take several forms and occur at a variety of scales. Examples include classic silled or intra-shelf basins; geographically restricted oceanic basins, such as the Gulf of Mexico and Arctic basins in Cretaceous times; and geographically enclosed epeiric seaways, such as the Cretaceous Western Interior Seaway, USA, the Jurassic of the North Sea and the Holocene Black Sea.

† Anoxia is used here to cover both truly anoxic and oxygen depleted, dysaerobic (<0.5 ml l^{-1} oxygen) bottom-waters under which organic matter preservation is enhanced.

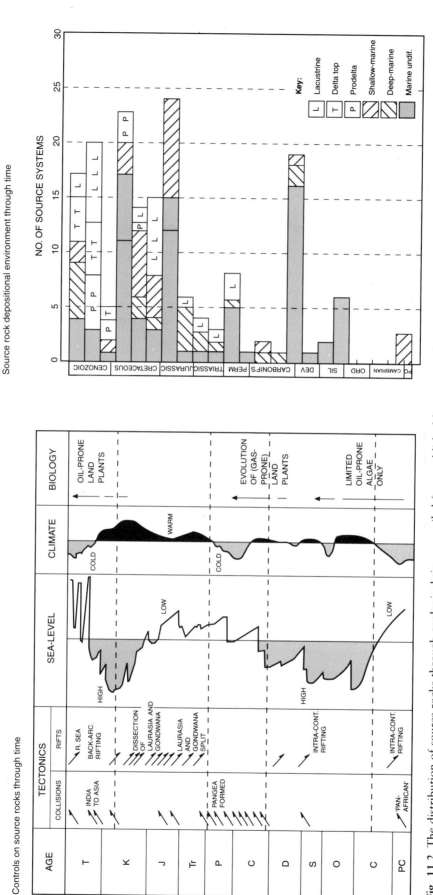

Fig. 11.2 The distribution of source rocks through geological time compiled from published literature and internal BP data bases. The column on the right contains source rocks that have been demonstrated to have generated a minimum of 500 million barrels of oil. These are classified into different environments of deposition. Peaks in marine source-rock deposition occur in the Late Devonian, Late Jurassic and Cretaceous, at times of continental break-up and rising first-order sea-level.

Water column stratification enhances the probability of anoxic conditions developing by inhibiting bottom-water replenishment. A positive water balance occurs where run-off from land and rainfall exceeds evaporation. A less dense freshwater cap forms over more saline marine water, and restricted circulation may lead to episodically anoxic conditions in basins of limited extent (see Fleet et al.'s (1987) discussion of the Lower Jurassic of Southern England). A negative water balance means that evaporation exceeds supply by runoff and rainfall. Dense saline water sinks to feed the bottom waters, which may become anoxic as oxygen is used by the degradation of organic matter. This is the model commonly used to explain anoxia in Jurassic-aged intrashelf basins in the Middle East (e.g. Droste, 1990). Stratification and hence anoxic conditions can persist only below the surface mixing layer, i.e. effectively below storm wave base. The surface mixing layer thickness varies with wind and tide energy in the basin, and usually is within the range 50–200 m.

Oceanic circulation

Water column oxygen profiles are key factors in predicting anoxic conditions and hence source potential in deep-water sediments of the continental slope and rise. In present-day oceans, the dissolved oxygen minimum is at depths of 100–1000 m because most oxidation of sinking organic matter occurs at these depths. However, at present the oxygen minimum is rarely sufficiently intense to give anoxic conditions and enhance organic matter preservation. Oxygen contents rise below the oxygen minimum zone because deep ocean water today is supplied by cold, oxygen-rich polar waters. The oxygen content of ocean bottom-water therefore decreases with increasing distance from the poles. The oxygen minimum (and hence anoxic conditions) is presently most intense in areas of high surface productivity, i.e. upwelling areas. Where the oxygen minimum is intense, anoxic conditions may occur on the upper slope and outer continental shelf.

Although it is understood that oceanic circulation patterns must have changed dramatically through geological time, prediction of ancient water-column oxygen profiles is difficult. For example, at times of high global sea-level and more extensive continental shelves, the oxygen minimum zone may have covered large areas of the continental shelf. In an ice-free Earth, sources of deep water may have been the warm and saline waters from low latitudes rather than the present-day cold and less saline polar waters. This could have resulted in an expanded oxygen minimum zone (Brass et al., 1982). Palaeogeography is a pointer to past areas of intense oxygen minima. Intense minima seem to have occurred in physiogeographically restricted areas of the oceans, distant from the main body of the oceans (e.g. Cretaceous Arctic Ocean, Cretaceous Gulf of Mexico).

Sedimentation rate

Changes in sedimentation rate and, in particular, intervals of condensed sedimentation can be predicted from sequence stratigraphic analysis. It has been suggested that condensed sections are likely candidates for source-rock intervals (Loutit et al., 1988). However, sedimentation can have either a positive or negative effect on organic carbon preservation (see review in Schwarzkopf, 1993). The relationship between organic preservation and sedimentation rate is complex because it is a delicate balance between enhanced preservation through rapid burial and dilution of organic matter by clastic material during rapid sedimentation (Schwartzkopf, 1993). Nevertheless, there are increasing numbers of case studies demonstrating the correlation between enhanced organic preservation and regional transgression and, in particular, anoxic condensed facies (Hallam and Bradshaw, 1979; Demaison and Moore, 1980; Jenkyns, 1980; Loutit et al., 1988; Leckie et al., 1990; Palsey et al., 1991; Creaney and Passey, 1993).

11.1.2 Source-rocks, tectonics and sea-level change

Figure 11.2 shows the distribution of the world's major source rocks for oil through geological time (based on an in-house compilation). Certain plate tectonic configurations in the past were favourable for source-rock deposition. Peaks of marine source-rock development in late Devonian and in the Late Jurassic to Cretaceous coincide with continental break-up and peaks in extensional activity in the 'Wilson Cycle' of global plate tectonic movements. These periods also coincide with peaks in the first-order sea-level cycle.

The vast majority of source-rocks form in extensional tectonic settings, particularly passive margins, and intracratonic and back-arc basins. A low in marine source-rock development occurs in the Permo-Carboniferous interval, at a time of predominantly contractual tectonic activity and widespread glaciation.

Note also the influence of stratigraphic age. Gas-prone, land-plant-derived, source rocks are present only since the Devonian. Delta top oil-prone source-rocks are restricted to the late Mesozoic and Cenozoic times, coincident with the evolution of the flowering angiosperms.

11.2 Delta/coastal plain organic-rich facies and source rocks

11.2.1 Sequence stratigraphic significance of coals

Coals can form in a variety of basin settings. For example, Tertiary coals in southeast Asia formed predominantly in extensional basins, whereas the widespread Upper Cretaceous coal deposits in western USA formed in a foreland basin setting. Autochthonous coals are an important component of delta/coastal plain deposits because they represent vertical accumulation of sediment and hence pure aggradation of the delta/coastal plain. Controls on coal formation are many and varied, but given favourable factors, such as a hot and wet climate, a stable, high groundwater table and no clastic influx to the peat forming mire, coals can accumulate rapidly on the delta/coastal plain (McCabe, 1984). Recently, it has been proposed that coals can be used for regional correlation in non-marine basins (Hamilton and Tadros, 1994).

Peat accumulation rates in modern Arctic climates are very slow compared with modern tropical climates, where rates of 2.5 m per 1000 years are typical and where rates up to 5 m per 1000 years have been recorded (Fig. 11.3). In ancient coal deposits it is worth considering the time required for the accumulation of thick coals. After allowing for compaction, a 3 m coal seam deposited at 1.8 m per 1000 years may have required favourable conditions to persist for perhaps 16 000 years. Rates of coal accumulation therefore can be rapid but ultimately are limited by the rate of base level (relative sea-level) rise. Modern coal swamps are unlikely to be able to keep pace with the most rapid rates of Holocene eustatic sea-level rise of 20 m per 1000 years (Fairbanks, 1989).

Coals accumulate at or close to base level, and it follows that for a thick coal to accumulate, aggradation of the delta/coastal plain must occur. In other words, the rate of coal accumulation and the rate of creation of sediment accommodation must balance. In addition the coal swamp must be effectively sheltered from clastic input in order for low-ash coals to accumulate. It is suggested that the development of raised mires on the coastal plain is necessary for this to occur (e.g. McCabe and Shanley, 1993). Raised mires occur where the build-up of waterlogged peat elevates the surface of the mire above regional ground-water tables. In southeast Asia, raised mires tens of kilometres across can be elevated up to 7 m above adjacent flood plains (Cameron et al., 1989).

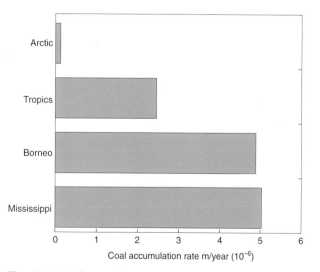

Fig. 11.3 Modern coal (peat) accumulation rates in different geographical settings

The development of coals during progradation and abandonment of a delta lobe will now be considered (Fig. 11.4). Coals can form in local depocentres on the delta plain during active delta progradation. These coals are likely to be thin owing to low delta-plain aggradation rates, and discontinuous owing to erosion by active alluvial channels and dilution by clastic sediment. Thick regionally extensive coals are more likely to form during abandonment of a delta lobe. The abandoned delta lobe will subside at a rate governed by subsidence and eustasy. If the abandonment phase persists, sediment starvation can ensue and a thick coal may accumulate. Eventually, either through a drop in coal accumulation rate or an increase in the rate of relative sea-level rise, the delta plain is flooded by transgressive marine sediments and the cycle begins again.

Workers in the Cretaceous Western Interior Seaway, USA have related the stratigraphic position of coals to parasequence stacking patterns (Ryer, 1984; Levey, 1985; Cross, 1988; Shanley and McCabe, 1993). Coals developed in coastal plain mires extending several tens of kilometres landward of coeval shorelines (Fig. 11.5). Coals are thickest landward of aggradational parasequence sets, i.e. where the shoreline stacks vertically. Coals associated with basinward and landward stepping parasequences are thinner (Cross, 1988). Ryer (1984) specifically relates the largest coal fields in Utah with extensive aggradational stacking of parasequences (fourth-order cycles) developed at the transgressive and regressive maxima of third-order cycles. The inference from these observations is that thick coals correlate with periods of rapid coastal plain aggradation. However, the fact that coals can form up to 45% of the coastal plain succession in these deposits, and that low-ash coals accumulate only 4 km from a clastic shoreline, requires further thought. Shanley and McCabe (1993) suggest that the formation of raised mires in the coastal plain can stabilize the shoreline during moderate rates of sea-level rise, inhibiting transgression and encouraging vertical stacking of facies belts in much the same way as a carbonate platform margin might behave (Chapter 10).

11.2.2 Geochemistry of delta plain organic-rich facies

The controls on the oil- or gas-prone nature of organic-rich delta/coastal plain sequences are still debated. It is known that some late Cretaceous and Tertiary delta/coastal plain source-rocks have sourced significant quantities of oil, but can oil-prone delta/coastal plain source rocks be predicted? This is a complex subject for which generalized rules-of-thumb can sometimes be misleading (Fig. 11.6).

Oil-prone terrigenous kerogen is derived from plant cuticles and resins. It is thought that only in post-Jurassic times have plants had sufficient foliage to yield significant amounts of oil-prone kerogen (there is uncertainty over late Triassic and Jurassic land plants). Pre-Jurassic land plants would yield gas-prone kerogen.

Hot, wet climates are conducive to the production of

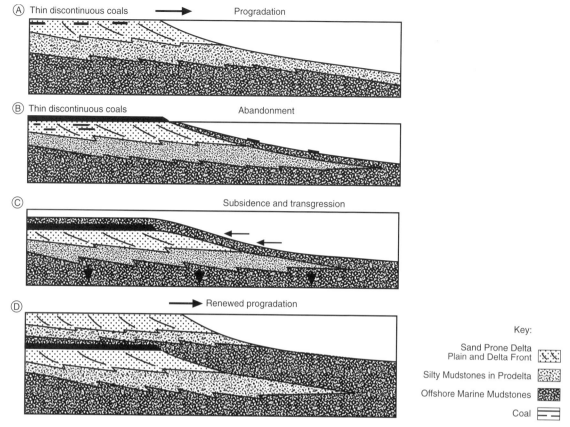

Fig. 11.4 Cartoons illustrating coal formation during progradation and abandonment of a delta lobe (after Allen and Mercier, 1988)

high foliage : wood ratio plant communities with potentially high yields of oil-prone kerogen. Cool, wet climates with gymnosperm (coniferous) dominated plant communities have yielded prolific resiniferous oil-prone kerogen in some Cretaceous delta/coastal plains in Australasia. Dry climates result in predominantly gas-prone kerogen.

Preservation of oil-prone terrigenous kerogen is thought to be enhanced in the brackish to saline conditions of the lower delta/coastal plain. Upper delta-plain (freshwater influenced) coals and coaly sediments are considered mainly to be gas-prone.

It sometimes may be possible to map coal distributions using techniques such as seismic attribute and seismic facies analysis given well calibration. However, the prediction of relative oil- or gas-prone potential of delta/coastal plain sediments in a seismic package is very difficult, if not impossible. Firstly, the differentiation of lower and upper delta plain settings within the top sets of a seismic sequence is unlikely. Secondly, the relative oil- and gas-prone potential of coals can vary laterally both within and between individual coal seams. Finally, there is presently no consensus over the relative potential of coals and associated carbonaceous shales.

11.3 Organic-rich facies and systems tracts in clastic systems

The following section attempts to integrate some of the controls on clastic source-rock development described earlier within a sequence stratigraphic framework. Source-rock development is discussed within a series of systems tracts developed during a cycle of changing relative sea-level. The systems tract block diagrams are adapted from those published by Posamentier and Vail (1988).

11.3.1 Lowstand systems tract

Significant coal accumulation on the coastal plain is unlikely in the early lowstand systems tract and terrestrial organic matter is likely to be highly oxidized. The shelf and upper slope are considered to be zones of sediment bypass and hence source potential there is negligible (Fig. 11.7). Terrestrial organic matter supply is limited to locally reworked material in basinal fan deposits. If anoxic conditions persist in the basin a condensed basinal source-rock facies may occur away from the sediment entry points. The lowstand prograding wedge has some potential for coal development during the aggradational phase but the areal extent of the

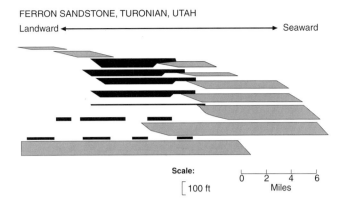

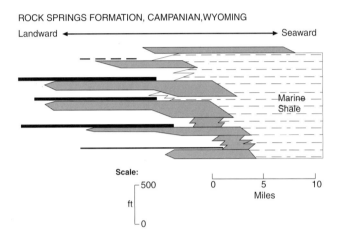

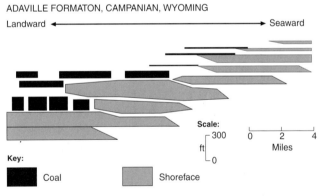

Fig. 11.5 Diagrammatic cross-sections illustrating the distribution of coal relative to the position of the shoreline in different stratigraphic intervals of the Cretaceous, Western Interior Seaway, USA. Shorelines stepping basinward indicate progradation, whereas shorelines stacking vertically indicate aggradation and shorelines stepping landward indicate retrogradation (from Cross, 1988)

organic-rich facies will be restricted to the incised valleys and associated canyons. Marine organic-rich slope and basinal facies are unlikely owing to the rapid sedimentation rates. Although marine source-rocks could occur, overall this appears to be the least prospective systems tract for source-rock development.

11.3.2 Transgressive systems tract

This is the most important systems tract for the development of marine oil-prone source rocks. Many authors have noted the correlation between the occurrence of organic-rich facies and regional transgression (Hallam and Bradshaw, 1979; Demaison and Moore, 1980; Jenkyns, 1980; Loutit *et al.*, 1988; and see Fig. 11.2). However, not all transgressive systems tracts result in deposition of organic-rich facies. The link is complex and is explored in some detail in this section.

General features of the transgressive systems tract

In the transgressive systems tract the shoreline retreats landward, resulting in a progressive increase in the geographical extent of shallow-marine shelf deposition, reaching a maximum at the maximum flooding surface. Increasing distance from the contemporary shoreline results in reduced clastic sediment supply and reduced terrestrial organic matter supply to the outer shelf and slope. Very low sedimentation rates on the shelf and slope can result in deposition of a condensed facies (Fig. 11.8).

However, reduced sedimentation rates alone will not result in deposition of organic rich facies. Enhanced preservation of organic matter in poorly oxygenated benthic environments is a common characteristic of transgressive black shales (e.g. Demaison and Moore, 1980; Oschmann, 1988; Miller, 1990; Wignall, 1991a,b). Shallow water depths on the shelf result in high organic fluxes and therefore high oxygen demand. The higher the surface productivity, the higher is oxygen demand in the bottom waters. In the surface mixing layer above storm wave base, storms and tidal currents can effectively mix the water column and reoxygenate the bottom waters. It follows that the sediment–water interface will have to remain below the surface mixing layer (H) for oxygen deficient/depleted conditions to persist for significant periods of time. The depth (H) will vary depending on storm and tide energy. For example, H may be shallower on wide shelves compared with narrow shelves owing to reduced tidal and wave activity (Hallam and Bradshaw, 1980). Development of a stratified water column and physiogeographic restriction will increase the chances of anoxic conditions and organic-rich facies development during transgression.

Models for source-rock development

In the model shown (Fig. 11.9), the basin is permanently anoxic below a certain water depth, marking the base of the surface mixing layer. A distal deep-water, relatively condensed, source-rock facies occurs in an anoxic basin centre in the distal toe-sets of both lowstand and highstand systems tracts, which would be manifested as a downlap surface on seismic sections. The transgressive systems tract results in landward translation of this facies belt until the

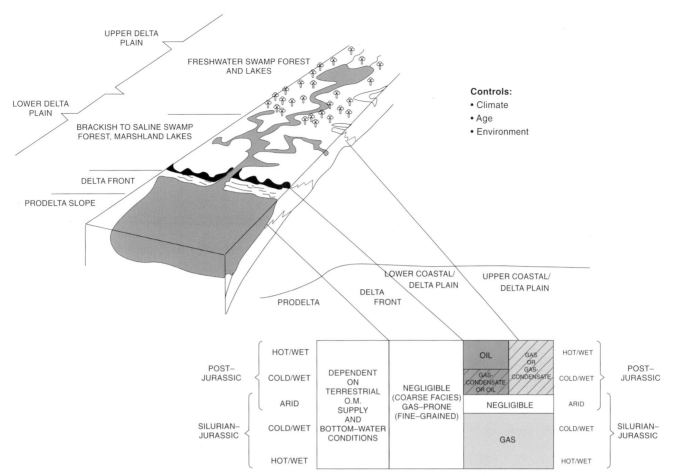

Fig. 11.6 Cartoon illustrating different types of source-rock setting on a typical modern delta plain from southeast Asia (based on the Klang Delta). Lower delta-plain environments are thought to be areas where the preservation of oil-prone (organic matter likely to generate oil) is enhanced in brackish alkaline environments. Hot humid climatic regimes in the post-Cretaceous era are the most favourable for generating high foliage-to-wood ratio plant communities, which, given favourable preservation, can result in deposition of oil-prone coal source rocks

deeper water condensed source-rock facies impinges on the transgressed shelf. In this model the greatest extent of source-rock facies will coincide with the time of maximum flooding. During subsequent highstand progradation, water depths decrease and the area of source-rock deposition gradually shrinks. Thus source rocks are not restricted to a particular systems tract but expand to a maximum during the transgressive systems tract. This model is similar to the 'expanding puddle' model of Wignall (1991a), derived from the study of black shale deposits in epicontinental shelf seas (Fig. 11.10). It is less applicable to passive continental margins, where the deep ocean may be well-oxygenated.

Wireline logs are particularly useful in characterizing source rocks (Passey *et al.*, 1990; Myers and Jenkyns, 1992). Creaney and Passey (1993) noted that many marine oil-prone source rocks are characterized by an initially abrupt upward increase in organic richness, against background values, and a subsequent gradual decrease in richness (Fig. 11.11). They attributed this pattern to the control of organic carbon contents by clastic sedimentation rate under anoxic bottom water conditions. The initially rapid increase in TOC (total organic carbon) results from a rapid decrease in the rate of clastic sediment supply to the shelf during transgression. The subsequent gradual decrease in TOC reflects increasing clastic sediment supply and dilution of organic carbon during highstand progradation. In this model, the source interval would thicken from the basin margin into a series of stacked source intervals in the basin centre. Wignall (1991a) interprets the Toarcian black shales of England to be of this type.

An alternative model is where the anoxic conditions develop only in the transgressive systems tract itself (Fig. 11.12). Basinal areas will have interbedded source facies deposited in the TST, with non-source facies deposited in the HST and LST. Shelfal source-rocks are restricted to the TST. This model implies that the palaeogeographic conditions necessary for anoxia to develop are unique to the transgressive systems tract. Leckie *et al.* (1990) document an example from the Cretaceous Shaftsbury Formation of Canada, where a nearshore organic-rich zone deposited during rapid transgression passes basinward to more normally oxygenated sediments. In this case, a nearshore zone

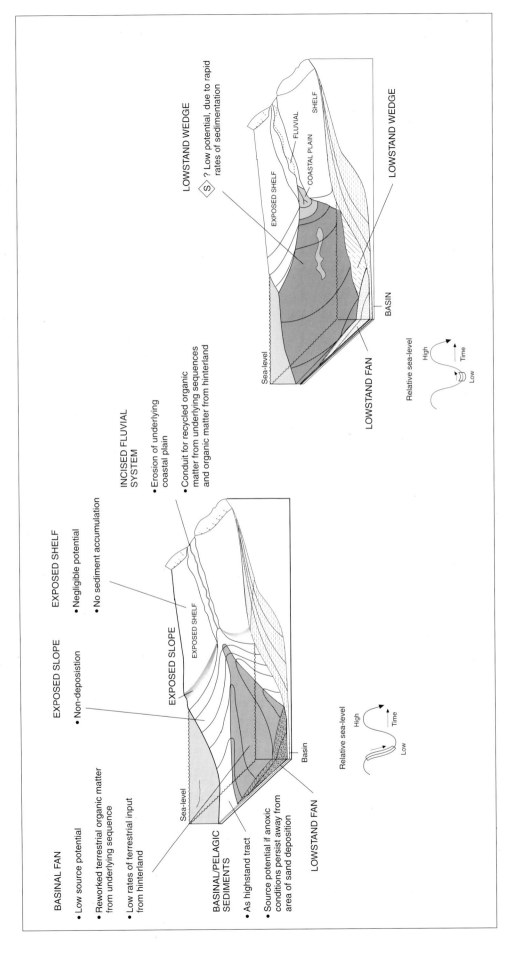

Fig. 11.7 Schematic cartoon of a lowstand systems tract on a shelf-break margin, adapted to show the potential for organic-rich facies development (after Posamentier and Vail, 1988)

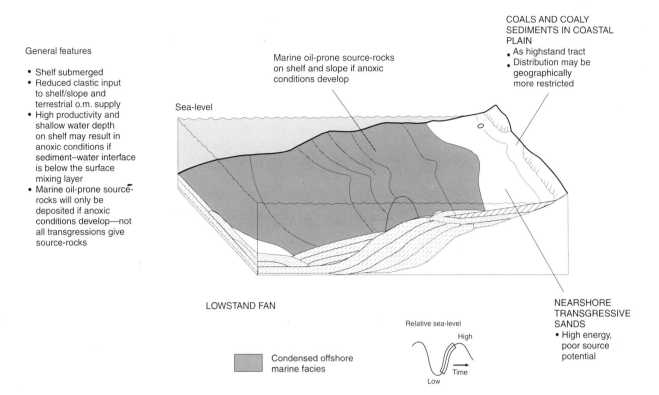

Fig. 11.8 Schematic cartoon of a transgressive systems tract on a shelf-break margin adapted to show the potential for organic-rich facies development (after Posamentier and Vail, 1988)

of high productivity is assumed to be localized in the transgressive systems tract.

It should be clear from the above that although there is a general relationship between source rocks and transgression, the stratigraphy of organic facies is complex in detail. This complexity is shown by the work of Curiale *et al.* (1991) and Palsey *et al.* (1991) in the Cretaceous Western Interior Seaway. Curiale *et al.* (1991) show maximum organic richness occurring in the early highstand systems tract *above* the condensed section associated with maximum flooding in the Cenomanian–Turonian interval they studied. Palsey *et al.* (1991), in contrast, show that in the overlying Coniacian strata, maximum organic richness occurs in the transgressive systems tract below the condensed section associated with maximum flooding.

Organic richness will vary from the proximal to distal portions of the systems tract and should not be considered to be uniform. A common problem for the petroleum geologist is to extrapolate a source-rock proven in the shelfal portion of a transgressive systems tract into the

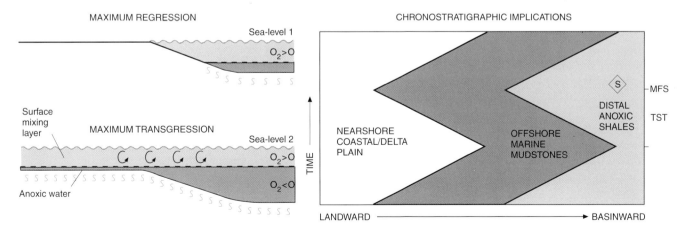

Fig. 11.9 Chronostratigraphic implications of a source-rock model where organic-rich rocks are developed in a distal facies in the basin at times of both maximum transgression and regression. Transgression serves only to spread the area covered by organic-rich facies landward. MFS, maximum flooding surface; TST, transgressive systems tract

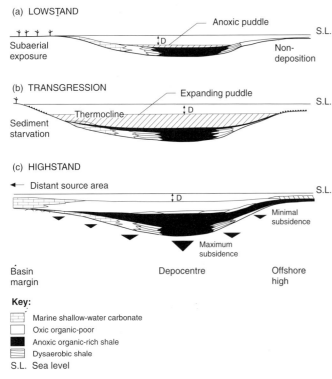

Fig. 11.10 The expanding puddle model proposed by Wignall (1991a) to explain the distribution of Lower Toarcian organic-rich facies in a shallow epeiric basin. D is the depth of the surface mixing layer below which anoxic bottom waters are trapped. (a) During lowstand organic facies accumulate only in the basin depocentres. (b) A combination of sediment starvation, sea-level rise and continuing subsidence during transgression causes marked expansion in the area of deep water and thus the areal extent of organic-rich facies deposition. (c) During highstand, progradation results in shallowing of the water column and a shrinkage in the area of organic-rich facies

basin. The first question to consider is whether anoxia is developed in the basin as well as on the shelf. Figure 11.13 shows models where (i) anoxia develops on both the shelf and in the basin, (ii) anoxia and source rocks develop on the slope and basin but not on the shelf, and (iii) high productivity and source rocks develop on the shelf but not in the basin. Many other permutations are possible.

The thickness of the anoxic organic-rich interval will be a function of sedimentation rate and the duration of the interval. Richness of the anoxic organic-rich interval (if only marine aquatic organic matter is considered) will be a function of surface productivity, water depth, sedimentation rate and bottom-water oxygenation.

11.3.3 Highstand systems tract

Thick coastal plain successions are most likely in the topsets of aggradational systems (Fig. 11.14). Coals and coaly sediments may occur if climate and other factors are favourable. Where rates of progradation are high, coastal plain aggradation rates are low, and thick delta/coastal-plain source rocks are less likely.

High sedimentation rates and oxygenated environments on actively prograding slopes will dilute organic carbon contents. Slope mudstones are usually, at best, sources for gas. However, where coal swamps are accumulating on an aggrading coastal plain, high rates of terrestrial organic matter may be supplied to interdistributary bay or upper slope mudrocks.

Basinal facies in the bottomsets of the prograding clinoforms may be organic-rich if the basin is anoxic at depth (Curiale *et al.*, 1991). Where submarine canyons tap into coastal plain sediments, terrestrial organic matter can be transported rapidly to deepwater areas by density currents.

11.4 Marine Carbonate Source Rocks

Many of the world's most prolific source rocks are developed in marine carbonate depositional systems. The depositional controls on organic carbon accumulation in carbonate systems are similar to clastic systems (Fig. 11.1), with the development of anoxic bottom-water conditions being critical, either as a result of restricted circulation or enhanced surface productivity. A terrigenous organic matter contribution tends to be less important given the arid cimates in which many carbonate systems develop.

Carbonate systems differ from clastic systems in that they can create the physiogeographic restriction necessary for the development of anoxia and enhanced preservation of organic matter by their response to rapid relative sea-level rise. The carbonate depositional geometries outlined here are discussed in detail in Chapter 10.

11.4.1 Genetic classification scheme

The following section will consider four genetic types of carbonate source rock defined on depositional geometry. The intercarbonate build-up source rock develops predominantly in areas of restricted circulation between carbonate platforms and platform margins. The intraplatform depression source rock develops when an isostatically sagged platform interior is drowned. These source rocks are characteristic of high carbonate productivity systems. The unrestricted carbonate-margin type develops on low carbonate productivity margins during periods of upwelling and/or oceanic anoxia. Finally, deep-ocean-basin source rocks develop in bathyal water depths in long-lived, tectonically silled, carbonate fringed anoxic basins.

11.4.2 Intercarbonate build-up

Rapid relative sea-level rise may result in the differentiation of an antecedent platform into a series of isolated build-ups (Fig. 11.15, based on Stoakes, 1980). Build-ups may nucleate on topographic highs such as the platform margin. Restric-

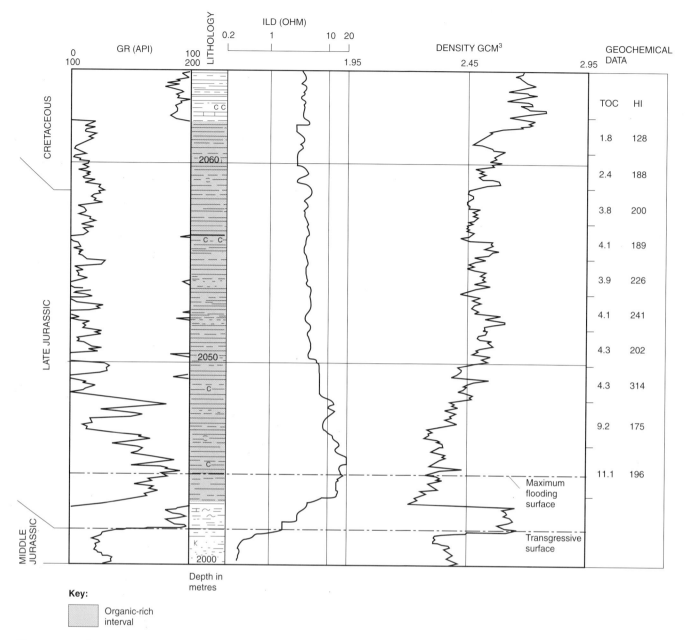

Fig. 11.11 A typical wireline log response of the organic-rich Kimmeridge Clay Formation in the North Sea. The log shows an abrupt upward increase in organic carbon content above a transgressive surface, as shown by the rapid upward decrease in density (Passey *et al.*, 1990). The high gamma radioactivity is due to enrichment of the shales in uranium, which indicates deposition under anoxic bottom-water conditions (Myers and Wignall, 1987). Organic carbon contents decrease upward more gradually, which can be explained by increasing clastic sedimentation rates and dilution of organic matter during highstand progradation. In this case organic-rich facies are deposited under anoxic bottom-water conditions, both in the transgressive and early highstand systems tracts, with the highest organic contents correlating with the lowest rates of clastic dilution

ted circulation in the inter-build-up area results in the development of anoxia below the surface mixing layer. Organic-rich carbonates are deposited in the transgressive systems tract at the time of maximum bathymetric relief, coeval with the rapidly aggrading carbonate build-ups. Organic carbon contents may be enhanced further by low rates of carbonate dilution. In the highstand systems tract, prograding systems gradually infill the topography developed during earlier transgression.

The Devonian Douvernay Formation is the type example of this type of source rock (Fig. 11.16). Organic-rich carbonates of the Douvernay Formation overlie the extensive antecedent Cooking Lake platform and are time equivalent to up-building pinnacle reefs of the Leduc Formation

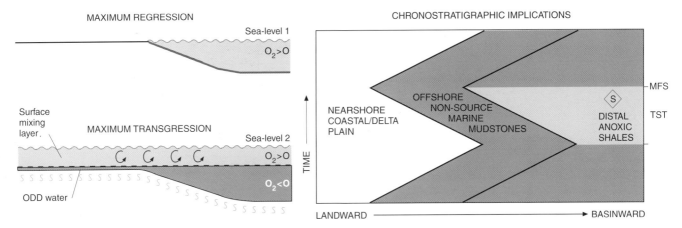

Fig. 11.12 Chronostratigraphic implications of a source-rock model where organic-rich rocks are developed in a distal facies in the basin only in the transgressive systems tract. During both lowstand and highstand systems tracts the basin is well aerated and there is a causal link between transgression (T) and the development of anoxic conditions, e.g. owing to the development of high surface productivity on the flooded shelf. ODD, oxygen deficient/depleted (conditions); MFS, maximum flooding surface; TST, transgressive systems tract

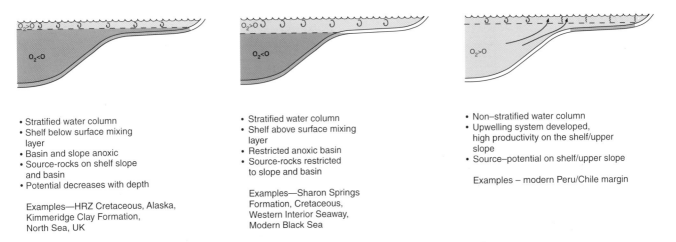

Fig. 11.13 Three models for transgressive systems tract variability and possible ancient analogues

(Stoakes, 1980; Fig. 11.15). Calcareous mudstones of the Ireton Formation form the prograding highstand deposits that infill the Leduc–Douvernay topography. Net thickness of the organic-rich facies varies from 10 to 20 m, with total organic carbon contents in the 4–7% range. The source rock is estimated to have been deposited over a period of 2–3 million years, during a 'second order' sea-level rise.

11.4.3 Intraplatform depression

In this case differential subsidence in the platform interior results in the formation of a long-lived depression during rapid sea-level rise (Fig. 11.17). No convincing mechanism for the isostatic sagging has been demonstrated but salt withdrawal has been suggested (Burchette and Britton, 1985; Aigner and Lawrence, 1990). The sequence strati-graphic model in Fig. 11.17 is based largely on Droste's (1990) work on the Late Jurassic Hanifa Formation in Saudi Arabia.

During highstand, when relative sea-level is rising slowly, the platform generates enough sediment to fill accommodation space across the platform and shallow water depths are maintained. During rapid sea-level rise in the transgressive systems tract the whole platform cannot keep pace and a shallow intraplatform depression forms. Restricted circulation results in anoxia and organic-rich sedimentation. Aggradation of the margins of the basin result in minimal carbonate dilution in the basin centre. In the following highstand systems tract the topography is wholly or partly infilled as the margins of the basin prograde into the depression. There also may be a close association with evaporites if the basin becomes isolated and drawdown

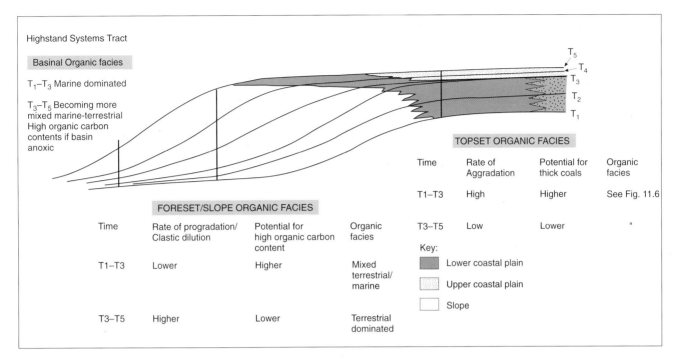

Fig. 11.14 The potential for source-rock development in a hypothetical highstand systems tract. It considers probability of organic-rich facies occurring in basinal, slope and topset settings. Times 1–3 represent aggradation during early highstand and times 3–6 represent progradation during late highstand

ensues. This can happen in any systems tract but may be more likely during lowstand (Chapter 10).

The Hanifa Formation described by Droste (1990) is one of a chain of intraplatform-depression source rocks deposited in the Gulf of Arabia area in Oxfordian–Kimmeridgian times (Murris, 1980). The thickness of the organic-rich interval reaches 30–50 m in the centre of the depression, with TOC of 5–10% (Droste, 1990). The Hanifa Formation was deposited in Oxfordian–Kimmeridgian times during a second-order sea-level rise and shows an internal cyclicity reflecting higher order sea-level changes (Fig. 11.18).

11.4.4 Unrestricted basin margin

The unrestricted-basin-margin source rock has a similar depositional geometry to the transgressive marine clastic source rocks described in section 11.3.2. This type of source rock occurs in low-productivity carbonate systems, when transgression results in sediment starvation of the outer shelf or deep ramp and deposition is dominated by pelagic carbonate (Fig. 11.19). Low carbonate productivity inhibits up-building and deep-water environments develop across the shelf or ramp. Source rocks will be deposited if anoxia and/or high phytoplankton productivity is developed. As the rate of sea-level rise decreases, highstand progradation of the clastic or carbonate ramp proceeds and source-rock deposition ceases.

There are a number of examples of this type of source rock: the Upper Jurassic Smackover Formation, Gulf Coast, the Upper Cretaceous La Luna Formation, South America and the Upper Cretaceous Brown Limestone Formation, Gulf of Suez. Figure 11.19 shows the typical stratal geometry of this type of source rock, with the organic-rich facies located in the transgressive systems tract underlying the low-angle clinoforms. Figure 11.20 is a typical well-log through the Brown Limestone Formation. The high radio-activity unit is an organic and phosphate-rich pelagic limestone interpreted as a distal ramp carbonate deposited following rapid transgression of the underlying clastic Matulla Formation. The organic-rich interval is 20–50 m thick and contains 3–6% TOC. Its precise age range is not well controlled, although it may represent 4–5 million years of deposition in the early Campanian.

11.4.5 Deep ocean basin

The deep-ocean-basin type carbonate source rock is controlled more by palaeo-oceanographic factors than relative sea-level changes because it is deposited at 100s to 1000s of metres water depth in silled basins. The type example of

251

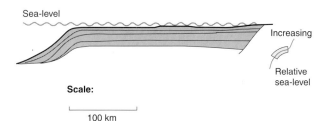
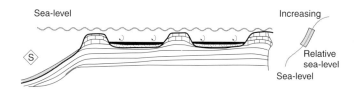
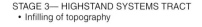
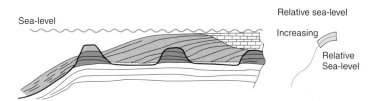

Fig. 11.15 Intercarbonate build-up source rock. The hypothetical two-dimensional cross-section illustrates depositional geometries at three stages: (1) prior to rapid relative sea-level rise with the platform at keep-up stage; (2) transgressive systems tract with platform at catch-up stage; (3) highstand systems tract with platform at keep-up stage. Cross-section is based on Stoakes (1980) interpretation of the Devonian Leduc–Douvernay Formations of the Western Canada Basin

this source rock is the early to middle Cretaceous pelagic carbonates of the deep water Gulf of Mexico (Katz, 1984). These were deposited periodically over 43 million years in water depths of around 1500–2000 m. In DSDP boreholes 535 and 540 they attain up to 250 m in thickness and contain an average of 1.4 ± 1.8% TOC (Katz, 1984).

11.5 Conclusions

Sequence stratigraphy clearly has an important role to play in the study of organic-rich facies. Prediction of source rocks in an undrilled frontier basin using seismic stratigraphy alone will not be possible, because factors such as climate, oceanic circulation and bottom-water anoxia cannot be interpreted from seismic data. However, by identifying likely transgressive systems tracts and condensed facies, the most prospective intervals of the basin fill can be highlighted. Sequence stratigraphy will be particularly useful in predicting the likely lateral distribution of organic-rich intervals away from well control, by constraining palaeobathymetry and lateral changes in sedimentation rate. Coals are important facies as they represent periods of coastal plain aggradation and often mark the early stages of transgression.

This chapter has presented very much an overview of a complex topic and to a large extent has considered organic richness only in terms of total organic carbon. However, the type of organic matter controls the potential of the organic facies to generate oil or gas on maturation. Published case studies integrating sequence stratigraphy and detailed geochemistry are rare at present (e.g. Palsey *et al.*, 1991; Curiale *et al.*, 1991) and future work must continue to study the link between organic facies and systems tracts.

Fig. 11.16 (opposite) A typical wireline-log profile through the Devonian Douvernay Formation, the type example of an intercarbonate build-up source rock. The organic-rich interval is shown by the shaded low-density (DRHO gm/cc) and high sonic-travel-time (DT) zone, which together with the high resistivity (ILD) indicates a mature source rock (e.g. see Passey *et al.*, 1990). The Douvernay overlies the highstand carbonates of the Cooking Lake Formation and is time equivalent to the Leduc pinnacle reefs (Stoakes, 1980). The calcareous mudstones of the Ireton Formation represent the overlying highstand systems tract (HST). TST, transgressive systems tract; S, indicates source rock potential; GR, gamma ray log; NPHI, neutron porosity (%)

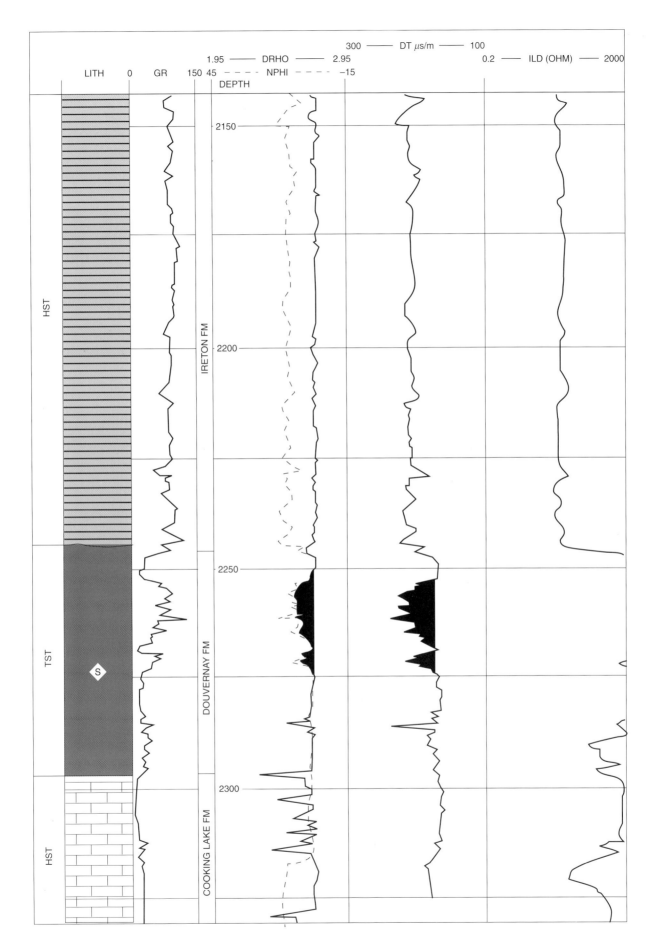

STAGE 1 – PRIOR TO RAPID RELATIVE SEA-LEVEL RISE

STAGE 2 – TRANSGRESSIVE SYSTEMS TRACT

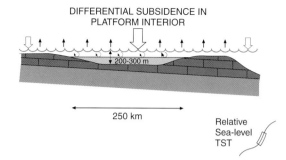

- Initial antecedent platform or gently-dipping ramp
- Shallow water depths preclude source-rock deposition

- An intra-platform (or intrashelf) basin forms during a rapid rise in sea-level which drowns an isostatically sagged platform interior (Aigner et al, 1990)
- Restricted circulation (enhanced by stratification in an arid climate) results in anoxia and organic-rich sedimentation
- Aggradation of the margins of the basin reduces carbonate dilution of organic matter in the centres of the depression

STAGE 3 – HIGHSTAND PROGRADATION

- Infilling of topography
- Either or both margins prograde into the topographic depression
- In arid environments, lowstands may result in the cut off of the intrashelf basin from the ocean, drawdown and evaporite deposition

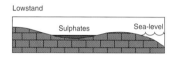

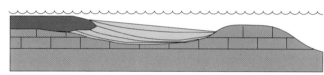

Fig. 11.17 Intraplatform depression source-rock. The hypothetical two-dimensional cross-section illustrates depositional geometries at three stages: (1) prior to rapid relative sea-level rise, with the platform or ramp at keep-up stage; (2) in the transgressive systems (TST) tract during rapid relative sea-level rise; (3) during highstand progradation (after Droste, 1990). HST, highstand systems tract

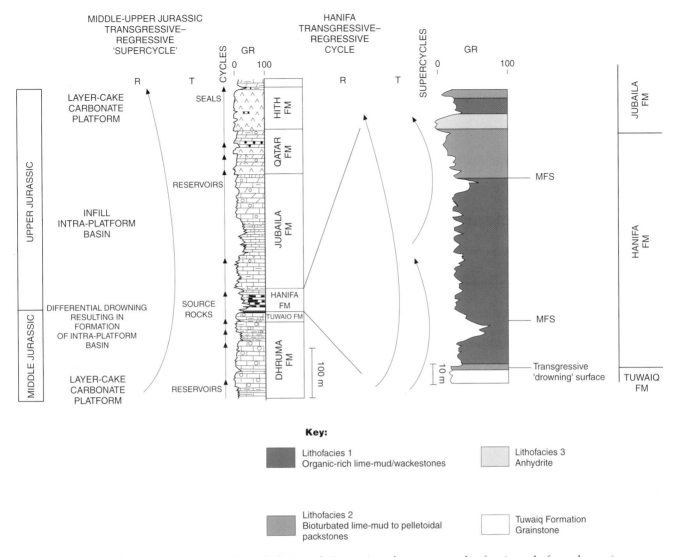

Fig. 11.18 A typical wireline log response through the Hanifa Formation, the type example of an intraplatform-depression source rock (from Droste, 1990). MFS, maximum flooding surface; R, regressive; T, transgressive; GR, Gamma-Ray log

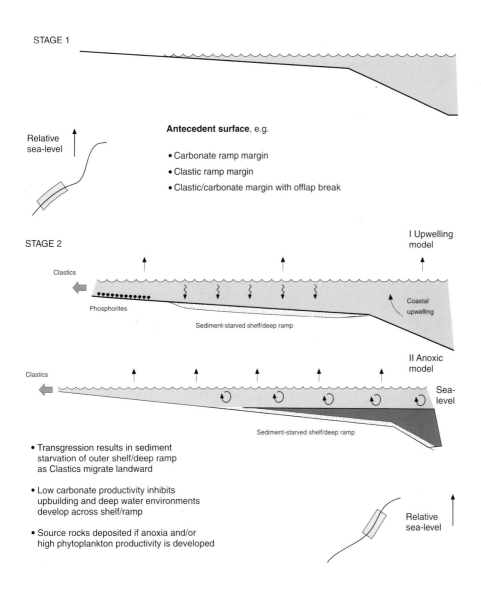

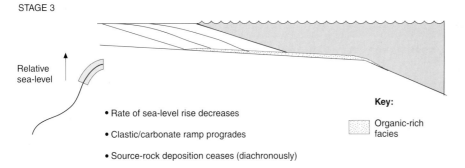

Fig. 11.19 Unrestricted-basin-margin source rock. The hypothetical two-dimensional cross-section illustrates typical depositional geometries at different stages of the sea-level cycle. Geometries are similar to clastic source rocks

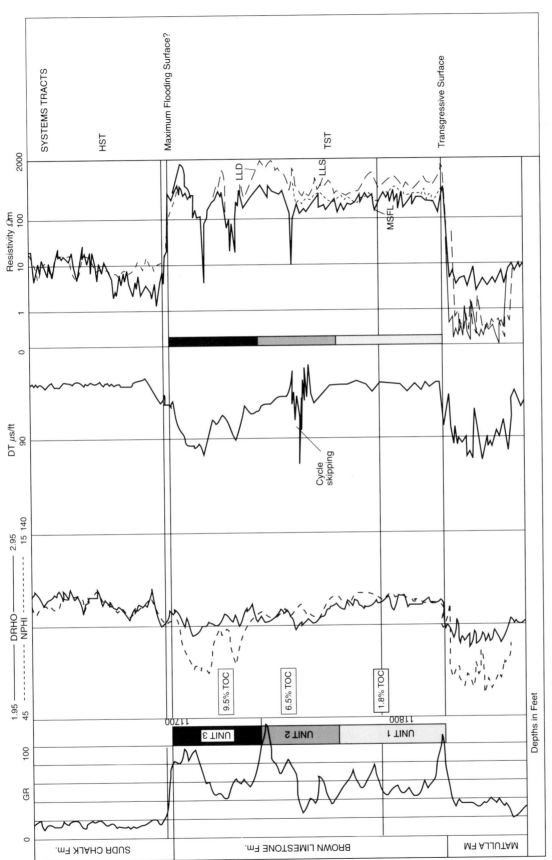

Fig. 11.20 A typical wireline log profile through the Upper Cretaceous Brown Limestone, Gulf of Suez, Egypt. The organic-rich carbonate is shown by the low density and high sonic travel times together with the high resistivity. Geochemical parameters refer to the results of geochemical analyses of cuttings samples. The Brown Limestone is separated from the underlying clastic Matulla Formation by a transgressive surface, and is an organic- and phosphate-rich pelagic limestone considered to be upwelling related. However, the high radioactivity is due to uranium enrichment, indicating that anoxic conditions also prevailed during deposition. The overlying Sudr Chalk Formation is organic-poor and probably represents the overlying highstand systems tract

CHAPTER TWELVE

Computer Modelling of Basin Fill

12.1 Introduction

12.2 Forward model types
    12.2.1 Hydraulic process–response models
    12.2.2 Diffusion or potential gradient models

12.2.3 Geometrical models
12.2.4 Discussion of model types

12.3 Examples of the stratigraphic use of computer modelling

12.4 Practical forward computer modelling
    12.4.1 Points to consider
    12.4.2 Worked example

12.5 Conclusions

12.1 Introduction

Computer simulation of basin fill is used increasingly in basin analysis, both as a tool for attempting to simulate known geometries within a basin, and to predict the stratigraphy and facies distributions in poorly constrained areas. Modelling packages have been developed at a variety of scales to suit a range of purposes, from simulation of the fill of entire basins, to replicating growth patterns of small carbonate systems or sand-body distribution in a submarine fan. This chapter discusses the range of model types and their application to understanding the stratigraphical and sedimentological development of basins.

Most natural systems are extremely complex and difficult to understand. A model of any natural system is an attempt to simplify the important parts while still being useful. We could say for example that the North Sea was a water-filled hole that is in the process of being filled by sediment. This would be a very simple model that could be used to convey some aspects of the 'North Sea' to a colleague (i.e. it is not a desert environment and we expect marine or lacustrine rather than aeolian processes to be acting). To be able to use this model to predict the location of a sand bar, or the distribution of organic-rich shales we need not only more information but also more quantitative information, such as the direction and perhaps strength of tidal and river currents, position of sediment entry points, and changes in water level. As we ask more detailed questions of the model, so we need more detailed information, and we quickly reach the stage where both the number of parameters and complexity of interaction between parameters are more than we can manage without mechanical assistance. Hence the need for computer modelling. If we need only the answers to simple, qualitative questions then simple mental models may often be enough, but if we need either simple answers to complex problems or quantitative answers to simple problems, then we can often gain by using a computer.

There is another benefit from computer modelling, which is that computers do not take kindly to ambiguity in interpretation, they force us to make decisions in a way that mental models do not. A mental model can carry uncertainty in an intuitive sense very far along the interpretation or prediction chain (in fact so far that we often forget that there was any uncertainty in our original data), but when the final map or section has to be drawn then we cannot easily express the doubt other than by stippling the boundaries. Quantifying uncertainty at an early stage and using mathematics to carry the uncertainty through all stages in interpretation and prediction leads to the development of alternative working hypotheses and better communication of our state of knowledge.

Modelling may be either 'forward' or 'inverse'. In the forward model we start with what we hope are the major processes acting on a system, wave direction and amplitude for example, and some boundary conditions, such as beach grain-size distribution, water depth, beach slope, etc. We let the model run for an appropriate length of time and look at the change in the system. If both the process model and the boundary conditions were specified sufficiently then the end result will be a prediction that can be tested at an appropriate scale. There will always be higher resolution features in reality that were not predicted by the simulation, and we have to hope that these are not significant for our proposed use. An important point to remember about any modelling is that a model is always a simplification in some aspect. The skill of the modeller is to ensure that the model is neither too simple nor too complex to answer the question being asked. Another way of looking at this is to say that if you do not know what question you are asking, then no model will ever satisfy your requirements. There is no such thing as a generally applicable model.

Inverse modelling takes the present state of a system and tries to derive both its initial state and the processes acting to produce the current state. In simple cases, where a single linear process has acted on a few variables, this may be possible, for example a rough estimate of the initial height above a given point of a steel ball rolling down an inclined plane may be derived by measuring the velocity of the ball at that point. If we wanted to predict the initial height to several significant digits, however, then we would have to include information about surface roughness, friction, material properties, etc. Note also that we cannot use an inversion approach to predict distance of the measurement point from the release point. In the inverse modelling case we have to understand the processes involved and their relation to the question that we wish to ask.

By modelling a system we can test our understanding of its fundamental properties. Such modelling is often called simulation. Simulation is a form of model building, where sufficient features of the system or object being modelled are included to enable predictions to be made (either qualitatively or quantitatively) concerning the behaviour of its real-life analogue. Simulation can be physical, for example flume-tank experiments, mathematical (where analytical solutions exist for predictive equations), algorithmic/computational (where numerical solutions exist), or heuristic/computational (heuristic = 'rule-of-thumb'), where knowledge exists in the form of general 'understanding' and logical rules rather than equations.

Both forward and inverse models have their weak points. A given present-day state can be derived from an initial set of conditions via several different routes, or a different route combined with different initial conditions can result in an identical present-day state. In other words it is relatively easy to prove sufficiency but almost impossible to prove necessity. A sedimentological example of this is the 'upward-coarsening unit', where it is relatively easy to show that a prograding marine unit can have produced that grain-size distribution, but impossible to show that only a prograding marine unit could have produced it.

Both forward and inverse modelling of natural systems suffer from the phenomena of 'non-uniqueness'. This is the property of being able to approach the same result by an infinite number of different paths. It is relatively easy to show that a given state can be reached by the process modelled, but it is often difficult if not impossible to prove that it can only have been reached that way.

The present discussion concerns only forward models, although mathematical inverse models of basin fill processes have been shown to be practicable, at least on synthetic data (Lessenger, 1991, 1993). Sedimentologists and stratigraphers usually work with intuitive inverse models, and forward models offer one way of testing the sufficiency of the inversion but not the necessity. In other words if a seismic stratigrapher interprets a given seismic feature as a lowstand wedge we can use a forward model to find out if indeed a lowstand at that particular time, in that location, may have produced a clastic wedge in that position given that our functional model of the way sediment packages respond to sea-level changes is accurate. However, we cannot test whether a lowstand did produce the wedge in question. Note that we are trapped into testing two things simultaneously. We are testing both our hypothesis concerning the response of sediment packages to sea-level change and the possibility of that model producing a clastic wedge at that location. The only defensible conclusion must be that if our hypothesis is correct then a clastic wedge can/cannot have formed under those conditions. It is important to realize that this is not a test of whether or not a clastic wedge formed under lowstand conditions, but only of the sufficiency of this particular model to produce a clastic wedge at a lowstand. A clastic wedge may also form under other conditions, or may not form under the specified conditions if our model is wrong. We have no way of separating the two. With this caveat we will look at the background to existing models, placing them in their developmental perspective.

Four books can be considered essential reading on this topic, of which one is out of print. Harbaugh and Bonham-Carter's (1970, out of print) book *Computer Simulation in Geology* and Schwarzacher's (1975) book *Sedimentation Models and Quantitative Stratigraphy* are classics and well worth the attempt to find a copy. Two more recent publications give an idea of the complexity and breadth of this topic today, they are Tetzlaff and Harbaugh's (1989) *Simulating Clastic Sedimentation*, and Cross' (1990) *Quantitative Dynamic Stratigraphy*.

12.2 Forward model types

12.2.1 Hydraulic process–response models

Hydraulic process–response models are those in which fundamental laws concerning hydraulic flow are used to model sediment transport and deposition in three dimensions. Most are based on solutions to simplified versions of the Navier–Stokes equations describing flow in two spatial dimensions for an isotropic Newtonian fluid. The book by Tetzlaff and Harbaugh (1989) discusses this type of modelling in detail, as well as providing a series of FORTRAN subroutines. Sediment transport and deposition within this flow regime is another, separate, problem. The model developed by the University of Stanford, 'SEDSIM3', simulates sediment transport as a function of bed roughness, flow depth and flow rate. The effective transport of sediment depends on the transport capacity of the fluid (a function of flow conditions), and the effective sediment concentration. Sediment transport and deposition pose the greatest problem for hydraulic process models because it is particularly difficult to scale-up the timing of processes and sediment transport without having to resort to heuristics. In their section on 'economizing on arithmetic', Tetzlaff and Harbaugh (1989, p. 81) discuss some of these scaling problems, solutions to which rely on judicious use of extrapolation, which in some cases may not be valid.

12.2.2 Diffusion or potential gradient models

Diffusion models are those in which the transport and deposition of sediment is a function of the potential gradient on the pre-existing (and continually changing) sediment mass/surface. In the following discussion mass concentration and topographic height (energy potential) are treated as equivalent and interchangeable. In other words a heap of sediment that will slide downhill (diffuse) is the same as a concentration of sodium ions that will diffuse throughout a block of porous sandstone from a high potential to a low potential. Sediment can be considered to follow the same rules, i.e. it needs more energy to get sediment to climb uphill.

12.2.3 Geometrical models

These are models that do not describe the process itself, but the geometric results of the process, usually filling or partially filling accommodation space. Many of these models are based on pioneering work by Sloss (1962). He developed a simplistic model for the relationship between sediment shape (S), volume (Q) of sediment supplied, rate of subsidence (R), rate of dispersal of sediment (D), and the 'nature' of the material supplied (M);

$$S = f(Q, R, D, M)$$

Sloss developed models for situations where the relationship between input sediment volume and subsidence rate varied, showing that both regressive and transgressive phases, and cyclical patterns could be generated by sediment supply variation. Sloss' subsidence term implicitly included relative sea-level change as a uniform subsidence. Sloss discussed the use of the model in passive margin and foreland settings, and also discussed turbidite development. His treatment was on a purely logical, conceptual basis. These ideas were taken up by Harbaugh and Bonham-Carter in Chapter 9 of their 1970 publication, which provides a mathematical treatment of Sloss' conceptual model. The following is taken from their chapter.

> A section through the basin topography is divided into a series of two-dimensional columns which provide the basis for sediment accumulation. Sediment enters the section from a predetermined point. Part of the sediment load is deposited on the first column encountered and the remainder is passed to the next column. The amount deposited on a particular column depends on (a) the amount of sediment available for deposition, and (b) the accommodation space in that column. Mass is conserved by accounting for all sediment as it moves throughout the system.

Harbaugh and Bonham-Carter (1970) mention three constraints.

1 Where the sediment–water interface is above or equal to base level (as specified for a given particle size class) then deposition is not allowed and all remaining load is passed to the next column.
2 Where the sediment–water interface is only slightly below base level, only part of the sediment load is deposited. Sediment is base-levelled and the remainder passed on to the next column.
3 The sediment–water interface is well below base level for all particle-size fractions.

Over the years additional constraints have been added by many packages, such as angle of repose for various grain-size mixtures, slope failure, etc. Subsidence can be either an instantaneous or a lagged (occurring after a defined time interval) function of sediment loading or external tectonics. There are in practice few limits to the algorithmic complexity that can be built into this type of model. A FORTRAN IV program to carry out this type of modelling is provided by Harbaugh and Bonham-Carter (1970, pp. 382–384), and this program formed the precursor to many of today's geometric models. Quite complex geometries and sedimentological situations can be built in. In 1978 Pitman developed a geometric model of clastic sedimentation that can be considered a development of Sloss', and Harbaugh and Bonham-Carter's, and the SEDPAK program produced by the University of South Carolina is also a good example of the geometric approach (Scaturo et al., 1987, 1989).

A fourth model type not discussed in detail here is an algorithmic and heuristic hybrid. In this case a pragmatic mixture of models is used under different circumstances. For example, large-scale clastic features may be filled geometrically, whereas carbonates use a partial process model, and coal is deposited heuristically.

12.2.4 Discussion of model types

The above model categories are to some extent arbitrary and there is a good deal of overlap in practice in many operational programs. There remains room for all types at appropriate scales and appropriate problems. For studying the entrainment, transport, distribution, and deposition of clastic sediments in restricted basins, such as harbours and estuaries, over short time intervals then pure hydraulic process models are of most interest. Practical problems that can be examined include pipeline scour studies, platform and loading-quay effects on sediment build-up, pollutant dispersal including heavy minerals from drilling operations, and placer deposit distribution in restricted regimes. In a stratigraphic context they may be of interest where basin/channel topography is complex but well defined.

The difficulty (impossibility) of analytical solutions to many of the equations also forces workers to finite-difference numerical solutions in which both the topography and time intervals are discretized. The size of the steps involved if several million years and several hundred kilometres of section are to be simulated in a 'reasonable' study period means that in practice there are minor differences only between a potential-gradient (diffusion) equation approach and a geometric model at geological scales.

12.3 Examples of stratigraphic use of computer modelling

The stratigraphic use of these models in practice has been rather limited, at least as far as published results are concerned. Papers by Lawrence and co-workers at Shell (Aigner *et al.*, 1990; Lawrence *et al.*, 1990) and by the University of South Carolina group (Scaturo *et al.*, 1987, 1989) show the type of modelling that is possible in clastic and carbonate environments, respectively.

Tetzlaff and Harbaugh (1989, p. 137) applied SEDSIM to the Golden Meadow field, Louisiana, where Miocene and Pliocene sands produce hydrocarbons, the goal was to predict sand distribution and quality throughout the field. The sands are laterally variable and a 300 ft-thick stratigraphic section around the '8900 ft sand' at 8900 ft depth was modelled over a square 65 km by 65 km. Information from 55 wells was used to both build and control the model (some wells being held in reserve for testing). The hydraulic model was run for 50 000 years, using a point source (palaeo-Mississippi) with discharge rates similar to the present river. It was found necessary to increase both flow rate and sediment concentration in the flow to achieve the observed sand percentage (initially too low). Results indicate that SEDSIM3 could be used to investigate reservoir-scale features and investigate net/gross variation given sufficient input information (Fig. 12.1). Perhaps most noteworthy is the very small time span needed to produce the reservoir sands given present-day Mississippian discharge rates. This indicates that for reservoir-scale work, a hydraulic model may be the most appropriate.

Using a geometric/heuristic approach, Royal Dutch/Shell has published applications for the Paris Basin (Fig. 12.2) and the Arabian Platform (Aigner *et al.*, 1990). These are much larger scale studies, aiming to understand the controls on basin stratigraphy. The paper emphasizes both the learning and predictive aspects of computer modelling of basin fill. They state that 'computer modelling is ideal for generating and quantitatively evaluating conceptual models'. In the case of the Arabian Platform the program was used to gain some feel for the way in which several orders of relative sea-level change affect carbonate development on the platform. They show the effects of hierarchical cyclicity up to 'fifth order'. Fourth- or fifth-order subcycles are stacked into almost symmetrical transgressive–regressive third-order cycles, which in turn are combined into the Akhdar Group 'supercycle'. Eustatic, sediment supply, and tectonic influences are modelled yielding different stratal patterns. The conclusion is that the cyclicity was more eustatically controlled in this area.

In a second example the Shell group discuss the modelling of the Paris Basin, a radially symmetric, intracratonic basin filled with Mesozoic and Tertiary sediment. Modelling in this case aimed at predicting/hindcasting the location of the known Triassic sand reservoirs, Callovian carbonate, and Lower Cretaceous sand reservoirs. The input parameters were not published in detail, and it is difficult to judge the sensitivity of the model, but Shell report that several subsidence curves were needed at different points in the basin, and successive runs were needed to match the observed features in the basin. The best match was obtained by allowing sediment influx from both sides of the basin. Data from only one well in the basin and the Haq *et al.* (1987) sea-level curve were sufficient to provide the input for a rough initial model with some estimates of sediment flux variation. An acceptable match required better control on subsidence history and sea-level variation across the basin (Fig. 12.2).

Bosence *et al.* (1991) have used a carbonate modelling program, 'CARBONATE', to simulate the geometries of a well-constrained outcrop section of Miocene carbonates of Cap Blanc, Mallorca. The computer program incorporates five processes; *in situ* carbonate production, pelagic carbonate production, chemical erosion, slope failure and wave-generated sediment erosion, transport and deposition (Bosence and Waltham, 1990). The rates of these processes were specified for the Late Miocene from empirical evidence, and a relative sea-level curve was derived from outcrop-section observations. The match obtained between the outcrop section and computer model was very good (Fig. 12.3), and indicated that the major control on the stratigraphic architecture of carbonate platforms of this size (kilometres across and tens of metres high) was relative sea-level, with compaction and differential subsidence having little effect on stratigraphy.

Scaturo *et al.* (1989) describe a simulation of the Devonian Judy Creek carbonate platform complex in Alberta, Canada (Fig. 12.4). They emphasized the fact that although the simulation results matched the observations reasonably well, this in no way provides proof that eustatic sea-level did in fact vary as modelled, only that it could have done. The study concluded that:

1 the progradation of the Judy Creek reef margin occurred when the rate of carbonate accumulation was greater than the rate of sea-level rise, and the slope of the marginal escarpment was less than 0.25°;
2 the reef margin aggraded vertically if the foreslope was too steep for stable deposition, or if the rate of carbonate accumulation equalled the rate of relative sea-level rise;
3 backstepping of the reef margin occurred as the relative sea-level rise exceeded carbonate growth provided the platform top remained in the photic zone (see Chapter 10);
4 clastic sediment input to the area of carbonate development reduced growth rates exponentially;
5 the platform complex could be modelled using carbonate growth rates of 3.0 m per thousand years and initial relative sea-level rises of 3.5–5.0 m per ten thousand years,

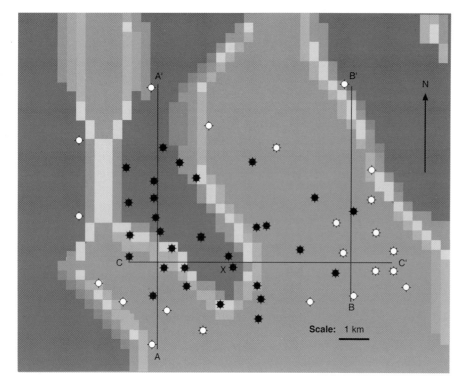

Fig. 12.1 Golden Meadow simulation results. (a) Plan view of the Golden Meadow SEDSIM simulation results showing the facies distribution. X denotes the location with the lobate sand body that was predicted to contain the maximum sand thickness. Darker shades represent increasing sand percentages

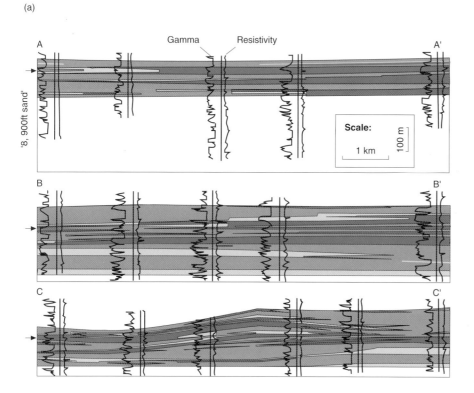

Fig. 12.1 (b) Simulated lithostratigraphic sections A–A′, B–B′ and C–C′ located on (a). Actual gamma and resistivity log traces are superimposed. Wells aligned with the A–A′ and B–B′ sections were used as a blind test (from Tetzlaff and Harbaugh, 1989)

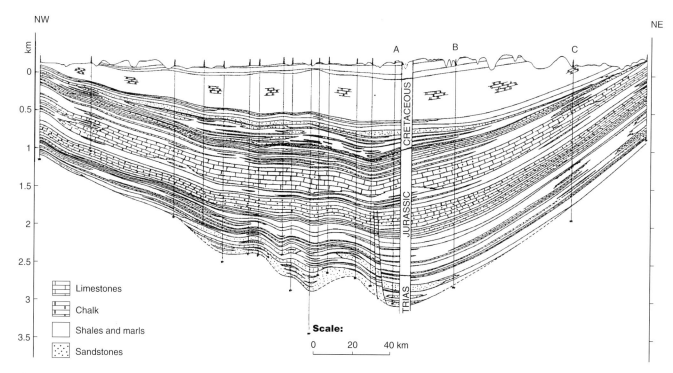

Fig. 12.2 Actual (*above*) and simulated (*overleaf*) stratigraphy of the Paris Basin, Rhaetian to Maastrichtian (from Aigner *et al.*, 1990)

followed by a more gradual relative sea-level rise of 0.5–1.5 m per ten thousand years.

12.4 Practical forward computer modelling

12.4.1 Points to consider

Before an appropriate model to simulate basin fill can be chosen, it is essential to consider the nature of the problem to be addressed, the scale of the problem in space and time, and our knowledge of the input parameters, such as sediment supply, subsidence rates and pre-depositional bathymetry.

Nature and scale of the problem

The types of problem that we may wish to address range from the simulation of the filling of an entire basin during a significant period of the Earth's history, to the prediction of small-scale reservoir heterogeneities, both of which are described in the preceding section. The nature of the problem will allow an appropriate scale to be chosen (e.g. 1000 km by 1000 km, or 100 m by 100 m), and this choice of scale will practically determine the resolution, owing to limitations in computer power and human patience. The larger the scale of the problem the coarser the answer. The greater the resolution needed, the greater the patience required. In addition to the size of the problem, the time resolution must be defined. This could be at intervals of hours if we are wanting to model scour around a pipe while tide levels change, or at intervals of 100 thousand years if we are interested in the Early Jurassic. It is important to ensure that the model time interval is less than half the time period of the event of interest (e.g. Milankovitch cycles) in order to avoid aliasing problems.

From a practical computing point of view, the grid size must be specified because it determines the spatial resolution of the model. A good way to approach this problem is with some idea of the spatial wavelength of interest, and to ensure that the grid size used is less than half the wavelength of the smallest feature of interest (e.g. point bar, clinoform, offshore bar, delta lobe, etc.).

Input parameters for simulation

The input parameters for the simulation can be determined from both empirical observations and theoretical values based on a model of a given process. Figure 12.5 can be used to place our knowledge of a system in a global perspective. At one end of the scale, models predicting (forecasting or hindcasting) large features at time-scales of 100 million years may start at the far left. On the other hand a highly detailed model of a present-day process can use direct observations from the centre column in Fig. 12.5.

Most available basin-scale and reservoir-scale stratigraphic models start somewhere towards the fourth column from the left, with estimates of sediment input, sea-level and subsidence rates. There are a few approaches that use climate modelling (second and third columns from the left

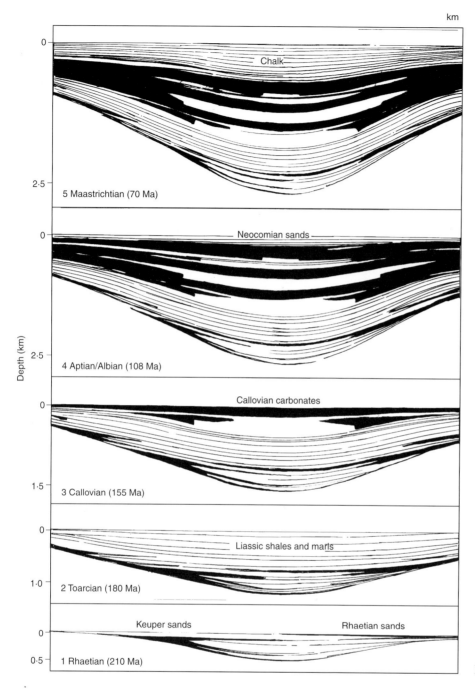

Fig. 12.2 (continued)

in Fig. 12.5) to predict sediment input variation (Koltermann and Gorelick, 1992), and some that have combined climate and Milankovitch models (Perlmutter and Matthews, 1990).

For simulation purposes, two broad choices are available. The first is to assume theoretical values for our input parameters, based on a model of the process (e.g. we can assume that Milankovitch cycles occurred in the Miocene, that sea-level changes are driven by glaciation, that a basin with a given β-value will subside at a given rate, etc.). The other possibility is to derive all input parameters from observations. By choosing the second route it is necessary to have a mechanism for deriving empirical values. To the right of Fig. 12.5 can be seen the measurements that are available from preserved strata, bearing in mind that post- and syn-depositional processes (third column from the right in Fig. 12.5) will have modified the original values. As can be seen there is uncertainty in both the direction of theoretical process, and in the interpretation from present-day observations. In this case estimates of global sea-level, subsidence, and sedimentation rate are not only empirical but are impossible to separate out from observations in one-dimensional sections. Only by working in two dimensions and three dimensions can we hope to separate these effects (Griffiths and Nordlund, 1993; Lessenger, 1993).

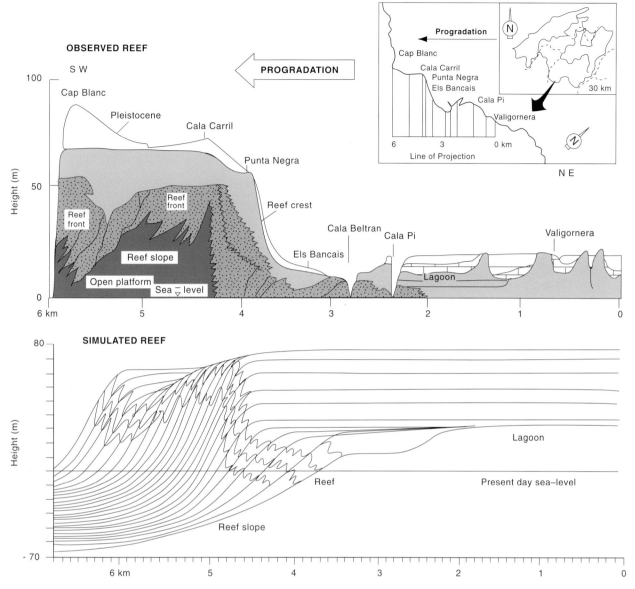

Fig. 12.3 Observed and simulated carbonate reef in Mallorca (from Bosence *et al.*, 1991)

Griffiths and Hadler-Jacobsen (1993) and Nordlund and Griffiths (1993a,b) discuss the derivation of input parameters for forward models from seismic observations. Table 12.1 lists a range of input parameters for simulation and points to consider prior to simulation.

12.4.2 Worked example

To illustrate the practical use of these tools we can look at how we may simulate a simple two-dimensional sequence stratigraphic cartoon. A problem that we may address is the prediction of the sand:mud ratio 15 km from the proximal end of the section shown on the cartoon.

The first stages are as described in Chapter 5. The seismic geometries are mapped and digitized (Fig. 12.6a). Given the assumption that each reflection is an isochron we convert the two-dimensional section to a Wheeler diagram (Wheeler, 1958) or chronostratigraphic chart (Fig. 12.6b). From this chart we calculate the area of each chronosome, which represents the area/volume of sediment deposited and preserved during the time interval represented by the chronosome. From the Wheeler diagram we can also estimate a coastal onlap curve, and make a first approximation to a relative sea-level (RSL) curve, again as described in Chapter 5.

We now have two options. We can choose either an RSL curve based on our observations (Fig. 12.6c,d) or one based on a theoretical curve (in Fig. 12.6e,f a sine wave has

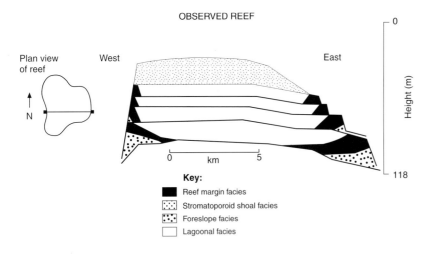

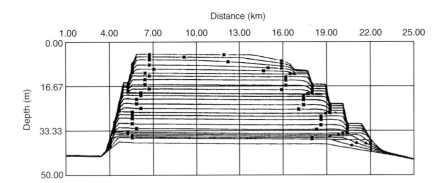

Fig. 12.4 SEDPAK simulation of the Judy Creek carbonate platform, Alberta, Canada (from Scaturo et al., 1989)

been used). The same possibilities exist for sediment input. Case 12.6c,d uses an observed input curve, whereas case 12.6e,f uses a constant sediment input throughout the experiment. The geological age of the reflectors also must be estimated and here we have used an assumption that the length : thickness ratio of the chronosome is proportional to elapsed time. This allocates more time to deposit long, thin chronosomes than short, thick chronosomes. We have assumed that the total package duration is 1 million years. The integrals of the sediment input curves both equal the observed section area (we are ignoring compaction in this example). Subsidence rates were the same in both cases.

The simulation was run using the Marco Polo program PHIL^{TM} version 1.5. The colour graphical output from PHIL has been modified. The results illustrate many of the features of the current state of two-dimensional modelling. Remember that we wanted the model to duplicate the observed reflection geometries and make predictions concerning sand/shale distribution within the chronosomes at a given point.

The first feature to note is that the basin-floor fan is missing from the simulations. This is because we need to specify special conditions to generate such fans. The computer model must be told that in the case of sea-level falling at a given rate then sediment will not be eroded and redeposited as basin-stepping units, but will be deposited on the basin floor. We could also build in a case where the oversteepened edge of a clinoform slumps during a highstand, also depositing basin-floor fans. The point being that we are not judging either case, only building alternative models to test hypotheses and their consequences.

Another feature is the distal extent of each chronosome. The computer model does not know about seismic resolution but does know about fine sediment being carried far out into the basin. Consequently, instead of chronosomes terminating neatly when they drop below seismic resolution, the existence of thin deep-marine sediment is predicted.

The predicted sand/shale distribution using the empirical RSL curve is shown in Fig. 12.7a. A coarse, clean sand body is predicted from 75 to 60 m. Using the sine-wave (Fig. 12.7b), a more clay-rich section is predicted, with thinner sands around 80 and 70 m. Which is right? Probably neither, although in this case the Wheeler diagram for the sine-wave case is a closer match to the observed Wheeler diagram and we may be tempted to choose that. In practice the model would be rerun using criteria that managed to reproduce the observed basin-floor fans, and give a better match to the clinoform geometries.

All this adjusting of variables takes time if done manually, but help is at hand in the form of mathematical inversion. It has been demonstrated (Lessenger, 1993) using synthetic

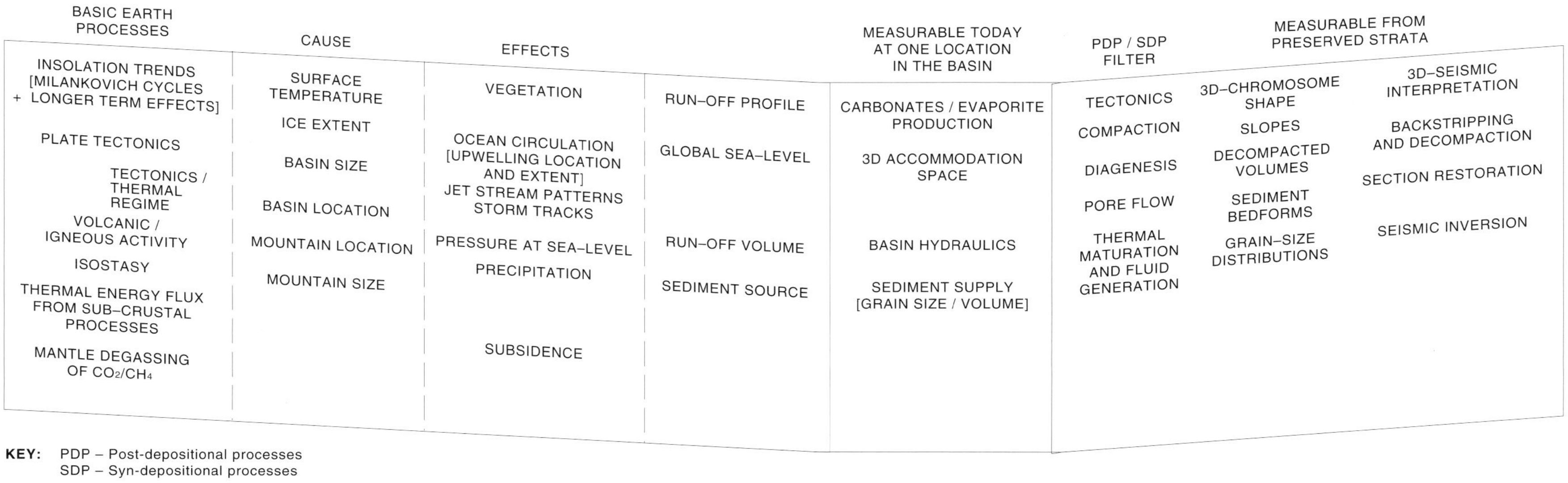

BASIC EARTH PROCESSES	CAUSE	EFFECTS	MEASURABLE TODAY AT ONE LOCATION IN THE BASIN	PDP / SDP FILTER	MEASURABLE FROM PRESERVED STRATA	
INSOLATION TRENDS [MILANKOVICH CYCLES + LONGER TERM EFFECTS]	SURFACE TEMPERATURE	VEGETATION	CARBONATES / EVAPORITE PRODUCTION	TECTONICS	3D-CHROMOSOME SHAPE	3D-SEISMIC INTERPRETATION
	ICE EXTENT	OCEAN CIRCULATION [UPWELLING LOCATION AND EXTENT]		COMPACTION	SLOPES	BACKSTRIPPING AND DECOMPACTION
PLATE TECTONICS	BASIN SIZE	JET STREAM PATTERNS STORM TRACKS	3D ACCOMMODATION SPACE	DIAGENESIS	DECOMPACTED VOLUMES	SECTION RESTORATION
TECTONICS / THERMAL REGIME	BASIN LOCATION			PORE FLOW	SEDIMENT BEDFORMS	SEISMIC INVERSION
VOLCANIC / IGNEOUS ACTIVITY	MOUNTAIN LOCATION	PRESSURE AT SEA-LEVEL PRECIPITATION	BASIN HYDRAULICS	THERMAL MATURATION AND FLUID GENERATION	GRAIN-SIZE DISTRIBUTIONS	
ISOSTASY	MOUNTAIN SIZE		SEDIMENT SUPPLY [GRAIN SIZE / VOLUME]			
THERMAL ENERGY FLUX FROM SUB-CRUSTAL PROCESSES		SUBSIDENCE				
MANTLE DEGASSING OF CO_2/CH_4						

Additional row entries (middle column): RUN-OFF PROFILE, GLOBAL SEA-LEVEL, RUN-OFF VOLUME, SEDIMENT SOURCE

KEY: PDP – Post-depositional processes
SDP – Syn-depositional processes

Fig. 12.5 Relationship between process variables and observations in computer modelling of basin fill. PDP, post-depositional processes; SDP, syn-depositional processes

Table 12.1 Input parameters and points to consider for computer modelling

Initial topography/bathymetry
This needs to be determined via backstripping and must be carried out in three dimensions. Even if the modelling is to be carried out in two dimensions, we need to know the three-dimensional surface in order to estimate sediment transport directions

Nature of depositional surface
Loose sand, clay or crystalline rock? The answer determines the degree to which clastic input is supplemented by erosion products

Climate: rainfall
The runoff volume and sediment concentration can be input either directly or predicted from a climate model

Climate: wave and storm magnitude and frequency
River discharge is extremely irregular, and most sediment is transported and deposited/redeposited during brief extreme events, hence the statistics of such behaviour are of interest to modellers. It will not be possible to predict exactly when a flood or storm event will happen, but much can be said about the probability of it happening in a given climatic regime. Such information can be built either directly or indirectly into forward models (Koltermann and Gorelick, 1992)

Currents
Is the model in an area where marine currents might have redistributed sediment? If so what were their characteristics?

Subsidence rates
Is, or was, subsidence being driven by sediment loading, or was it a function of plate stress or thermal recovery happening independently of sediment load. The sediment geometries produced are very different in the two cases. See Chapter 2 for a discussion of subsidence rates in different tectonic settings

Sea-level
One of the most controversial variables, discussed in Chapter 2. The simulation may use the Haq et al. (1987) sea-level curves, or a local relative sea-level (RSL) curve, the construction of which is described in Chapter 5. If many such curves are available around a basin then a regional curve may be extracted. One practical decision that has to be made in computer modelling is how to treat the RSL curve. One option is to treat it as the sum of all changes to accommodation space, including global sea-level, tectonic subsidence, compaction, thermal subsidence, etc. This may be acceptable in a one-dimensional simulation but not in two or three dimensions. The three-dimensional shape of accommodation space creation varies with the different causes. An attempt should be made to break out the component parts

Sediment supply
Sediment supply can be treated either as a function of climatic process, or as an observation. Measurements of the volume of sediment input to the basin during a given time period can be made. Integrating the sediment supply rate over this time period must give the observed sediment volume after decompaction. The sediment supply curve within the simulated period can take a number of forms, such as constant, cyclic, or matched to chronosome areas

Post-depositional processes
In order to match the modelled geometries to observations there have to be some post-depositional processes, such as compaction, burial, faulting, etc. The timing of these processes in relation to the depositional event is often critical. Present-day steady-state compaction concepts are not adequate when used in conjunction with forward modelling. The rate of compaction largely determines the successive location of delta lobes and shale drapes, for example

---

two-dimensional data that one can converge on unique values for sea-level curve, sediment input rates, initial topography and subsidence. If this also proves to be possible on real data then computer modelling will have taken a major step forward.

12.5 Conclusions

The various types of computer models available probably should be treated as being complementary rather than as being competitive. We know that reservoir sands themselves are deposited in very short time periods (from seconds to years), therefore hydraulic models are a reasonable way of investigating, and possibly even predicting, lateral (three dimensional) variation in sediment net/gross and architecture. The larger scale (longer time) features are, however, not managed adequately by hydraulic models for various reasons, including the need for heuristics to reduce computational times. The question then remains as to the role of the larger scale models (geometric, diffusion, and mixed heuristic models). The pure geometric model aims to duplicate the stratal geometry observed in seismic sections based on external controls. Because we already have these geometries in three dimensions, and can quantify them (Griffiths and Nordlund, 1993; Nordlund and Griffiths, 1993a,b), it would seem reasonable (if the goal is prediction of sand : mud ratio or seal potential for example) to combine detailed three-dimensional backstripping with short time-scale modelling for the intrachronosome detail. The uncertainty lies in the contents of the three-dimensional packages rather than their location.

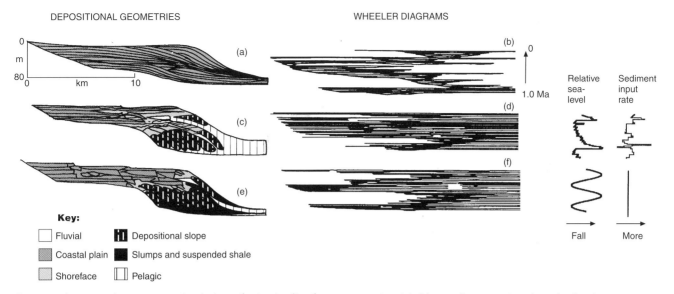

Fig. 12.6 An example computer simulation of seismic clinoform geometries. (a) Observed geometries. (b) Wheeler diagram derived from the observed geometries. (c) Simulated geometries using empirical relative sea-level curve. (d) Wheeler diagram derived from the above simulation. (e) Simulated geometries using theoretical relative sea-level curve. (f) Wheeler diagram derived from the above simulation

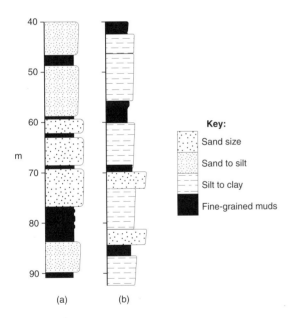

Fig. 12.7 Predicted lithologies from two simulations. (a) Using the empirical sea-level curve. (b) Using a sine-wave relative sea-level curve

References

Abbots, I.L. (1991) *United Kingdom Oil and Gas Fields 25 Years Commemorative Volume.* Geological Society of London, Bath, 573 pp.

Agassiz, L. (1840) *Etudes sur les Glaciers.* Privately published, Neuchatel, 346 pp.

Agassiz, L. (1842) The glacial theory and its recent progress. *Edinb. New Philos. J.*, 33, 271–283.

Ahr, W.M. (1973) The carbonate ramp: an alternative to the shelf model. *Trans. Gulf-Coast Assoc. geol. Soc.*, 23, 221–225.

Aigner, T. (1984) Dynamic stratigraphy of epicontinental carbonates. Upper Muschelkalk (M. Triassic), South German Basin. *Neues Jahrb. Geol. Palaeontol. Abh.*, 169, 127–159.

Aigner, T. & Reineck, H.E. (1982) Proximality trends in modern storm sands from the Helegoland Bight (North Sea) and their implications for basin analysis. *Senckenbergiana Marit.*, 14, 183–215.

Aigner, T., Brandenburg, A., Van Vliet, A., Doyle, M., Lawrence, D.T. & Westrich, J. (1990) Stratigraphic modeling of epicontinental basins — 2. Applications. *Sediment. Geol.*, 69, 167–190.

Ainsworth, R.B. & Pattison, S.A.J. (1994) Where have all the lowstands gone? Evidence for attached lowstand systems tracts in the Western Interior of North America. *Geology*, 22, 415–418.

Alexander, J. & Leeder, M.R. (1987) Active tectonic control on alluvial architecture. In: *Recent Developments in Fluvial Sedimentology* (ed. by F.G. Ethridge, R.M. Flores & M.D. Harvey). Special Publication, Society of Economic Paleontologists and Mineralogists, Tulsa, 39, 243–252.

Allen, G.P. (1984) Tidal processes in estuaries: a key to interpreting fluvial–tidal facies transitions. In: *International Association of Sedimentologists, 5th European Meeting*, pp. 23–24.

Allen, G.P. (1991) Sedimentary processes and facies in the Gironde Estuary: a recent model for macrotidal estuarine systems. In: *Clastic Tidal Sedimentology* (ed. by R.A. Rahmani, D.G. Smith, G.E. Reinson & B.A. Zaitlan). Memoir of the Canadian Society of Petroleum Geologists, Calgary, 16, 29–40.

Allen, G. & Mercier, F. (1988) Subsurface sedimentology of deltaic systems. *P.E.S.A.J.*, **March**, 30–44.

Allen, G.P. & Posamentier, H.W. (1993) Sequence stratigraphy and facies model of an incised valley-fill: the Gironde Estuary, France. *J. sediment. Petrol.*, 63, 378–391.

Allen, J.R.L. (1965) A review of the origin and characteristics of recent alluvial sediments. *Sedimentology*, 5, 98–191.

Allen, J.R.L. (1978) Studies in fluviatile sedimentation: an exploratory quantitative model for the architecture of avulsion controlled alluvial suites. *Sediment. Geol.*, 21, 129–147.

Allen, J.R.L. (1979) Studies in fluvatile sedimentation: an elementary geometrical model for the connectedness of avulsion controlled alluvial suites. *Sediment. Geol.*, 24, 253–267.

Allen, J.R.L. & Williams, B.J.P. (1982) The architecture of an alluvial suite: rocks between the Townsend Tuff and Pickard Bay Tuff Beds (Early Devonian), southwest Wales. *Philos. Trans. R. Soc. Lond., Ser. B*, 297, 51–89.

Allen, S., Coterill, K., Eisner, P., Perez-Cruz, G., Wornardt, W.W. & Vail, P.R. (1991) Micropaleontology, well log and seismic sequence stratigraphy of the Plio-Pleistocene depositional sequences — offshore Texas. In: *Sequence Stratigraphy as an Exploration Tool: Concepts and Practices in the Gulf Coast* (ed. by J.M. Armentrout & B.F. Perkins), Eleventh Annual Conference, Gulf Coast Section, Society of Economic Paleontologists and Mineralogists, June 2–5, Texas pp. 11–13.

Alonso, B. & Maldonado, A. (1990) Late Quaternary sedimentation patterns of the Ebro turbidite systems (northwestern Mediterranean): two styles of deep sea deposition. In: *The Ebro Continental Margin, Northwestern Mediterranean Sea* (ed. by C.H. Nelson & A. Maldonado). *Mar. Geol.*, 95, 353–377.

Arche, A. (1983) Coarse grained meander lobe deposits in the Jarama River, Madrid, Spain. In: *Modern and Ancient Fluvial Systems* (ed. by J.D. Collinson & J. Lewin). Special Publication, International Association of Sedimentologists, 6, 313–321, Blackwell Scientific Publications, Oxford.

Armentrout, J.M. (1987) Integration of biostratigraphy and seismic stratigraphy: Pliocene–Pleistocene, Gulf of Mexico. In: *Innovative Biostratigraphic Approaches to Sequence Analysis: New Exploration Opportunities.* 8th Annual Research Conference. Gulf Coast Section, Society of Economic Paleontologists and Mineralogists, pp. 6–14.

Armentrout, J.M. & Clement, J.F. (1991) Biostratigraphic calibration of depositional cycles: a case study in High Island–Galveston–East Breaks areas, offshore Texas. In: *Sequence Stratigraphy as an Exploration Tool: Concepts and Practices* (ed. by J.M. Armentrout & B.F. Perkins). Eleventh Annual Conference, June 2–5, Texas Gulf Coast Section, Society of Economic Paleontologists and Mineralogists, pp. 21–51.

Armentrout, J.M., Echols, R.J. & Lee, T.D. (1991) Patterns of foraminiferal abundance and diversity: implications for sequence stratigraphic analysis. In: *Sequence Stratigraphy as an Exploration Tool: Concepts and Practices* (ed. by J.M. Armentrout & B.F. Perkins). Eleventh Annual Conference, June 2–5, Texas Gulf Coast Section, Society of Economic Paleontologists and Mineralogists, pp. 53–58.

Armstrong, L.A., Ten Have, A. & Johnson, H.D. (1987) The geology of the Gannet Fields, Central North Sea, UK Sector. In: *Petroleum Geology of N.W. Europe* (ed. by J. Brooks & K. Glennie), Graham & Trotman, London, pp. 533–548.

Ashley, G.A. (1990) Classification of large scale subaqueous bedforms: a new look at an old problem. *J. Sediment. Petrol.*, 60, 160–172.

Ashton, M.A. (1980) The stratigraphy of the Lincolnshire Limestone Formation (Bajocian) in Lincolnshire and Leicestershire.

Asquith, D.O. (1970) Depositional topography and major marine environments, Late Cretaceous, Wyoming. *Am. Assoc. Petrol. Geol. Bull.*, **63**, 324–340.

Atkinson, C.D. (1983) Comparative sequences of ancient fluviatile deposits in the Tertiary South Pyrenean Basin, Northern Spain. Unpublished PhD thesis, University of Wales, Cardiff 360 pp.

Atkinson, C.D. (1986) Tectonic control on alluvial sedimentation as revealed by an ancient catena in the Capella Formation (Eocene) of northern Spain. In: *Paleosols, their Recognition and Interpretation* (ed. by V.P. Wright). Princeton University Press, Princeton, NJ, pp. 139–179.

Atkinson, C.D., Trumbley, P.N. & Kremer, M.C. (1988) Sedimentology and depositional environments of the Ivishak sandstone, Prudhoe Bay Field, North Slope, Alaska. In: *Giant Oil and Gas Fields, a Core Workshop* (ed. by A.J. Lomando & P.M. Harris). Society of Economic Paleontologists and Mineralogists, Tulsa, Core Workshop **12**, 561–613.

Atkinson, C.D., McGowen, J.C., Bloch, S., Lundell, L.L. & Trumbly, P.N. (1990) Braidplain and Deltaic Reservoir, Prudhoe Bay Field, Alaska. In: *Sandstone Petroleum Reservoirs* (ed. by J.H. Barwis, J.G. McPherson and J.R.J. Studlick). *Casebooks in Earth Sciences*. Springer-Verlag, New York, pp. 7–28.

Aurell, M. (1991) Identification of systems tracts in low-angle carbonate ramps; examples from the Upper Jurassic of the Iberian Chain (Spain). *Sediment. Geol.*, **73**, 101–115.

Autin, W.J.S., Burns, S.F., Miller, B.J., Saucier, R.T. & Snead, J.I. (1991) Quaternary geology of the lower Mississippi Valley. In: *Quaternary Nonglacial Geology; Conterminous U.S.* Colorado Geology Society of America, **K2**, 547–582.

Back, W., Hanshaw, B.B. & van Driel, J.N. (1984) Role of groundwater in shaping the eastern coastline of the Yucatan Peninsula, Mexico. In: *Groundwater as a Geomorphic Agent* (ed. by R.G. LaFleur). Allen and Unwin, Boston, pp. 157–172.

Banner, F.T. & Blow, W.H. (1965) Progress in the planktonic Foraminiferal biostratigraphy of the Neogene. *Nature*, **208**, 1164–1166.

Barron, J.A. (1985) Miocene to Holocene planktic diatoms. In: *Plankton Stratigraphy* (ed. by H.M. Bolli, J.B. Saunders & K. Perch-Nielsen). Cambridge Earth Science Series, Cambridge University Press, Cambridge, UK, pp. 763–809.

Bathurst, R.G.C. (1975) *Carbonate Sediments and Their Diagenesis*. Elsevier, Amsterdam, 658 pp.

Batten, D.J. (1974) Wealden palaeoecology from the distribution of plant fossils. *Proc. Geol. Assoc.*, **85**, 433–458.

Begin, Z.B., Meyer, D.F. & Schumm, S.A. (1981) Development of longitudinal profiles of alluvial channels in response to base-level lowering. *Earth Surf. Process. Land.*, **6**, 49–68.

Behrensmeyer, A.K. (1987) Miocene fluvial facies and vertebrate taphonomy in northern Pakistan. In: *Recent Developments in Fluvial Sedimentology* (ed. by F.G. Ethridge, R.M. Flores & M.D. Harvey). Special Publication, Society of Economic Paleontologists and Mineralogists, Tulsa, **39**, 169–176.

Behrensmeyer, A.K. & Tauxe, L. (1982) Isochronous fluvial system in Miocene deposits of northern Pakistan. *Sedimentology*, **29**, 331–352.

Bellaiche, G., Droz, L., Aloisi, J., Bouye, C., Monaco, A., Maldanado, A., Serra-Raventos, J. & Mirabile, L. (1981) The Ebro and Rhone deep sea fans: first comparative study. *Mar. Geol.*, **43**, 75–85.

Berg, R.R. & Royo, G.R. (1990) Channel-fill turbidite reservoir, Yowlumne Field, California. In: *Sandstone Petroleum Reservoirs* (ed. by J.H. Barwis, J.G. McPherson & R.J. Studlick). Springer-Verlag, New York, pp. 467–487.

Berggren, W.A. & Van Couvering, J.A. (1974) The Late Neogene biostratigraphy, geochronology and palaeoclimatology of the last 15 million years in marine and continental sequences. *Palaeogeogr. Palaeoclimatol. Palaeoecol.*, **16**, 1–215.

Bergman, K.M. & Walker, R.G. (1988) Formation of Cardium erosion surface E5, and associated deposition of conglomerate: Carrot Creek Field, Cretaceous Western Interior Seaway, Alberta. In: *Sequences, Stratigraphy, Sedimentology: Surface and Subsurface* (ed. by D.P. James and D.A. Leckie). Canadian Society Petroleum Geologists, Calgary, Memoir **15**, 15–24.

Bernard, H.A. & LeBlanc, R.J. (1965) Resume of the Quaternary geology of the northwestern Gulf of Mexico province. In: *The Quaternary Geology of the United States* (ed. by H.E. Wright, Jr. & D.G. Frey). Princeton University Press, Princeton, NJ, pp. 137–185.

Bernard, H.A. & Major, C.J. (1963) Recent meander belt deposits of the Brazos river, an alluvial 'sand' model. *Am. Assoc. petrol. Geol. Bull.* (abs.), **367**, 350.

Betzer, P.R., Showers, W.J., Laws, E.A., Winn, C.D., Ditulluo, G.R. & Kroopnick, P.M. (1984) Primary productivity and particle fluxes on a transect of the equator at 153°W in the Pacific Ocean. *Deep Sea Res.*, **31**, 1–11.

Bhattacharya, J.P. (1988) Autocyclic and allocyclic sequences in river- and wave-dominated deltaic sediments of the Upper Cretaceous Dunvegan Formation, Alberta: core examples. In: *Sequences, Stratigraphy, Sedimentology: Surface and Subsurface* (ed. by D.P. James & D.A. Leckie). Canadian Society of Petroleum Geologists, Calgary, Memoir **15**, 25–32.

Bhattacharya, J.P. (1991) Regional to subregional facies architecture of river-dominated deltas in the Alberta subsurface, Upper Cretaceous Dunvegan Formation. In: *The Three-Dimensional Facies Architecture of Terrigenous Clastic Sediments and its Implications for Hydrocarbon Discovery and Recovery* (ed. by A.D. Miall & N. Tyler). Society of Economic Paleontologists and Mineralogists (Society for Sedimentary Geology), Tulsa, OK, Concepts in Sedimentology and Paleontology, **3**, 189–206.

Biddle, K.T., Schlager, W., Rudolph, K.W. & Bush, T.L. (1992) Seismic model of a prograding carbonate platform, Picco de Vallandro, the Dolomites, Northern Italy. *Am. Assoc. petrol. Geol. Bull.*, **76**, 14–30.

Blakey, R.C. & Gubitosa, R. (1984) Controls on sandstone geometry and architecture of the Chinle Formation (Upper Triassic) Colorado Plateau. *Sediment. Geol.*, 51–86.

Blum, M.D. (1990) Climatic and eustatic controls on Gulf Coastal Plain fluvial sedimentation: an example from the Late Quaternary of the Colorado River, Texas. Extended abstract in: *Sequence Stratigraphy as an Exploration Tool: Concepts and Practices in the Gulf Coast* (ed. by J.M. Armentrout & B.F. Perkins), Eleventh Annual Research Conference, Gulf Coast Section, Society of Economic Paleontologists and Mineralogists, pp. 71–83.

Bolli, H.M. & Saunders, J.B. (1985) Oligocene to Holocene low latitude planktic foraminifera. In: *Plankton Stratigraphy* (ed. by H.M. Bolli, J.B. Saunders & K. Perch-Nielsen). Cambridge Earth Science Series, pp. 155–262.

Bolli, H.M., Saunders, J.B. & Perch-Nielsen, K. (1985) *Plankton Stratigraphy*. Cambridge Earth Science Series, Cambridge University Press, Cambridge, UK, 1006 pp.

Boreen, T. & Walker, R.G.W. (1991) Definition of allomembers and their facies assemblages in the Viking Formation, Willesden Green area, Alberta. *Bull. Can. Petrol. Geol.*, **39**, 123–144.

Boselini, A. (1984) Progradation geometries of carbonate platforms: examples from the Triassic of the Dolomites, northern Italy. *Sedimentology*, **31**, 1–24.

Bosence, D.W.J. (1989) Biogenic carbonate production in Florida Bay. *Bull. mar. Sci.*, **44**, 419–433,

Bosence, D.W.J. & Waltham, D. (1990) Computer modelling the internal architecture of carbonate platforms. *Geology*, **18**, 26–30.

Bosence, D.W.J., Pomer, L. & Waltham, D. (1991) Computer modelling Late Miocene carbonate platforms, Spain. In: *Dolomieu Conference on Carbonate Platforms and Dolomitization*, Abstract Volume (ed. by A. Bosellini, R. Brandner, E. Flugel, B. Purser, W. Schlager, M.E. Tucker & B. Zenger), Ortisei, Italy, pp. 30–31.

Bosscher, H. & Southam, J. (1992) CARBPLAT – a computer model to simulate the development of carbonate platforms. *Geology*, **20**, 235–238.

Bouma, A.H. & Hollister, C.D. (1973) Deep ocean sedimentation. In: *Turbidites and Deepwater Sedimentation* (ed. by G.V. Middleton & A.H. Bouma). Society of Economic Paleontologists and Mineralogists Pacific Section Short Course, pp. 79–118.

Bouma, A.H., Coleman, J.M., Stelting, C.E. & Kohl, B. (1989) Influence of relative sea level changes on the construction of the Mississippi fan. *Geo. Mar. Lett.*, **9**, 161–170.

Bouvier, J.D., Kaars-Sijpesteijn, C.H. & Kluesner, D.F., Onyejekwe, C.C. and Van Der Pal, R.C. (1989) Three-dimensional seismic interpretation and fault sealing investigations, Nun River field, Nigeria. *Am. Assoc. Petrol. Geol. Bull.*, **73**, 1397–1414.

Bown, T.M. & Kraus, M.J. (1981) Lower Eocene alluvial paleosols (Willwood Formation), Northwest Wyoming, U.S.A. and their significance for paleoecology, paleoclimatology and basin analysis. *Palaeogeogr. Palaeoclimatol. Palaeocol.*, **34**, 1–30.

Bown, T.M. & Kraus, M.J. (1987) Integration of channel and floodplain suites, I. Developmental sequence and lateral relations of alluvial paleosols. *J. Sediment. Petrol.*, **57**, 587–601.

Boyd, R., Suter, J. & Penland, S. (1989) Relation of sequence stratigraphy to modern sedimentary environments. *Geology*, **17**, 926–929.

Braga, J.C., Martin, J.M. and Alcala, B. (1990) Coral reefs in coarse-terrigenous sedimentary environments (Upper Tortonian, Granada Basin, southern Spain). *Sediment. Geol.*, **66**, 135–150.

Brass, G.W., Southam, J.R. and Peteroson, W.H. (1982) Warm saline bottom water in the ancient ocean. *Geology*, **12**, 614–618.

Bravais, A. (1840) Sur les lignes d'ancien niveau de la mer dans le Finnmark. *Compte Rendu Academie des Sciences de Paris*, **10**, 691.

Bridge, J.S. (1985) Paleochannel patterns inferred from alluvial deposits: a critical evaluation. *J. Sediment. Petrol.*, **55**, 579–589.

Bridge, J.S. & Leeder, M.R. (1979) A simulation model of alluvial stratigraphy. *Sedimentology*, **26**, 617–644.

Bristow, C.S. (1987) Brahmaputra River: channel migration and deposition. In: *Recent Developments in Fluvial Sedimentology* (ed. by F.G. Ethridge, R.M. Flores & M.D. Harvey). Special Publication, Society of Economic Paleontologists and Mineralogists, Tulsa, **39**, 63–74.

Brown, L.F. & Fisher, W.L. (1977) Seismic stratigraphic interpretation of depositional systems: examples from Brazil rift and pull-apart basins. *Am. Assoc. petrol. Geol. Memoir*, **26**, 213–248.

Bruun, P. (1962) Sea level rise as a cause of shore erosion. *J. Waterways and Harbour Div., Am. Soc. Civil Engineers*, **88**, 117–130.

Buckland, W. (1823) *Reliquiae Diluvianae; or observations on the organic remains contained in caves, fissures and diluvial gravel, and on other geological phenomena attesting the action of a universal deluge.* John Murray, London, 303 pp.

Bukry, D. (1973) Low-latitude coccolith biostratigraphic zonation. *Initial Rep. Deep Sea Drill. Proj.*, **15**, 685–703.

Burchette, T.P. (1987) Carbonate barrier shorelines during the basal Carboniferous transgression: the Lower Limestone Shale Group, South Wales and western England. In: *European Dinantian Environments* (ed. by J. Miller, A.E. Adams & V.P. Wright) John Wiley, Chichester, pp. 239–263.

Burchette, T.P. & Britton, S.R. (1985) Carbonate facies analysis in the exploration for hydrocarbons; a case-study from the Cretaceous of the Middle East. In: *Sedimentology; Recent Developments and Applied Aspects* (ed. by P.J. Brenchley & B.P.J. Williams). Blackwell Scientific Publications, Oxford.

Burke, K. (1972) Longshore drift, submarine canyons and submarine fans in the development of Niger Delta. *Am. Assoc. Petrol. Geol. Bull.*, **56**, 1975–1983.

Burnett, A.W. & Schumm, S.A. (1983) Alluvial-river response to neotectonic deformation in Louisiana and Mississippi. *Science*, **222**, 49–50.

Busby-Spera, C. (1985) A sand-rich submarine fan in the Lower Mesozoic Mineral King Caldera Complex, Sierra Nevada, California. *J. Sediment. Petrol.*, **55**, 376–391.

Butler, P.R. (1984) Fluvial response to ongoing tectonism and base level change, Amargosa River, Death Valley, California. *Sediment. Geol.*, **38**, 107–125.

Cairncross, B., Stanistreet, L.G., McCarthy, T.S., Ellery, W.N., Ellery, K. & Grobicki, T.S.A. (1988) Palaeochannels (stone rolls) in coal seams: modern analogues from fluvial deposits of the Okavango Delta, Botswana, southern Africa. *Sediment. Geol.*, **57**, 107–118.

Calvert, S.E. (1987) Oceanographic controls on the accumulation of organic matter in marine sediments. In: *Marine Petroleum Source Rocks* (ed. by J. Brooks & A. Fleet). Special Publication, Geological Society of London, **26**, 137–151.

Calvet, F. & Tucker, M.E. (1988) Outer ramp carbonate cycles in the Upper Muschelkalk, Catalan Basin, NE Spain. *Sediment. Geol.*, **57**, 185–198.

Calvet, F., Tucker, M.E. & Henton, J.M. (1990) Middle Triassic carbonate ramp systems in the Catalan Basin, northeast Spain: facies, systems tracts, sequences and controls. In: *Carbonate Platforms: Facies Sequences and Evolution* (ed. by M.E. Tucker, J.L. Wilson, P.D. Crevello, J.R. Sarg & J.F. Read). Special Publication, International Association of Sedimentologists, **9**, 79–108. Blackwell Scientific Publications, Oxford.

Cameron, C.C., Esterle, J.S. & Palmer, C.A. (1989) The geology, botany and chemistry of selected peat-forming environments from temperate and tropical latitudes. *Int. J. Coal Geol.*, **12**, 105–156.

Campbell, A.E. (1992) Unconformities in seismic records and outcrops. PhD Thesis, Vrije Universiteit, Amsterdam.

Campbell, C.V. (1976) Reservoir geometry of a fluvial sheet

sandstone. *Am. Assoc. Petrol. Geol. Bull.*, **60**, 1009–1020.

Campbell, J.E. & Hendry, H.E. (1987) Anatomy of a gravelly meander lobe in the Saskatchewan River, near Nipawin, Canada. In: *Recent Developments in Fluvial Sedimentology* (ed. by F.G. Ethridge, R.M. Flores & M.D. Harvey). Special Publication, Society of Economic Paleontologists and Mineralogists, Tulsa, **39**, 179–189.

Cant, D.J. (1978a) Development of a facies model for sandy braided river sedimentation: comparison of the South Saskatchewan River and the Battery Point Formation. In: *Fluvial Sedimentology* (ed. by A.D. Miall). Memoir of the Canadian Society of Petroleum Geologists, Calgary, **5**, 627–639.

Cant, D.J. (1978b) Fluvial facies models. In: *Sandstone Depositional Environments* (ed. by P.A. Scholle & D. Spearing). Memoir of the American Association of Petroleum Geologists, Tulsa, **31**, 115–137.

Cant, D.J. (1984) Subsurface facies analysis. In: *Facies Models* (ed. by R.G. Walker). Geoscience, Canada.

Cant, D.J. & Walker, R.G. (1976) Development of a braided fluvial facies model for the Devonian Battery Point Sandstone, Quebec, Canada. *J. Earth Sci.*, **13**, 102–119.

Cant, D.J. & Walker, R.G. (1978) Fluvial processes and facies sequences in the sandy braided South Saskatchewan River, Canada. *Sedimentology*, **25**, 625–648.

Caron, M. (1985) Cretaceous planktonic Foraminifera. In: *Plankton Stratigraphy* (ed. by H.M. Bolli, J.B. Saunders & K. Perch-Nielsen). Cambridge Earth Science Series, Cambridge University Press, Cambridge, UK, pp. 17–86.

Carozzi, A.W. (1992) De Maillet's Telliamed (1748): the diminution of the sea or the fall portion of a complete cosmic eustatic cycle. In: *Eustasy: The Ups and Downs of a Major Geological Concept* (ed. by R.H. Dott, Jr.). *Geol. Soc. Am. Mem.*, **180**, 17–24.

Chamberlin, T.C. (1898) The ulterior basis of time divisions and the classification of geological history. *J. Geol.*, **6**, 449–462.

Chamberlin, T.C. (1909) Diastrophism as the ultimate basis of correlation. *J. Geol.*, **17**, 685–693.

Chann, M.A. & Dott, R.H., Jr. (1983) Shelf and deep sea sedimentation in Eocene forearc basin, Western Oregon — fan or nonfan? *Am. Assoc. Petrol. Geol. Bull.*, **67**, 2100–2116.

Church, M. & Slaymaker, O. (1989) Disequilibrium of Holocene sediment yield in glaciated British Columbia. *Nature*, **337**, 452–454.

Clifton, H.E., Hunter, R.E. & Phillips, R.L. (1971) Depositional structures and processes in the non-barred, high energy nearshore. *J. Sediment. Petrol.*, **41**, 651–670.

Cloetingh, S. (1988) Intraplate stress: a tectonic cause for third order cycles in apparent sea level. In: *Sea Level Changes — An Integrated Approach* (ed. by C.K. Wilgus, B.S. Hastings, C.G.C. Kendall, H.W. Posamentier, C.A. Ross & J.C. Van Wagoner). Special Publication, Society of Economic Paleontologists and Mineralogists, Tulsa, **42**, pp. 19–30.

Cloetingh, S., McQueen, H. & Lambeck, K. (1985) On a tectonic mechanism for regional sea level variations. *Earth Planet. Sci. Lett.*, **75**, 157–166.

Colella, A. (1988) Pliocene–Holocene fan deltas and braid deltas in the Crati basin, Southern Italy: a consequence of varying tectonic conditions. In: *Fan Deltas* (ed. by W. Nemec & R.J. Steel). Blackie, Glasgow, pp. 50–74.

Coleman, J.M. (1969) Brahmaputra River: channel processes and sedimentation. *Sediment. Geol.*, **13**, 129–239.

Coleman, J.M. & Prior, D.B. (1982) Deltaic environments of deposition. In: *Sandstone Depositional Environments* (ed. by P.A. Scholle & D. Spearing). Memoir of the American Association of Petroleum Geologists, Tulsa, **31**, 410.

Coleman, J.M., Prior, D.B. & Lindsay, J.F. (1983) Deltaic influences on shelf-edge instability processes. In: *The Shelf-break; Critical Interface on Continental Margins* (ed. by D.J. Stanley & G.T. Moore). Special Publication, Society of Economic Paleontologists and Mineralogists, Tulsa, **33**, 121–137.

Collinson, J.D. (1970) Bedforms of the Tana River, Norway. *Geografiska Annaler*, **52A**, 31–55.

Collinson, J.D. (1986) Alluvial sediments. In: *Sedimentary Environments and Facies*, 2nd edn (ed. by H.G. Reading). Blackwell Scientific Publications, Oxford, pp. 20–62.

Collinson, J.D. & Lewin, J. (1983) *Modern and Ancient Fluvial Systems*. Special Publication, International Association of Sedimentologists, **6**, 575 pp. Blackwell Scientific Publications, Oxford.

Cook, H.E., Field, M.E. and Gardner, J.V. (1982) Characteristics of sediments on modern and ancient continental slopes. In: *Sandstone Depositional Environments* (ed. by P.A. Scholle & D. Spearing). Memoir of the American Association of Petroleum Geologists, Tulsa, **31**, 329–364.

Corso, W., Buffler, R.T. & Austin, J.A. (1988) Erosion of the Southern Florida Escarpment. In: *Atlas of Seismic Stratigraphy* (ed. by A.W. Bally). American Association of Petroleum Geologists, Tulsa, Studies in Geology, **27**, 149–151.

Craig, D.H. (1988) Caves and other features of Permian Karst in San Andres Dolomite, Yates Field Reservoir, West Texas. In: *Paleokarst* (ed. by N.P. James & P.W. Choquette). Springer-Verlag, New York, pp. 342–363.

Craig, D.H. (1990) Yates and other Guadalupian (Kazanian) oil fields, US Permian Basin. In: *Classic Petroleum Provinces* (ed. by J. Brooks). Special Publication, Geological Society of London, **50**, 249–264.

Crawley and Fleet (1986)

Creaney, S. & Passey, Q.R. (1993) Recurring patterns of total organic carbon and source rock quality within a sequence stratigraphic framework. *Am. Assoc. Petrol. Geol. Bull.*, **77**, 386–401.

Croll, J. (1864) On the physical cause of the change of climate during geological epochs. *Philos. Mag.*, **28**, 435–436.

Cross, T.A. (1988) Controls on coal distribution in transgressive–regressive cycles, Upper Cretaceous, Western Interior, USA. In: *Sea Level Changes — An Integrated Approach* (ed. by C.K. Wilgus, B.S. Hastings, C.G. Kendall, H.W. Posamentier, C.A. Ross & J.C. Van Wagoner). Special Publication, Society of Economic Paleontologists and Mineralogists, Tulsa, **42**, 371–380.

Cross, T.A. (1990) *Quantitative Dynamic Stratigraphy*. Prentice Hall, Princeton, New Jersey.

Curiale, J.A., Cole, R.D. & Witmer, R.J. (1991) Application of organic geochemistry to sequence stratigraphic analysis: Four Corners area, New Mexico, U.S.A. *Org. Geochem.*, **19**, 53–75.

Dalrymple, R.W., Knight, R.J., Zaitlin, B.A. & Middleton, G.V. (1990) Dynamics and facies of a macrotidal sand-bar complex, Cobequid Bay — Salmon River Estuary (Bay of Fundy). *Sedimentology*, **37**, 577–612.

Dalrymple, R.W., Zaitlan, B.A. & Boyd, R. (1992) Estuarine facies models: conceptual basics and stratigraphic implications.

Damuth, J.R. (1977) Late Quaternary sedimentation in the western equatorial Atlantic. *Am. Assoc. Petrol. Geol. Bull.*, **88**, 695–710.

Damuth, J.R. (1980) Use of high frequency (3.5–12 kHz) echograms in the study of bottom sedimentation processes in the deep sea: a review. *Mar. Geol.*, **38**, 51–75.

Damuth, J.R. & Kumar, N. (1975) Amazon Cone: morphology, sediments, age and growth pattern. *Bull. Geol. Soc. Am.*, **86**, 863–878.

Damuth, J.R., Kolla, V., Flood, R.D., Kowsmann, R.O., Monteiro, M.C., Gorini, M.A., Palma, J.J.C. & Belderson, R.H. (1983) Distributary channel meandering and bifurcation patterns on the Amazon deep sea fan as revealed by long range side-scan sonar (GLORIA). *Geology*, **11**, 94–98.

Damuth, J.R., Flood, R.G., Kowsmann, R.O., Belderson, R.H. & Gorini, M. (1988) Anatomy and growth pattern of Amazon Deep-Sea Fan as revealed by long range side-scan sonar (Gloria) and high resolution seismic studies. *Am. Assoc. Petrol. Geol. Bull.*, **72**, 885–911.

Davidson-Arnott, R.G.D. & Greenwood, B. (1976) Facies relationships on a barred coast, Kouchibouguac bay, New Brunswick, Canada. In: *Beach and Nearshore Sedimentation* (ed. by R.A. Davis Jr. & R.L. Ethington). Special Publication, Society of Economic Paleontologists and Mineralogists, Tulsa, OK, **24**, 149–168.

Davies, P.J., Symonds, P.A., Feary, D.A. & Pigram, C.J. (1989) The evolution of the carbonate platforms of northeast Australia. In: *Controls on Carbonate Platform and Basin Development* (ed. by P.D. Crevello, J.L. Wilson, J.F. Sarg & J.F. Read). Special Publication, Society of Economic Paleontologists and Mineralogists, Tulsa, **44**, 233–258.

De Graciansky, P.C., Dardeau, G., Dumont, T., Jacquin, T., Marchand, D., Mouterde, R. & Vail, P. (1993) Depositional sequence cycles, transgressive–regressive facies cycles and extensional tectonics: example from the southern subalpine Jurassic Basin, France. *Bull. Soc. Geol. Fr.*, **164**, P709–718.

De Maillet, B. (1748) Telliamed: ou entretiens d'un philosophe indien avec un missionaire francais sur la diminution de la mer, la formation de la terre, l'origine de l'homme & Mis en ordere sur les memoires de feu M. de Maillet par J.A.G. Amsterdam, Chez L'Honore & Fils, 201 pp.

De'Ath, N.G. & Schuyleman, S.F. (1981) The geology of the Magnus Oilfield. In: *Petroleum Geology of the Continental Shelf of North-west Europe* (ed. by L.V. Illing & G.D. Hobson). Heyden, London, pp. 342–351.

Deegan, C.E. & Scull, B.J. (1977) A standard lithostratigraphic nomenclature for the central and northern North Sea. *Inst. geol. Sci. Rept.*, 77/25; *Bull. Norw. Pet. Dir. Bull.*, 1.

Defant, A. (1961) *Physical Oceanography*. Pergamon Press, Oxford, 598 pp.

Demaison, G. & Moore, G.T. (1980) Anoxic environments of oil source bed genesis. *Am. Assoc. Petrol. Geol. Bull.*, **64**, 1979–2109.

Den Hartog Jager, D., Giles, M.R. & Griffiths, G.R. (1993) Evolution of Palaeogene submarine fans of the North Sea in space and time. In: *Petroleum Geology of Northwest Europe: Proceedings of the 4th Conference* (ed. by J.R. Parker). Geological Society of London, Bath, pp. 59–72.

Denison, C.N. & Fowler, R. (1980) *Palynological Identification of Facies in a Deltaic Environment*. Norwegian Petroleum Society, 22 pp.

Devine (1991) Transgressive origin of channelled estuarine deposits in the Point Lookout Sandstone, northwestern New Mexico, a model for Upper Cretaceous, cyclic regressive parasequences of the U.S. Western Interior. *Am. Assoc. Petrol. Geol. Bull.*, **75**, 1039–1063.

Dingus, W.F. & Galloway, W.E. (1990) Morphology, paleogeographic setting and origin of the Middle Wilcox Yoakum Canyon, Texas coastal plain. *Am. Assoc. Petrol. Geol. Bull.*, **74**, 1055–1076.

Dixon, J. (1979) The Lower Cretaceous Atkinson Point Formation on the Tuktoyaktuk Peninsula, NWT: a coastal fan-delta to marine sequence. *Bull. Can. Petrol. Geol.*, **27**, 163–182.

Dixon, R.J., Schofield, K., Anderton, R., Reynolds, A.D., Alexander, R.W.S., Williams, M.C. & Davies, K.G. (1993) Sandstone diapirism and clastic intrusion in the Tertiary fans of the Bruce-Beryl Embayment, Quadrant 9, UKCS. In: *Proceedings of the Conference of Reservoir Characterization of Deep Marine Clastic Systems*.

Dolson, J., Muller, D., Evetts, M.J. & Stein, J.A. (1991) Regional paleotopographic trends and production, Muddy Sandstone (Lower Cretaceous), Central and Northern Rocky Mountains. *Am. Assoc. Petrol. Geol. Bull.*, **75**, 409–435.

Dott, R.H., Jr. (1992a) Introduction to the ups and downs of eustasy. In: *Eustasy: The Ups and Downs of a Major Geological Concept* (ed. R.H. Dott, Jr.). *Geol. Soc. Am. Mem.*, **180**, 1–16.

Dott, R.H., Jr. (1992b) T.C. Chamberlin's hypothesis of diastrophic control of worldwide changes of sea level: a precursor of sequence stratigraphy. In: *Eustasy: The Ups and Downs of a Major Geological Concept* (ed. R.H. Dott, Jr.). *Geol. Soc. Am. Mem.*, **180**, 31–41.

Driscoll, N.W., Weissel, J.K., Karner, G.D. & Mountain, G.S. (1991) Stratigraphic response of a carbonate platform to relative sea level changes: Broken Ridge, southeast Indian Ocean. *Bull. Am. Petrol. Geol.*, **75**, 808–831.

Droste, M. (1990) Depositional cycles and source rock development in an epeiric intra-platform basin; the Hanifa Formation of the Arabian Peninsula. *Sediment. Geol.*, **69**, 281–296.

Droxler, A.W. & Schlager, W. (1985) Glacial versus interglacial sedimentation rates and turbidite frequency in the Bahamas. *Geology*, **13**, 799–802.

Droxler, A.W., Schlager, W. & Whallon, C.C. (1983) Quaternary aragonite cycles and oxygen-isotope record in Bahamian carbonate ooze. *Geology*, **11**, 235–239.

Droz, L. (1983) L'eventail sous marin profond du Rhone (Golfe de Lion): graind traits morphologiues et structure semi-profonde. Theses 3e Cycle, Paris, 195 pp.

Droz, L. & Bellaiche, G. (1985) Rhone Deep Sea Fan: morphostructure and growth pattern. *Am. Assoc. petrol. Geol. Bull.*, **69**, 460–479.

Dutton, S.P. (1982) Pennsylvanian fan-delta and carbonate deposition. Moobeetie field, Texas Panhandle. *AAPG Bull.*, **66**, 389–407.

Duval, B., Cramez, C. & Vail, P.R. (1992) Types and hierarchy of stratigraphic cycles. In: *Mesozoic and Cenozoic Sequence Stratigraphy of European Basins International Symposium (Dijon France)* (ed. by Centre Nat. Rech. Sci. et al.). Abstract, pp. 44–45.

Ebanks, W.J., Jr. (1987) Flow unit concept — integrated approach to reservoir description for engineering projects. *Am. Assoc. Petrol. Geol. Bull.*, **71**, 551–552.

Eberli, G.P. & Ginsburg, R.N. (1987) Segmentation and coalescence of Cenozoic carbonate platforms, northwestern Great Bahama Bank. *Geology*, **15**, 75–79.

Edwards, M.B. (1981) Upper Wilcox Rosita delta system of south Texas: growth-faulted shelf-edge deltas. *Am. Assoc. Petrol Geol. Bull.*, **65**, 54–73.

Elliott, T. (1986a) Deltas. In: *Sedimentary Environments and Facies*, 2nd edn (ed. by H.G. Reading). Blackwell Scientific Publications, Oxford, pp. 113–154.

Elliott, T. (1986b) Siliciclastic shorelines. In: *Sedimentary Environments and Facies*, 2nd edn (ed. by H.G. Reading). Blackwell Scientific Publications, Oxford, pp. 155–188.

Elliott, T. (1989) Deltaic systems and their contribution to an understanding of basin-fill successions. In: Deltas: Sites and Traps for Fossil Fuels (ed. by M.K. Whateley & Pickering K.T.) *Geol. Soc. Spec. Publ.*, **41**, 3–10.

Emery, D. and Dickson, J.A.D. (1991) The subsurface correlation of the Lincolnshire Limestone Formation in Lincolnshire. *Proc. Geol. Assoc.*, **102**, 109–122.

Emery, D. & Robinson, A.G. (1993) *Inorganic Geochemistry: Applications to Petroleum Geology*. Blackwell Scientific Publications.

Enachescu, M.E. (1993) Amplitude interpretation of 3-D reflection data. *The Leading Edge*, **12**, 678–685.

Epting, M. (1980) Sedimentology of Miocene carbonate buildups, Central Luconia, offshore Sarawak. *Bull. Geol. Soc. Malaysia*, **12**, 17–30.

Epting, M. (1989) Miocene carbonate buildups of central Luconia, offshore Sarawak. In: *Atlas of Seismic Stratigraphy Vol. 3* (ed. by A.W. Bally). American Association of Petroleum Geologists, Tulsa, Studies in Geology **27**, 168–173.

Erlich, R.N., Barrett, S.F. & Guo Bai Ju (1990) Seismic and geological characteristics of drowning events on carbonate platforms. *Am. Assoc. Petrol. Geol. Bull.*, **74**, 1523–1537.

Eschard, R. (1989) *Geometry and dynamics of depositional sequences in a deltaic system, Middle Jurassic, Cleveland Basin, UK*. Unpubl. PhD Thesis, University of Strasbourg.

Eschard, R., Ravenne, C., Houel, P. & Knox, R. (1991) Three dimensional reservoir architecture of a valley fill sequence and a deltaic aggradational sequence: influences of minor relative sea-level variations (Scalby Formation, England). In: *The Three-Dimensional Facies Architecture of Terrigenous Clastic Sediments and its Implications for Hydrocarbon Discovery and Recovery* (ed. by A.D. Miall & N. Tyler). Society of Economic Paleontologists and Mineralogists (Society for Sedimentary Geology), Tulsa, OK, Concepts in Sedimentology and Paleontology, **3**, 133–147.

Ethridge, F.G. & Flores, R.M. (1981) *Recent and Ancient Non-Marine Depositional Environments: Models for Exploration*. Special Publication, Society of Economic Paleontologists and Mineralogists, Tulsa, **31**, 349 pp.

Ethridge, F.G. & Westcott, W.A. (1984) Tectonic setting, recognition and hydrocarbon reservoir potential of fan-delta deposits. In: *Sedimentology of Gravels and Conglomerates* (ed. by E.H. Koster & R.J. Steel). Canadian Society of Petroleum Geologists, Calgary, Memoir **10**, 217–235.

Ethridge, F.G., Flores, R.M. & Harvey, M.D. (1987) *Recent Developments in Fluvial Sedimentology*. Special Publication, Society of Economic Paleontologists and Mineralogists, Tulsa, **39**, 389 pp.

Ewing, M., Ericson, D.B. & Heezen, B.C. (1958) Sediment and topography of the Gulf of Mexico. In: *Habitat of Oil* (ed. by L.G. Weeks). American Association of Petroleum Geologists, Tulsa, 995–1053.

Eyles, C.H. & Walker, R.G. (1988) 'Geometry' and facies characteristics of stacked shallow marine sandier-upwards sequences in the Cardium Formation at Willesden Green, Alberta. In: *Sequences, Stratigraphy, Sedimentology: Surface and Subsurface* (ed. by D.P. James & D.A. Leckie). Canadian Society of Petroleum Geologists, Calgary, Memoir **15**, 85–96.

Fairbanks, R.G. (1989) A 17 000-year glacio-eustatic sea level record: influence of glacial melting rates on the younger Dryas event and deep-ocean circulation. *Nature*, **342**, 637–642.

Fenner, J. (1985) Late Cretaceous to Oligocene planktic diatoms. In: *Plankton Stratigraphy* (ed. by H.M. Bolli, J.B. Saunders & K. Perch-Nielsen). Cambridge Earth Science Series, Cambridge University Press, Cambridge, UK, pp. 713–762.

Ferguson, R.J. (1987) Hydraulic and sedimentary controls on channel patterns. In: *Rivers: Environment, Form and Process* (ed. by K.S. Richards). Blackwell, Oxford, pp. 129–158.

Ferrentinos, G.G., Papatheodorou, G. & Collins, M.B. (1988) Submarine transport processes on an active submarine fault scarp: Gulf of Corinth, Greece. *Mar. Geol.*, **83**, 43–61.

Fischer, A.G. (1964) The Lofer cyclothems of the Alpine Triassic. In: *Symposium on Cyclic Sedimentation* (ed. by D.F. Merriam). *Bull. Geol. Surv. Kansas*, **169**, 107–149.

Fischer, A.G. (1961) Stratigraphic record of transgressing seas in the light of sedimentation on the Atlantic coast of New Jersey. *Am. Assoc. Petrol. Geol. Bull.*, **45**, 1656–1666.

Fisher, W.L. & McGowen, J.H. (1967) Depositional systems in the Wilcox Group of Texas and their relationship to the occurrence of oil and gas. *Gulf Coast Assoc. Geol. Soc. Trans.*, **17**, 105–125.

Fitzsimmons, R. (1994) Identification of high order sequence boundaries and land attached shoreface regressions. In *High Resolution Sequence Stratigraphy: Innovations and Applications*, Abstract Volume, University of Liverpool, pp. 332–333.

Fleet, A.J., Clayton, C.J., Jenkyns, H. & Parkinson, D.N. (1987) Liassic source-rock deposition in Western Europe. In: *Petroleum Geology of North West Europe* (ed. by J. Brooks & K.W. Glennie). Graham & Trotman, London, pp. 59–70.

Flood, R.C., Manley, P.L., Kowsmann, R.O., Appi, C.J. & Pirmez, C. (1991) Seismic facies and late Quaternary growth of the Amazon submarine fan. In: *Seismic Facies and Sedimentary Processes of Modern and Ancient Submarine Fans and Turbidite Systems* (ed. by P. Weimer & M.H. Link) Springer-Verlag, New York, pp. 415–434.

Folk, R.L. (1959) Practical petrographic classification of limestones. *Am. Assoc. Petrol. Geol. Bull.*, **43**, 1–38.

Forbes, D.L. (1983) Morphology and sedimentology of a sinuous gravel-bed channel system, lower Babbage River, Yukon coastal, Canada. In: *Modern and Ancient Fluvial Systems* (ed. by J.D. Collinson & J. Lewin). Special Publication, International Association of Sedimentologists, **6**, 165–206. Blackwell Scientific Publications, Oxford.

Frasier, D.E. & Osanik, A. (1967) Recent peat deposits – Louisiana coastal plain. In: *Environments of Coal Deposition* (ed. by E.C. Dapples & M.E. Hopkins). Geological Society of America, Boulder, Colorado Special Paper **114**, 85 pp.

Frazier, D.E. (1974) Depositional episodes: their relationship to the Quaternary stratigraphic framework in the northwestern portion of the Gulf Basin. University of Texas at Austin, Bureau of Economic Geology. *Geol. Circ.* **4**(1), 28 pp.

Friedman, G.M. & Sanders, J.E. (1978) *Principles of Sedimentology*. John Wiley and Sons, New York, pp. 792.

Friend, P.F. (1983) Towards the field classification of alluvial architecture or sequence. In: *Modern and Ancient Fluvial Systems* (ed. by J.D. Collinson & J. Lewin). Special Publication, International Association of Sedimentologists, 6, 345–354. Blackwell Scientific Publications, Oxford.

Friend, P.F., Slate, M.J. & Williams, R.C. (1978) Vertical and lateral building of river, sandstones bodies, Ebro Basin, Spain. *J. Geol. Soc. Lond.*, **136**, 39–46.

Galloway, W.E. (1975) Process framework for describing the morphologic and stratigraphic evolution of deltaic depositional systems. In: *Deltas* (ed. by M.L. Broussard). Houston Geological Society, pp. 87–98.

Galloway, W.E. (1981) Depositional architecture of Cenozoic Gulf Coast Plain fluvial systems. In: *Recent and Ancient Nonmarine Depositional Environments* (ed. by F.G. Ethridge & R.M. Flores). Special Publication, Society of Economic Paleontologists and Mineralogists, Tulsa, **31**, 127–155.

Galloway, W.E. (1989) Genetic stratigraphic sequences in basin analysis: architecture and genesis of flooding surface bounded depositional units. *Am. Assoc. Petrol. Geol. Bull.*, **73**, 125–142.

Galloway, W.E. & Brown, L.F. Jr (1972) *Depositional Systems and Shelf-slope Relationships in Upper Pennsylvanian Rocks, North Central Texas*. Texas University Bureau of Economic Geology Report. No. 75–62 pp.

Galloway, W.E. & Hobday (1983) *Terrigenous Clastic Depositional Systems*. Springer-Verlag, New York.

Garcia-Gil, S. (1993) The fluvial architecture of the upper Buntsandstein in the Iberian Basin, central Spain. *Sedimentology*, **36**, 125–143.

Garland, C.R. (1993) Miller Field: reservoir stratigraphy and its impact on development. In: *Geology of New Europe, Proceedings of the 4th Conference* (ed. by J.D. Parker). Geological Society of London, Bath, 401–414.

Gaskell, B.A. (1991) Extinction patterns in Paleogene benthic foraminiferal faunas: relationship to climate and sea level. *Palaios*, **6**, 2–16.

Germanoski, D. & Schumm, S.A. (1993) Changes in braided river morphology resulting from aggradation and degradation. *J. Geol.*, **101**, 451–466.

Gilluly, J. (1949) Distribution of mountain building in geologic time. *Geol. Soc. Am. Bull.*, **60**, 561–590.

Ginsburg, R.N. & James, N.P. (1974) Holocene carbonate sediments of continental margins. In: *The Geology of Continental Margins* (ed. by C.A. Burk & C.L. Drake). Springer-Verlag, Berlin, pp. 137–155.

Ginsburg, R.N., McNeill, D.F., Eberli, G.P., Swart, P.K. & Kenter, J.A. (1991) Transformation of morphology and facies of Great Bahama Bank by Plio-Pleistocene progradation. In: *Dolomieu Conference on Carbonate Platforms and Dolomitization*, Abstract Volume (ed. by A. Bosellini, R. Brandner, E. Flugel, B. Purser, W. Schlager, M.E. Tucker & B. Zenger). Ortisei, Italy, pp. 88–89.

Glaser, K.S. & Droxler, A. (1991) High production and highstand shedding from deeply submerged carbonate banks, northern Nicaragua Rise. *J. Sediment. Petrol.*, **61**, 128–142.

Gloppen, T.G. & Steel, R.J. (1981) The deposits, internal structure and geometry in six alluvial fan–fan delta bodies (Devonian-Norway) – a study in the significance of bedding sequence in conglomerates. In: *Recent and Ancient Nonmarine Depositional Environments* (ed. by F.G. Ethridge & R.M. Flores). Special Publication, Society of Economic Paleontologists and Mineralogists, Tulsa, **31**, 49–69.

Goldhammer, R.K., Dunn, P.A. & Hardie, L.A. (1987) High frequency glacio-eustatic sea level oscillations with Milankovitch characteristics recorded in Middle Triassic cyclic platform carbonates, northern Italy. *Am. J. Sci.*, **287**, 853–892.

Goldhammer, R.K., Oswald, E.J. & Dunn, P.A. (1989) The hierarchy of stratigraphic forcing – an example from Middle Pennsylvanian shelf carbonates of the southwestern Paradox Basin, Honaker Trail, Utah. In: *Sedimentary Modeling: Computer Simulation of Depositional Sequences* (ed. by E.K. Franseen & W.L. Watney). Subsurface Geology Series, 12, Kansas Geological Survey, pp. 27–30.

Goldhammer, R.K., Dunn, P.A. & Hardie, L.A. (1990) Depositional cycles, composite sea level changes, cycle stacking patterns, and the hierarchy of stratigraphic forcing: examples from platform carbonates of the Alpine Triassic. *Geol. Soc. Am. Bull.*, **102**, 535–562.

Goldhammer, R.K., Oswald, E.J. & Dunn, P.A. (1991) High frequency glacio-eustatic cyclicity in the Middle Pennsylvanian of the Paradox Basin: an evaluation of Milankovitch forcing. In: *Dolomieu Conference on Carbonate Platforms and Dolomitization*, Abstract Volume (ed. by A. Bosellini, R. Brandner, E. Flugel, B. Purser, W. Schlager, M.E. Tucker & B. Zenger). Ortisei, Italy, p. 91.

Goldhammer, R.K., Lehmann, P.J. & Dunn, P.A. (1993) The origin of high frequency platform carbonate cycles and third-order sequences (Lower Ordovician El Paso Gp., West Texas): constraints from outcrop data and stratigraphic modeling. *J. Sediment. Petrol.*, **63**, 318–359.

Gorsline, D.S. & Emery, K.O. (1959) Turbidity current deposits in San Pedro and Santa Monica basins of southern California. *Geol. Soc. Am. Bull.*, **70**, 279–290.

Gould, H.R. (1970) The Mississippi delta complex. In: *Deltaic Sedimentation: Modern and Ancient* (ed. by J.P. Morgan). Special Publication, Society of Economic Paleontologists and Mineralogists, Tulsa, **15**, 3–30.

Gowland, S. & Riding, J.B. (1991) Stratigraphy, sedimentology and palaeontology of the Scarborough Formation (Middle Jurassic) at Hundale Point, North Yorkshire. *Proc. Yorks. Geol. Soc.*, **48**, 375–392.

Grabau, A.W. (1940) *The Rhythm of the Ages*. Henri Vetch, Peking, 561 pp.

Grainge, A.M. & Davies, K.G. (1985) Reef exploration in the East Sengkang Basin, Sulawesi, Indonesia. *Mar. Petrol. Geol.*, **2**, 142–155.

Grammer, G.M. (1991) Formation and evolution of Quaternary carbonate foreslopes, Tongue of the Ocean, Bahamas. PhD Thesis, University of Miami.

Grammer, G.M., Ginsburg, R.N. & McNeill, D.F. (1990) Morphology and development of modern carbonate foreslopes, Tongue of the Ocean, Bahamas. In: *Transactions 12. Caribbean Geological Conference, St Croix, Miami* (ed. by D.K. Larue & G. Draper). Miami Geological Society, Miami pp. 27–32.

Grammer, G.M. & Ginsburg, R.N. (1992) Highstand vs. lowstand deposition on carbonate platform margins: insight from Quaternary foreslopes in the Bahamas. *Mar. Geol.*, **103**, 125–136.

Gray, D.I. (1981) Lower Carboniferous shelf carbonate environments in North Wales. PhD Thesis, University of Newcastle upon Tyne.

Griffiths, C.M. & Nordlund, U. (1993) Chronosomes and Quantitative Stratigraphy. *Geoinformatics* (Japanese Society of Geoinformatics), 4(3).

Griffiths, C.M. & Hadler-Jacobsen, F. (1993) Practical dynamic modelling of sequence stratigraphy. Presented at the Norwegian Petroleum Society Conference on *Sequence Stratigraphy: Advances and Applications for Exploration and Production in North West Europe*, Stavanger, 1–3 February 1993.

Grotzinger, J.P. (1986) Cyclicity and paleoenvironmental dynamics, Rocknest platform, northwest Canada. *Geol. Soc. Am. Bull.*, 97, 1208–1231.

Guardado, L.R., Camboa, L.A.P. & Lucchesi, C.F. (1989) Petroleum geology of the Campos Basin, Brazil, a model for producing an Atlantic type basin. In: *Divergent/Passive Margins* (ed. J.D. Edwards & P.A. Santagrossi). Memoir of the American Association of Petroleum Geologists, Tulsa, 48, 3–79.

Gustavson, T.C. (1978) Bedforms and stratification types of modern gravel meander lobes, Nueces River, Texas. *Sedimentology*, 25, 401–426.

Hallam, A. & Bradshaw, M.J. (1979) Bituminous shales and oolitic ironstones as indicators of transgression and regression. *J. Geol. Soc. Lond.*, 136, 157–164.

Hamblin, A.P. & Walker, R.G. (1979) Storm-dominated shallow marine deposits — the Fernie–Kootenay (Jurassic) transition, southern Rocky Mountains. *Can. J. Earth Sci.*, 16, 1673–1690.

Hamilton, D.S. & Tadros, N.Z. (1994) Utility of coals as genetic stratigraphic sequence boundaries in nonmarine basins: an example from the Gunnedah Basin Australia. *Am. Assoc. petrol. Geol. Bull.*, 78, 267–286.

Hancock, N.J. & Fisher, M.J. (1981) Middle Jurassic North Sea deltas with particular reference to Yorkshire. In: *Petroleum Geology of the Continental Shelf of Northwest Europe* (ed. by G.V. Illing & G.D. Hobson). Heyden, London, pp. 186–195.

Handford, C.R. (1982) Sedimentology and evaporite genesis in a Holocene continental sabkha playa basin — Bristol Dry Lake, California. *Sedimentology*, 29, 239–253.

Haq, B.U., Hardenbol, J. & Vail, P.R. (1987) Chronology of fluctuating sea-levels since the Triassic. *Science*, 235, 1153–1165.

Haq, B.U., Hardenbol, J. & Vail, P.R. (1988) Mesozoic and Cenozoic chronostratigraphy and cycles of sea-level change. In: *Sea-level Changes: an Integrated Approach* (ed. by C.K. Wilgus, B.S. Hastings, C.G. St Kendall, H.W. Posamentier, C.A. Ross & J.C. Van Wagoner). Special Publication, Society of Economic Paleontologists and Mineralogists, Tulsa, 42, 40–45.

Harbaugh, J.W. & Bonham-Carter, G. (1970) *Computer Simulation in Geology*. John Wiley & Sons, New York, 575 pp.

Hardie, L.A., Bosellini, A. & Goldhammer, R.K. (1986) Repeated subaerial exposure of subtidal carbonate platforms, Triassic, northern Italy: evidence for high frequency sea level oscillations on a 10^4 year scale. *Paleoceanography*, 1, 447–457.

Harms, J.C., Tackenberg, P., Pickles, E. & Pollock, R.E. (1981) The Brae-Oilfield area. In: *Petroleum Geology of the Continental Shelf of North-west Europe* (ed. by L.V. Illing & G.D. Hobson). Heyden, London, pp. 352–357.

Hart B.S. & Plint A.G. (1989) Gravelly shoreface deposits; a comparison of modern and ancient facies sequences. *Sedimentology*, 36, 551–557.

Hazeldine, R.S. (1983) Fluvial bars reconstructed from a deep straight channel, Upper Carboniferous coalfield of northeast England. *J. Sediment Petrol.*, 53, 1233–1248.

Heezen, B.C., Menzies, R.J., Schneider, E.D., Ewing, W.M. & Granelli, C.L. (1964) Congo Submarine Canyon. *Am. Assoc. Petrol. Geol. Bull.*, 48, 1126–1149.

Heezen, B.C., Hollister, C.D. & Ruddiman, W.F. (1966) Shaping of the continental rise by geostrophic countour currents. *Science*, 152, 502–508.

Heller, P.L. & Dickinson, W.R. (1985) Submarine ramp facies model for delta-fed sand-rich turbidite systems. *Am. Assoc. Petrol. Geol. Bull.*, 69, 960–976.

Heretier, F.E., Lossel, P. & Wathne, E. (1979) Frigg Field — large submarine fan trap in the Lower Eocene rocks of the North Sea Viking Graben. *Am. Assoc. Petrol. Geol. Bull.*, 63, 1999–2020.

Heward, A.P. (1978) Alluvial fan sequence and megasequence models: with examples from the Westphalian D-Stephanian Coalfields, northern Spain. In: *Fluvial Sedimentology* (ed. by A.D. Miall). Memoir of the Canadian Society of Petroleum Geologists, Calgary, 5, 669–702.

Hill, P.R. (1984) Sedimentary facies of the Nova Scotian upper and middle continental slope, offshore eastern Canada. *Sedimentology*, 31, 293–309.

Hine, A.C., Wilber, R.J. & Neumann, A.C. (1981) Carbonate sand-bodies along contrasting shallow-bank margins facing open seaways, northern Bahamas. *Am. Assoc. Petrol. Geol. Bull.*, 65, 261–290.

Howarth, M.J. (1982) Tidal currents of the continental shelf. In: *Offshore Tidal Sands* (ed. by A.H. Stride). London, Chapman & Hall, pp. 10–26.

Hubbard, R.J. (1988) Age and significance of sequence boundaries on Jurassic and Early Cretaceous rifted continental margins. *Am. Assoc. Petrol. Geol. Bull.*, 72, 49–72.

Hubbard, R.J., Pape, J. & Roberts, D.G. (1985) Depositional sequence mapping as a technique to establish tectonic and stratigraphic framework and evaluate hydrocarbon potential on a passive continental margin. In: *Seismic Stratigraphy II* (ed. by O.R. Berg & D. Woolverton). Memoir of the American Association of Petroleum Geologists, Tulsa, 39, 79–91.

Hubbard, R.J., Edrich, S.P. & Rattey, R.P. (1987) Geologic evolution and hydrocarbon habitat of the 'Arctic Alaska Microplate'. *Mar. Petrol. Geol.*, 4, 2–34.

Hunt, D. & Tucker, M.E. (1991) Responses of rimmed shelves to relative sea level rises; a proposed sequence stratigraphic classification. In: *Dolomieu Conference on Carbonate Platforms and Dolomitization*, Abstract Volume (ed. by A. Bosellini, R. Brandner, E. Flugel, B. Purser, W. Schlager, M.E. Tucker & B. Zenger). Ortisei, Italy, pp. 114–115.

Hunt, D. & Tucker, M.E. (1992) Stranded parasequences and the forced regressive wedge systems tract: deposition during base-level fall. *Sediment. Geol.*, 81, 1–9.

Hunter, R.E., Clifton, H.E. & Phillips, R.L. (1979) Depositional processes, sedimentary structures, and predicted vertical sequences in barred nearshore systems, southern Oregon coast. *J. Sediment. Petrol.*, 49, 711–726.

Hutton, J. (1788) Theory of the Earth. *Proc. Royal Soc. Edinburg*.

Imperato, D.G. & Nilsen, T.H. (1990) Deep-sea-fan channel-levee complexes, Arbuckle Field, Sacramento Basin, California. In: *Sandstone Petroleum Reservoirs* (ed. by J.H. Barwis & J.H.R. Studlick). Springer-Verlag, New York, pp. 535–555.

Ineson, J.R. (1989) Coarse-grained submarine fan and slope apron deposits in a Cretaceous back-arc basin, Antarctica. *Sedimentology*, 36, 793–820.

Jackson, R.G., II (1976) Depositional model of a point bar in the Lower Wabash River. *J. Sediment Petrol.*, 46, 579–594.

Jackson, R.G., II (1978) Preliminary evaluation of lithofacies models for meandering alluvial streams. In: *Fluvial Sedimentology* (ed. by A.D. Miall). Memoir of the Canadian Society of Petroleum Geologists, Calgary, 5, 543–576.

Jacquin, T., Arnaud-Vanneau, A. & Arnaud, H. (1991) Systems tracts and depositional sequences in a carbonate setting: a study of continuous outcrops from platform to basin setting at the scale of seismic lines. *Mar. Petrol. Geol.*, 8, 122–139.

James, N.P. (1983) Reefs. In: *Carbonate Depositional Environments* (ed. by P.A. Scholle, D.G. Bebout & C.H. Moore). Memoir of the American Association of Petroleum Geologists, Tulsa, 33, 345–462.

James, N.P. (1984) Shallowing-upward sequences in carbonates. In: *Facies Models* (ed. by R.G. Walker). Geoscience Canada, pp. 213–228.

James, N.P. & Ginsburg, R.N. (1979) The seaward margin of Belize barrier and atoll reefs. *Spec. Publ. Int. Ass. Sedimentol.*, 3, 191. Blackwell Scientific Publications, Oxford.

James, N.P. & Choquette, P.W. (1988) *Paleokarst*. Springer-Verlag, New York, 146 pp.

Jamieson, H.C., Brocket, L.D. & McIntosh, R.A. (1980) Prudhoe Bay – a 10 year perspective. In: *Giant Oil and Gas Fields of the Decade* (ed. by M.T. Halbouty). Memoir of the American Association of Petroleum Geologists, Tulsa, 30, 289–310.

Jenkins, D.G. (1985) Southern mid-latitude Paleocene to Holocene planktonic Foraminifera. In: *Plankton Stratigraphy* (ed by H.M. Bolli, J.B. Saunders & K. Perch-Nielsen). Cambridge Earth Science Series, Cambridge University Press, Cambridge, UK pp. 263–282.

Jenkyns, H.C. (1980) Cretaceous anoxic events from continents to oceans. *J. Geol. Soc. Lond.*, 1387, 171–188.

Jennette, D.D., Jones, C.R., Van Wagoner, J.C. & Larsen, J.E. (1992) High resolution sequence stratigraphy of the Upper Cretaceous Tocito sandstone: the relationship between incised valleys and hydrocarbon accumulation, San Juan basin, New Mexico. In: *Sequence Stratigraphy. Applications to Shelf Sandstone Reservoirs* (ed. by J.C. Van Wagoner, C.R. Jones, D.R. Taylor, D. Nummedal, D.C. Jennette & G.W. Riley). American Association of Petroleum Geologists Field Conference, September 1991, pp. 21–28.

Jervey, M.T. (1988) Quantitative geological modelling of siliciclastic rock sequences and their seismic expressions. In: *Sea Level Changes: an Integrated Approach* (ed. by C.K. Wilgus, B.S. Hastings, C.G. St Kendall, H.W. Posamentier, C.A. Ross & J.C. Van Wagoner). Special Publication, Society of Economic Paleontologists and Mineralogists, Tulsa, 42, 47–69.

Jev, B.I., Kaars-Spijpesteijn, C.H., Peters, M.P.A.M., Watts, N.L. & Wilkie, J.T. (1993) Akaso field, Nigeria: use of integrated 3-D seismic, fault slicing, clay smearing, and RFT pressure data on fault trapping and dynamic leakage. *Am. Assoc. Petrol. Geol. Bull.*, 77, 1389–1404.

Johnson, G.D. (1977) Paleopedology of *Ramapoithecus* bearing sediments. *Geol. Rundsch.*, 66, 192–216.

Johnson, H.D. & Baldwin, C.T. (1986) Shallow siliciclastic seas. In: *Sedimentary Environments and Facies*, 2nd edn (ed. by H.G. Reading). Blackwell Scientific Publications, Oxford, pp. 229–282.

Johnson, H.D. & Stewart, D.J. (1985) Role of clastic sedimentology in the exploration and production of oil and gas in the North Sea. In: *Sedimentology, Recent Developments and Applied Aspects* (ed. by P.J. Brenchley & B.P.J. Williams). Special Publication, Geological Society of London, 18, 249–310.

Johnson, M.A. & Belderson, R.H. (1969) The tidal origin of some vertical sedimentary changes in Epicontinental seas. *J. Geol.*, 77, 353–357.

Johnson, N.M., Stix, J., Tauxe, L., Cerveny, P.F. & Tahirkheki, R.A.K. (1985) Palaeomagnetic chronology, fluvial processes and implications of the Siwalik deposits near Chinji Village, Pakistan. *J. Geol.*, 93, 27–40.

Johnson, N.M., Sheikh, K.A., Dawson-Saunders, E. & McCrae, L.E. (1988) The use of magnetic reversal time lines in stratigraphic analysis; a case study in measuring variability in sedimentation rates. In: *New Perspectives in Basin Analysis* (ed. by K. Kleinspehn & C. Paola). Springer-Verlag, New York, pp. 189–200.

Jolley, E.J., Turner, P., Williams, G.D., Hartley, A.D. & Flint, S.D. (1990) Sedimentological response of an alluvial system to Neogene thrust tectonics. Atacama Desert, northern Chile. *J. Geol. Soc. Lond.*, 147, 769–784.

Jones, H.P. & Speers, R.G. (1976) Permo-Triassic reservoirs of Prudhoe Bay field, North Slope, Alaska. In: *North American Oil and Gas Fields* (ed. by J. Braunstein). Memoir of the American Association of Petroleum Geologists, Tulsa, 24, 23–50.

Jones, R.W. & Milton, N.J. (1994) Sequence development during uplift: Palaeogene stratigraphy and relative sea level history of the Outer Moray Firth, UK North Sea. *Mar. Petrol. Geol.*, 11, 57–165.

Jones, R.W., Ventris, P.A., Wonders, A.A.H., Lowe, S., Rutherford, H.M., Simmons, M.D., Varney, T.D., Athersuch, J., Sturrock, S.J., Boyd, R. & Brenner, W. (1993) Sequence stratigraphy of the Barrow group (Berriasian–Valanginian) siliciclastics, Northwest Shelf, Australia, with emphasis on the sedimentological and palaeontological characterization of systems tracts. In: *Applied Micropalaeontology* (ed. by D.G. Jenkins). pp. 193–223.

Katz, B.J. (1984) Source quality and richness of Deep Sea Drilling Project Site 535 sediments, southeastern Gulf of Mexico. In: (ed. by R.T. Buffler, W. Schlager, J.L. Bowdler, et al.). *Init. Rep. Deep-Sea Drill. Proj.*, 77, 445–450.

Kennett, J.P. & Srinivasan, S. (1983) *Neogene Planktonic Foraminifera*. Hutchinson Ross. 265 pp.

Kenter, J.A.M. (1990) Carbonate platform flanks: slope angle and sediment fabric. *Sedimentology*, 37, 777–794.

Kessler, L.G. & Moorhouse, K. (1984) Depositional processes and fluid mechanics of Upper Jurassic conglomerate accumulations, British North Sea. In: *Sedimentology of Gravels and Conglomerates* (ed. by E.H. Koster & R.J. Steel). Memoir of the Canadian Society of Petroleum Geologists, Calgary, 10, 383–397.

Kessler, L.G., Zany, R.D., Engelhorn, J.A. & Eger, J.D. (1980) Stratigraphy and sedimentology of a Palaeocene fan complex, Cod Field, Norwegian North Sea. In: *The Sedimentation of North Sea Reservoir Rocks*. Norsk Petroleumsforening, Geilo.

Kirschbaum, M.A. & McCabe, P.J. (1992) Controls on the accumulation of coal and on the development of anastomosed fluvial systems in the Cretaceous Dakota Formation of southern Utah. *Sedimentology*, 39, 581–598.

Kleverlaan, K. (1989) Three distinctive feed-lobe systems within one time slide of the Tortonian Tabernas fan, S.E. Spain. *Sedimentology*, **36**, 25–46.

Knighton, A.D. (1984) *Fluvial Forms and Process*. Edward Arnold, London, 218 pp.

Kooi, H. & Cloetingh, S. (1991) Regional isostasy and stress induced vertical motions of extensional basins. Sixth Meeting of the European Union of Geosciences. *Terra-Abstracts*, **3**, 263.

Kolla, V. (1993) Lowstand deep-water siliciclastic depositional systems: characteristics and terminologies in sequence stratigraphy and sedimentology. *Bull. Centres Rech. Explor-Produ. Elf Aquitaine*, **17**, 67–78.

Kolla, V. & Coumes, F. (1987) Morphology, internal structure, seismic stratigraphy and sedimentation of the Indus Fan. *Am. Assoc. Petrol. Geol. Bull.*, **71**, 650–677.

Kolla, V. & Macurda, D.B. (1988) Sea level changes and timing of turbidity-current events in deep sea fan systems. In: *Sea Level Changes — An Integrated Approach* (ed. by C.K. Wilgus, B.S. Hastings, C.G.C. St. Kendall, H.W. Posamentier, C.A. Ross & J.C. Van Wagoner). Special Publication, Society of Economic Paleontologists and Mineralogists, Tulsa, **42**, 381–392.

Kolla, V. & Perimutter, M.A. (1993) Timing of turbidite sedimentation on the Mississippi fan. *Am. Assoc. Petrol. Geol. Bull.*, **77**, 1129–1141.

Kolla, V., Kostecki, J.A., Henderson, L. & Hess, L. (1980) Morphology and Quaternary sedimentation of the Mozambique Fan and environs, southwestern Indian Ocean. *Sedimentology*, **27**, 357–378.

Kolla, V., Buffler, R.T. & Ladd, J.W. (1984) Seismic stratigraphy and sedimentation of Magdelana Fan, Southern Colombian Basin, Caribbean Sea. *Am. Assoc. Petrol. Geol. Bull.*, **68**, 316–332.

Kolla, V., Martin, R. & Weimer, P. (1990) Lowstand deepwater clastic fans and related depositional systems: terminology, characteristics, process & variability. In: *Sequence Stratigraphy as an Exploration Tool: Concepts and Practices from the Gulf Coast* (ed. by J.M. Armentrout & B.F. Perkins). Eleventh Annual Research Conference, Gulf Coast Section, Society of Economic Paleontologists and Mineralogists, pp. 213–215.

Koltermann, C.E. & Gorelick, S.M. (1992) Paleoclimatic signature in terrestrial flood deposits. *Science*, **256**, 1775–1782.

Koss, J.E., Ethridge, F.G. & Schumm, S.A. (1994) An experimental study of the effects of base-level change on fluvial, coastal plain and shelf systems. *J. Sediment. Petrol.*, **64**, 90–98.

Kraus, M.J. (1987) Integration of channel and flood plain suites II. Vertical relations of alluvial paleosols. *J. Sediment. Petrol.*, **57**, 602–612.

Kraus, M.J. & Middleton, L.T. (1987) Contrasting architecture of two alluvial suites in different structural settings. In: *Recent Developments in Fluvial Sedimentology* (ed. by F.G. Ethridge, R.M. Flores & M.D. Harvey). Special Publication, Society of Economic Paleontologists and Mineralogists, Tulsa, **39**, 253–262.

Kulpecz, A.A. & Van Geuns, L.C. (1990) Geological modelling of a turbidite reservoir, Forties Field, North Sea. In: *Sandstone Petroleum Reservoirs* (ed. by J.H. Barwis, J.G. McPherson & R.J. Studlick). Springer-Verlag, New York, pp. 489–507.

Kumar, N. & Slatt, R.M. (1984) Submarine fan and slope facies of Tonkawa (Missourian–Virgilian) Sandstone in Deep Anadarko Basin. *Am. Assoc. Petrol. Geol. Bull.*, **68**, 1839–1856.

Lankester, T. (1993) PhD Thesis. University of London.

Lawrence, D.A. & Williams, B.P.J. (1987) Evolution of drainage systems in response to Arcadian deformation: the Devonian Battery Point Formation, Eastern Canada. In: *Recent Developments in Fluvial Sedimentology* (ed. by F.G. Ethridge, R.M. Flores & M.D. Harvey). Special Publication, Society of Economic Paleontologists and Mineralogists, Tulsa, **39**, 288–300.

Lawrence, D.T., Doyle, M. & Aigner, T. (1989) Application of stratigraphic forward models in exploration settings. *Bull. Houston Geol. Soc.*, **32**, 7.

Lawrence, D.T., Doyle, M. & Aigner, T. (1990) Stratigraphic simulation of sedimentary basins; concepts and calibration. *Am. Assoc. Petrol. Geol. Bull.*, **74**, 273–295.

Lawton, T.F., Geehan, G.W. & Voorhees, B.J. (1987) Lithofacies and depositional environments of the Ivishak Formation, Prudhoe Bay Field. In: *Alaska North Slope Geology* (ed. by I. Tailleur & P. Weimer). Society of Economic Paleontologists and Mineralogists (Pacific Section) and Alaskan Geological Society, **50**, 61–76.

LeBlanc, R.J., O'Brien, G.D., Barwis, J.H. & Buettner, C.E.J. (1989) Book Cliffs Utah. In: *Atlas of Seismic Stratigraphy*, Vol. 3 (ed. by A.W. Bally). pp. 130–133.

Leckie, D.A. (1994) Canterbury Plains, New Zealand — implications for sequence stratigraphic models. *Am. Assoc. Petrol. Geol. Bull.*, **78**, 1240–1256.

Leckie, D.A., Singh, C., Goodarzi, F. & Wall, J.H. (1990) Organic-rich, radioactive marine shale: a case study of a shallow-water condensed section, Cretaceous Shaftsbury Formation, Alberta, Canada. *J. Sediment. Petrol.*, **60**, 101–117.

Leeder, M.R. & Strudwick, A.E. (1987) Delta–marine interactions: a discussion of sedimentary models for Yoredale-type cyclicity in the Dinantian of northern England. In: *European Dinantian Environments* (ed. by J. Miller, A.E. Adams & V.P. Wright). John Wiley, Chichester, pp. 115–130.

Leg 96 Scientific Party (1984) *Challenger* drills Mississippi fan. *Geotimes*, **29**, 15–18.

Leopold, L.B. & Wolman, M.G. (1957) River channel patterns: braided, meandering and straight. Physiographic and hydraulic studies of rivers. *US Geol. Surv. Prof. Pap.*, **282B**, 39–85.

Leopold, L.B., Wolman, M.G. & Miller, J.P. (1964) *Fluvial Processes in Geomorphology*. W.H. Freeman, San Francisco, 522 pp.

Lessenger, M. (1991) A stratigraphic inverse simulation model. *Am. Assoc. Petrol. Geol. Bull.* (abs.), **5**, 620.

Lessenger, M. (1993) *A stratigraphic inverse simulation model*. PhD Thesis, Colorado School of Mines.

Levey, R.A. (1975) Bedform distribution and internal stratification of coarse grained point bars, Upper Congaree River, South Carolina. In: *Fluvial Sedimentology* (ed. by A.D. Miall). Memoir of the Canadian Society of Petroleum Geologists, Calgary, **5**, 105–127.

Levey, R.A. (1985) Depositional model for understanding geometry of Cretaceous coals: major coal seams, Rock Springs Formation, Green River Basin, Wyoming. *Am. Assoc. Petrol. Geol. Bull.*, **69**, 1359–1380.

Link, M.H. & Nilsen, T.H. (1980) The Rocks Sandstone, an Eocene sand-rich deep sea fan deposit, Northern Santa Lucia Range, California. *J. Sediment. Petrol.*, **50**, 583–602.

Link, M.H. & Welton, J.E. (1982) Sedimentology and reservoir potential of Matilija Sandstone: an Eocene sand-rich deep sea fan and shallow marine complex, California. *Am. Assoc. Petrol. Geol. Bull.*, **66**, 1514–1534.

Longman, M.W. (1980) Carbonate diagenetic textures from near-shore diagenetic environments. *Am. Assoc. Petrol. Geol. Bull.*, **64**, 461–487.

Loucks, R.G. & Anderson, J.H. (1985) Depositional facies, diagenetic terranes and porosity development in lower Ordovician Ellenburger dolomite, Puckett Field, West Texas. In: *Carbonate Petroleum Reservoirs* (ed. by P.O. Roehl & P.W. Choquette). Springer-Verlag, New York, pp. 19–37.

Loutit, T.S., Hardenbol, J., Vail, P.R. & Baum, G.R. (1988) Condensed sections: the key to age determination and correlation of continental margin sequences. In: *Sea Level Changes — An Integrated Approach* (ed. by C.K. Wilgus, B.S. Hastings, C.G. Kendall, H.W. Posamentier, C.A. Ross & J.C. Van Wagoner). Special Publication, Society of Economic Paleontologists and Mineralogists, Tulsa, **42**, 183–213.

Lowe, D.R. (1979) Sediment gravity flows: their classification and some problems of application to natural flows and deposits. In: *Geology of Continental Slopes* (ed. by L.J. Doyle & O.H. Pilkey). Special Publication, Society of Economic Paleontologists and Mineralogists, Tulsa, **27**, 75–82.

Lowe, D.R. (1982) Sediment gravity flows II: depositional models with special reference to the deposits of high density turbidity currents. *J. Sediment. Petrol.*, **52**, 279–297.

Lyell, C. (1835) The Bakerian Lecture — on the proofs of a gradual rising of land in certain parts of Sweden. *Philos. Trans. R. Soc. Lond.*, **125**, 1–38.

MacEachern, J.A., Raychaudhuri, I. & Pemberton, S.G. (1992) Stratigraphic applications of the *Glossifungites* ichnofacies: delineating discontinuities in the rock record. In: *Applications of Ichnology to Petroleum Exploration* (ed. by S.G. Pemberton) SEPM Core Workshop No 17, 169–198.

Macdonald, D.I.M. (1986) Proximal to distal sedimentological variation in a linear turbidite trough: implications for the fan model. *Sedimentology*, **33**, 243–260.

Mackin, J.H. (1948) Concept of the graded river. *Bull. Geol. Soc. Am.*, **59**, 463–512.

Macpherson, B.A. (1978) Sedimentation and trapping mechanisms in Upper Miocene Stevens and older turbidite fans of the southeastern San Joaquin Valley, California. *Am. Assoc. Petrol. Geol. Bull.*, **62**, 2243–2274.

Mandl, G. & Crans, W. (1981) Gravitational gliding in deltas. In: *Thrust and Nappe Tectonics* (ed. by K.R. McClay & N.J. Price). Special Publication, Geological Society of London, 9, pp. 41–54.

Manley, P.L. & Flood, R.D. (1988) Cyclic sediment deposition within Amazon deep-sea fan. *Am. Assoc. Petrol. Geol. Bull.*, **72**, 912–925.

Markello, J.R. & Read, J.F. (1981) Carbonate ramp to deeper shale shelf transitions of an upper Cambrian intrashelf basin, Nolichucky Formation, southwest Virgina Appalachians. *Sedimentology*, **28**, 573–597.

Martini, E. (1971) Standard Tertiary and Quaternary calcareous nanoplankton zonation. In: *Proceedings II Planktonic Conference Roma, 1970* (ed. by A. Farinacci). **2**, 739–785.

Marzo, M., Nijman, W. & Puigdefabregas, C. (1988) Architecture of the Castissent fluvial sheet sandstones, Eocene, South Pyrenees, Spain. *Sedimentology*, **35**, 719–738.

Massari F. & Parea G.C. (1988) Progradational gravel beach sequences in a moderate to high-energy microtidal marine environment. *Sedimentology*, **35**, 881–913.

Matthews, R.K. (1984) *Dynamic Stratigraphy*. Prentice-Hall, Englewood Cliffs, NJ, 489 pp.

Mazzullo, S.J. & Reid, A.M. (1989) Lower platform and basin depositional systems, Northern Midland Basin, Texas. In: *Controls on Carbonate Platform and Basin Development* (ed. by P.D. Crevello, J.L. Wilson, J.F. Sarg & J.F. Read). Special Publication, Society of Economic Paleontologists and Mineralogists, Tulsa, **44**, 305–322.

McCabe, P.J. (1984) Depositional models of coal and coal-bearing strata. In: *Sedimentology of Coal and Coal Bearing Strata* (ed. by R.R. Rahmani & R.M. Flores). Special Publication, International Association of Sedimentologists, 7, 13–42. Blackwell Scientific Publications, Oxford.

McCabe, P.J. & Shanley, K.W. (1993) Organic control on shoreface stacking pattern: bogged down in the mire. *Geology*, **20**, 741–744.

McCave, I.N. (1985) Recent shelf clastic sediments. In: *Sedimentology, Recent Developments and Applied Aspects* (ed. by P.J. Brenchley & B.P.J. Williams). Special Publication, Geological Society of London, **18**, 49–66.

McCrory, V.L.C. & Walker, R.G. (1986) A storm and tidally-influenced prograding shoreline — Upper Cretaceous Milk River Formation of Southern Alberta. *Sedimentology*, **33**, 47–60.

McGovney, J.E. & Radovitch, B.J. (1985) Seismic stratigraphy of the Frigg Fan Complex. In: *Seismic Stratigraphy II: An Integrated Approach* (ed. by O.R. Berg & D.G. Wolverton). Memoir, Association of American Petroleum Geologists, Tulsa, **39**, 139–156.

McGowen, J.H. & Bloch, S. (1985) Depositional facies, diagenesis and reservoir quality of Ivishak Sandstone (Sadlerochit Group), Prudhoe Bay Field. *Am. Assoc. Petrol. Geol. Bull.* (abs.), **69**, 286.

McGowen, J.H. & Garner, L.E. (1970) Physiographic features and stratification types of coarse-grained point bars: modern and ancient examples. *Sedimentology*, **14**, 77–111.

McHargue, T. & Webb, J.E. (1986) Internal geometry, seismic facies and petroleum potential of canyons and inner fan channels of the Indus submarine fan. *Am. Assoc. Petrol. Geol. Bull.*, **70**, 161–180.

McKenzie, D.P. (1978) Some remarks on the development of sedimentary basins. *Earth Planet. Sci. Lett.*, **40**, 25–32.

McMillen, K.J. & Winn, R.D. (1989) Seismic stratigraphy of Lewis Shale deltaic and deep-water clastics, Red Desert–Washakie basins, Wyoming. In: *Atlas of Seismic Stratigraphy*, Vol. 3 (ed. by A.W. Bally). American Association of Petroleum Geologists, Tulsa, USA.

McNeil, D.H., Dietrich, J.R. & Dixon, J. (1990) Foraminiferal biostratigraphy and seismic sequences: examples from the Cenozoic of the Beaufort-Mackenzie Basin, Arctic Canada. In: *Palaeoecology, Biostratigraphy, Palaeoceanography and Taxonomy of Agglutinated Foraminifera* (ed. by Ch. Hemelben, M.A. Kaminski, W. Kuhnt & D.B. Scott). Kluwer Academic Publishers, Dordrecht, pp. 859–882.

McQuillin, R., Bacon, M. & Barclay, W. (1984) *An Introduction to Seismic Interpretation*. Graham & Trotman, London, 287 pp.

Meckel, L.D. (1975) Holocene sand bodies in the Colorado delta, Salton Sea, Imperial County, California. In: *Deltas* (ed. by M.L. Broussard). Houston Geological Society, Houston, TX, pp. 239–266.

Meckel, L.D., Jr. & Nath, A.K. (1977) Geologic considerations for stratigraphic modelling and interpretation. In: *Seismic Stra-*

tigraphy — *Applications to Hydrocarbon Exploration* (ed. by C.E. Payton). American Association of Petroleum Geologists, Tulsa, OK, Memoir **26**, 417–438.

Melvin, J. & Knight, A.S. (1984) Lithofacies, diagenesis and porosity of the Ivishak Formation, Prudhoe Bay Area, Alaska. In: *Clastic Diagenesis* (ed. by D.A. MacDonald & R.C. Surdam). Memoir of the Association of Petroleum Geologists, Tulsa, **37**, 347–365.

Miall, A.D. (1977) A review of the braided river depositional environment. *Earth. Sci. Rev.*, **13**, 1–62.

Miall, A.D. (ed.) (1978) *Fluvial Sedimentology*. Memoir of the Canadian Society of Petroleum Geologists, **5**, Calgary, 859 pp.

Miall, A.D. (1981) Alluvial sedimentary basins: tectonic setting and basin architecture. In: *Sedimentation and Tectonics in Alluvial Basins* (ed. by A.D. Miall). Special Paper, Geological Association of Canada, Waterloo, Ontario, **23**, 1–33.

Miall, A.D. (1983) Basin analysis of fluvial sediments. In: *Modern and Ancient Fluvial Systems* (ed. by J.D. Collinson & J. Lewin). Special Publication, International Association of Sedimentologists, **6**, 279–286. Blackwell Scientific Publications, Oxford.

Miall, A.D. (1985) Architectural element analysis: a new method of facies analysis applied to fluvial deposits. *Earth. Sci. Rev.*, **22**, 261–308.

Miall, A.D. (1986) Eustatic sea-level changes interpreted from seismic stratigraphy: a critique of the methodology with particular reference to the North Sea Jurassic record. *Am. Assoc. Petrol. Geol. Bull.*, **70**, 131–137.

Miall, A.D. (1987) Recent developments in the study of fluvial facies models. In: *Recent Developments in Fluvial Sedimentology* (ed. by F.G. Ethridge, R.M. Flores & M.D. Harvey). Special Publication, Society of Economic Paleontologists and Mineralogists, Tulsa, **39**, 1–9.

Miall, A.D. (1988) Reservoir heterogeneities in fluvial sandstones: lessons from outcrop studies. *Am. Assoc. Petrol. Geol. Bull.*, **72**, 682–694.

Miall, A.D. (1991) Stratigraphic sequences and their chronostratigraphic correlation. *J. Sediment. Petrol.*, **61**, 497–505.

Miall, A.D. (1992) Alluvial deposits. In: *Facies Models — Response to Sea Level Change* (ed. by R.G. Walker & N.P. James). Geological Association of Canada, Waterloo, Ontario, pp. 119–142.

Middleton, G.V. & Hampton, M.A. (1973) Sediment gravity flows: mechanics of flow and deposition. In: *Turbidites and Deepwater Sedimentation* (ed. by G.V. Middleton & A.H. Bouma). Society of Economic Paleontologists and Mineralogists, Los Angeles, pp. 1–38.

Middleton, G.V. & Hampton, M.A. (1976) Subaqueous sediment transport and deposition by sediment gravity flows. In: *Marine Sediment Transport and Environmental Management* (ed. by D.J. Stanley & D.J.P. Swift). John Wiley, New York, pp. 197–218.

Middleton, G.V. & Southard, J.B. (1984) *Mechanics of Sediment Movement*. Society of Economic Paleontologists and Mineralogists, Tulsa, Short Course. Revised Notes.

Milankovitch, M. (1920) *Théorie mathématique des Phenomenes Thermiques Produit par la Radiation Solaire*. Gauthier-Villars, Paris, 338 pp.

Miller, K.G.J., Wright, J.D. & Fairbanks, R.G. (1991) Unlocking the ice-house: Oligocene–Miocene oxygen isotopes, eustacy and margin erosion. *J. Geophys. Res.*, **96**, 6829–6848.

Miller, K.G.J., Thompson, P.R. & Kent, D.V. (1993) Integrated Late Eocene–Miocene stratigraphy of the Alabama Coastal Plain: correlation of hiatuses and stratal surfaces to glacio-eustaic lowerings. *Palaeoceanography*, **8**, 313–331.

Miller, R.G. (1990) A palaeooceanographic approach to the Kimmeridge Clay Formation. In: *Deposition of Organic-rich Facies*. American Association of Petroleum Geologists, Tulsa, Studies in Geology, **30**, pp. 13–26.

Milton, N.J. (1993) Evolving depositional geometries in the North Sea Jurassic Rift. In: *Geology of Northwest Europe, Proceedings of the 4th Conference* (ed. by J.R. Parker). Geological Society of London, Bath, pp. 425–442.

Milton, N.J. & Bertram, G.B. (in press) Tectonic controls on systems tract development: implications for hydrocarbon exploration. In: *Sequence Stratigraphy and its Application to the British Stratigraphic Record* (ed. by S.P. Hesselbo & D.N. Parkinson). Special Publication, Geological Society of London.

Milton, N.J. & Dyce, M. (1995) Systems tract geometries associated with Early Eocene lowstands, imaged on a 3D seismic dataset from the Bruce Area, UK North Sea. In: *Sequence Stratigraphy on the Northwest European Margin* (ed. by R.J. Steel, V.L. Felt, E.P. Johannesson & C. Mathieu). Special Publication, Norwegian Petroleum Society (NPF), **5**, 429–442.

Milton, N.J., Bertram, G.T. & Vann, I.R. (1990) Early Palaeogene tectonics and sedimentation in the Central North Sea. In: *Tectonic Events Responsible for Britain's Oil and Gas Reserves* (ed. by R.P.F. Hardman & J. Brooks). Special Publication, Geological Society of London, **55**, 339–351.

Mitchener, B.C., Lawrence, D.A., Partington, M.A., Bowman, M.B.J. & Gluyas, J. (1992) Brent Group: sequence stratigraphy and regional implications. In: *Geology of the Brent Group* (ed. by A.C. Morton, R. Haszeldine, M.R. Giles & S. Brown.) Special Publication, Geological Society of London, **61**, 45–80.

Mitchum, R.M., Jr. (1985) Seismic stratigraphic expression of submarine fans. In: *Seismic Stratigraphy II — An Integrated Approach* (ed. by O.R. Berg & D.G. Woolverton). Memoir of the American Association of Petroleum Geologists, Tulsa, **39**, 116–136.

Mitchum, R.M., Jr. & Vail, P.R. (1977) Seismic stratigraphy and global changes of sea-level, part 7: stratigraphic interpretation of seismic reflection patterns in depositional sequences. In: *Seismic Stratigraphy — Applications to Hydrocarbon Exploration* (ed. by C.E. Payton). Memoir of the American Association of Petroleum Geologists, Tulsa, **26**, 135–144.

Mitchum, R.M., Jr. & Van Wagoner, J.C. (1991) High frequency sequences and their stacking patterns: sequence stratigraphic evidence of high frequency eustatic cycles. *Sediment. Geol.*, **70**, 135–144.

Mitchum, R.M., Vail, P.R., Todd, R.G. & Sangree, J.B. (1976) Regional seismic interpretations using sequences and eustatic cycles. *Am. Assoc. Petrol. Geol. Bull.* (abs.), **60**, 669.

Mitchum, R.M., Jr., Vail, P.R. & Thompson, S., III (1977a) Seismic stratigraphy and global changes of sea-level, part 2: the depositional sequence as a basic unit for stratigraphic analysis. In: *Seismic Stratigraphy — Applications to Hydrocarbon Exploration* (ed. by C.E. Payton). Memoir of the American Association of Petroleum Geologists, Tulsa, **26**, 53–62.

Mitchum, R.M., Jr., Vail, P.R. & Sangree, J.B. (1977b) Seismic stratigraphy and global changes of sea-level part 6: seismic stratigraphic interpretation procedure. In: *Seismic Stratigraphy — Applications to Hydrocarbon Exploration* (ed. by C.E. Payton).

Memoir of the American Association of Petroleum Geologists, Tulsa, **26**, 117–134.

Mitchum, R.M., Jr., Sangree, J.B., Vail, P.R. & Wornardt, W.W. (1990) Sequence stratigraphy in Late Cenozoic expanded sections, Gulf of Mexico. In: *Sequence Stratigraphy as an Exploration Tool: Concepts and Practices from the Gulf Coast* (ed. by J.M. Armentrout & B.F. Perkins). Eleventh Annual Research Conference, Gulf Coast Section, Society of Economic Paleontologists and Mineralogists, pp. 237–256.

Molina, E. (1992) The Stratotypic Ilerdian revisited: integrated stratigraphy across the Paleocene/Eocene boundary. *Rev. Micropaleontol.*, **35**, 143–156.

Montanez, I.P. & Osleger, D.A. (1993) Parasequence stacking patterns, third order accommodation events and sequence stratigraphy of Middle to Late Cambrian platform carbonates, Bonanza King Formation, southern Great Basin. In: *Recent Advances and Applications of Carbonate Sequence Stratigraphy* (ed. by B. Loucks & J.F. Sarg). Memoir of the American Association of Petroleum Geologists, Tulsa, **57**.

Moody-Stuart, M. (1966) High and low sinuosity stream deposits, with examples from the Devonian of Spitsbergen. *J. Sediment. Petrol.*, **36**, 1102–1117.

Moore, C.H. (1984) The Upper Smackover of the Gulf Rim: depositional systems, diagenesis, porosity, evolution and hydrocarbon production. In: *Jurassic of the Gulf Rim* (ed. by W.P.S. Ventress, D.G. Bebout, B.F. Perkins & C.H. Moore). Special Publication, Gulf Coast Section, Society of Economic Paleontologists and Mineralogists, pp. 283–308.

Morgridge, D.L. & Smith, W.B., Jr. (1972) Geology and discovery of Prudhoe Bay field, eastern Arctic Slope, Alaska. In: *Stratigraphic Oil and Gas Fields — Classification, Exploration Methods and Case Histories*. Memoir of the American Association of Petroleum Geologists, Tulsa, **16**, 499–501.

Morley, R. (1991) Tertiary stratigraphic palynology in Southeast Asia: current status and new directions. *Geol. Soc. Malaysia Bull.*, **28**, 1–36.

Mullins, H.T. (1983) Modern carbonate slopes and basins of the Bahamas. In: *Platform Margin and Deepwater Carbonates* (ed. by H.E. Cook, A.C. Hine & H.T. Mullins). Society of Economic Paleontologists and Mineralogists, Tulsa, Short Course **12**, 4.1–4.138.

Murray, J.W. (1973) *Distribution and Ecology of Living Benthic Foraminiferids*. Heinemann, London, 274 pp.

Murray, J.W. (1976) A method of determining proximity of marginal seas to an ocean. *Mar. Geol.*, **22**, 103–119.

Murray, J.W. (1992) *Ecology and Palaeoecology of Benthic Foraminifera*. Longman Scientific & Technical, New York.

Mutti, E. (1985) Turbidite systems and their relations to depositional sequences. In: *Provenance of Arenites* (ed. by G.G. Zuffa). NATO-ASI Series, D. Reidel Publishing Co. Dordrecht, pp. 65–93.

Mutti, E. (1992) *Turbidite Sandstones*. Istituto di Geologia Universita di Parma. Agip S.p.a. Publications, 275 pp.

Mutti, E. & Normark, W.R. (1987) Comparing examples of Modern and ancient turbidite systems: problems and concepts. In: *Marine Clastic Depositional Environments: Concepts and Case Studies* (ed. by J.K. Leggett & G.G. Zuffa). Graham & Trotman, London, pp. 1–38.

Mutti, E. & Ricci-Lucchi, F. (1972) Le torbiditi dell'Apennino Settentrionale: introduzione all'analisi di facies. *Memorie Soc. Geoligica Italiana*, **11**, 161–199.

Mutti, E., Remacha, E., Sgavetti, E., Rosell, M., Valloni, R. & Zamorano, M. (1985a) Stratigraphy of the Eocene Hecho Group turbidite systems, South-central Pyrenees. In: *Excursion Guidebook* (ed. by M.D. Mila & J. Rosell). 6th European Meeting of the International Association of Sedimentologists, Lleida, Spain, pp. 521–576.

Mutti, E., Rosell, J., Allen, G., Fonnesu, F. & Sgaretti, M. (1985b) The Eocene Baronia tide dominated delta/shelfsystem in the Ager basin. In: *Excursion Guidebook* (ed. by M.D. Mila & J. Rosell). 6th Regional Meeting of the International Association of Sedimentologists, Lleida, Spain, pp. 577–600.

Mutti, E. Seguret, M. & Sgavetti, M. (1988) *Sedimentation and Deformation in the Tertiary Sequences of the Southern Pyrenees*. Special Publication of the Institute of Geology of the University of Parma, Italy. American Association of Petroleum Geologists Mediterranean Basins Conference, Field Trip 7.

Myers, K.J. & Jenkyns, K.F. (1992) Determining total organic carbon contents from well logs: an intercomparison of GST data and a new density log method. In: *Geological Applications of Wireline Logs, II* (ed. by A. Hurst, C.M. Griffiths & P.F. Worthington). Special Publication, Geological Society of London, **65**, 369–376.

Myers, K.J. & Wignall, P.B. (1987) Understanding Jurassic organic-rich mudrocks: new concepts using gamma-ray spectrometry and palaeoecology: examples from the Kimmeridge Clay of Dorset and the Jet Rock of Yorkshire. In: *Marine Clastic Depositional Environments: Concepts and Case Studies* (ed. by J.K. Leggett & G.G. Zuffa). Graham & Trotman, London, pp. 175–182.

Nagy, J., Dypvik, H. & Bjaerke, T. (1984) Sedimentological and paleontological analyses of Jurassic North Sea deposits from deltaic environments. *J. Petrol. Geol.*, **7**, 169–188.

Nami, M. & Leeder, M.R. (1978) Changing channel morphology and magnitude in the Scalby Formation (M. Jurassic) of Yorkshire, England. In: *Fluvial Sedimentology* (ed. by A.D. Miall). Memoir of the Canadian Society of Petroleum Geologists, **5**, Calgary, pp. 431–440.

Nanson, G.C., Rust, B.R. & Taylor, G. (1986) Coexistant mud braids and anastomosing channels in an arid zone river: Cooper Creek, central Australia. *Geology*, **14**, 175–178.

Nardin, T.R., Hein, F.J., Gorsline, D.S. & Edwards, B.D. (1979) A review of mass movement processes, sediment and acoustical characteristics, and contrasts in slope and base of slope systems versus canyon fed basin floor systems. In: *Geology of Continental Slopes* (ed. by L.J. Doyle & O.H. Pilkey). Special Publication, Society of Economic Paleontologists and Mineralogists, Tulsa, **27**, 61–73.

Nelson, H.C. & Maldonado, A. (1988) *Factors Controlling Depositional Patterns of Ebro Turbidite Systems, Mediterranean Sea*. American Association of Petroleum Geologists, Memoir, Tulsa, **72**, 698–716.

Nelson, H.C. & Nilsen, T. (1984) Modern and ancient deep sea fan sedimentation. Society of Economic Paleontologists and Mineralogists, Tulsa, Short Course **14**, 404 pp.

Nelson, H.C., Twichell, D.C., Schwab, W.C., Lee, H.J. & Kenyon, N.H. (1992) Upper Pleistocene turbidite sand beds and chaotic silt beds in the channelized, distal, outer-fan lobes of the Mississippi fan. *Geology*, **20**, 693–696.

Neumann, A.C. & Land, L.S. (1975) Lime mud deposition and calcareous algae in the Bight of Abaco, Bahamas: a budget. *J. Sediment. Petrol.*, **45**, 763–786.

Neumann, A.C. & Macintyre, I. (1985) Reef response to sea level rise: keep-up, catch-up or give-up. In: *Proceedings of the Fifth International Coral Reef Congress*, 1985 Tahiti, 3, 105–110.

Nichols, G.J. (1987) Structural controls on fluvial; distributary systems – the Luna System, North Spain. In: *Recent Developments in Fluvial Sedimentology* (ed. by F.G. Ethridge, R.M. Flores & M.D. Harvey). Special Publication, Society of Economic Paleontologists and Mineralogists, Tulsa, 39, 269–277.

Nijman, W. & Nio, S.D. (1975) The Eocene Montanana Delta, Tremp Graus Basin, Southern Pyrenees. In: *Sedimentary Evolution of the Palaeogene South Pyrenean Basin, Excursion 19* (ed. by J. Rosell & C. Puidgefabregas). IXth International Sedimentologists Congress, Nice, France, 56 pp.

Nordlund, U. & Griffiths, C.M. (1993a) An example of the practical use of chronosomes. *Geoinformatics* (Japanese Society of Geoinformatics), 4(3).

Nordlund, U. & Griffiths, C.M. (1993b) Automatic construction of two- and three-dimensional chronostratigraphic sections. *Computers Geosci.*, 19.

Nordlund, U. & Griffiths, C.M. (in press). An example of the practical use of chronosomes in quantitative stratigraphy. In: *Quantitative Stratigraphy: New developments in Litho- and Bio-stratigraphy.*

Normark, W.R. (1978) Fan valleys, channels & depositional lobes on Modern submarine fans: characters for the recognition of sandy turbidite environments. *Am. Assoc. petrol. Geol. Bull.*, 62, 912–931.

Normark, W.R., Piper, D.J.W. & Hess, G.R. (1979) Distributary channels, sand lobes, and mesotopography of Navy Submarine Fan, California Borderland, with applications to ancient fan sediments. *Sedimentology*, 26, 749–774.

O'Connell, S. (1986) Anatomy of modern submarine fan depositional and distributary systems, Thesis. Columbia University.

Ori, G.G. (1982) Braided to meandering channel patterns in humid region alluvial fan deposits, River Reno, Po Plain (Northern Italy). *Sediment. Geol.*, 31, 231–248.

Ori, G.G. (1987) Plio-Pleistocene fan-delta deposition on listric normal faults in the Gulf of Corinth (Greece). In: *Fan Deltas – Sedimentology and Tectonic Settings* (ed. by W. Nemec). Abstracts, International Symposium, University of Bergen, 49.

Ori, G.G. & Friend, P.F. (1984) Sedimentary basins formed and carried piggy-back on active thrust sheets. *Geology*, 12, 475–478.

Orton, G.J. (1988) A spectrum of middle Ordovician fan deltas and braidplian deltas, North Wales: a consequence of varying fluvial clastic input. In: *Fan Deltas* (ed. by W. Nemec & R.J. Steel). Blackie, Glasgow, pp. 23–49.

Orton, G.J. & Reading, H.G. (1993) Variability of deltaic processes in terms of sediment supply, with particular emphasis on grain size. *Sedimentology*, 40, 475–512.

Oschmann, W. (1988) Kimmeridge Clay sedimentation – a new cyclic model. *Palaeogeogr. Palaeoclimatol. Palaeoecol.*, 65, 217–251.

Ouchi, S. (1985) Response of alluvial rivers to slow active tectonic movement. *Am. Assoc. Petrol. Geol. Bull.*, 96, 504–515.

Pacht, J.A., Bowen, B.E., Beard, J.H. & Shafer, B.L. (1990) Sequence stratigraphy of Plio-Pleistocene depositional facies in the offshore Lousiana South Additions. *Trans. Gulf Coast. Assoc. Geol. Soc.*, VXL, 643–659.

Palsey, M.A., Gregory, W.A. & Hart, G.F. (1991) Organic matter variations in transgressive and regressive shales. *Org. Geochem.*, 17, 483–501.

Parry, C.C., Whitley, P.K.J. & Simpson, R.D.H. (1981) Integration of palynological and sedimentological methods in facies analysis of the Brent Formation. In: *Petroleum Geology of the Continental Shelf of Northwest Europe* (ed. by G.V. Illing & G.D. Hobson). pp. 205–215. Heyden, London.

Partington, M.A., Mitchener, B.C., Milton, N.J. & Fraser, A.J. (1993a) Genetic sequence stratigraphy for the North Sea Late Jurassic and Early Cretaceous; distribution and prediction of Kimmeridgian–Late Ryazanian reservoirs in the North Sea and adjacent areas. In: *Petroleum Geology of Northwest Europe: Proceedings of the 4th Conference* (ed. by J.R. Parker). Geological Society of London, Bath, pp. 347–370.

Partington, M.A., Copestake, P., Mitchener, B.C. & Underhill, J.R. (1993b) Biostratigraphic calibration of genetic stratigraphic sequences in the Jurassic–lowermost Cretaceous (Hettangian to Ryazanian) of the North Sea and adjacent areas. In: *Petroleum Geology of Northwest Europe: Proceedings of the 4th Conference* (ed. by J.R. Parker). Geological Society of London, Bath, pp. 371–386.

Passey, Q.R., Creaney, J.B., Kulla, J.B., Moretti, F.J. & Stroud, J.D. (1990) A practical model for organic richness from porosity and resistivity logs. *Am. Assoc. Petrol. Geol. Bull.*, 74, 1777–1794.

Pattison, S.A.J. (1988) Transgressive, incised shoreface deposits of the Burnstick Member (Cardium "B" sandstone) at Caroline, Crossfield, Garrington and Lochend; Cretaceous Western Interior Seaway, Alberta, Canada. In: *Sequences, Stratigraphy, Sedimentology: Surface and Subsurface* (ed. by D.P. James & D.A. Leckie). Canadian Society of Petroleum Geologists, Calgary, Memoir 15, 155–166.

Pelet, R. (1987) A model for organic sedimentation on present day continental margins. In: *Marine Petroleum Source Rocks* (ed. by J. Brooks & A. Fleet). Special Publication, Geological Society of London, 26, 167–180.

Pequegnat, W.E. (1972) A deep bottom current on the Mississippi Cone. In: *Contributions on the Physical Oceanography of the Gulf of Mexico*. Texas A&M University Oceanographic Studies 2, Houston Gulf Publishing Co., pp. 65–87.

Perlmutter, A. & Matthews, D. (1990) Global cyclostratigraphy; a model. In: *Quantitative Dynamic Stratigraphy* (ed. by T.A. Cross), Prentice Hall, Englewood Cliffs, NJ, pp. 233–260.

Pickering, K.T. (1981) Two types of outer fan lobe sequence, from the late Precambrian Kongsfjord Formation submarine fan, Finnmark, North Norway. *J. Sediment. Petrol.*, 51, 1277–1286.

Pickering, K.T. (1983) Transitional submarine fan deposits from the late Precambrian Kongsfjord Formation submarine fan, Finnmark, N. Norway. *Sedimentology*, 30, 181–199.

Pickering, K.T., Stow, D.A.V., Watson, M. & Hiscott, R. (1986) Deep water facies, processes and models: a reservoir and classification scheme for modern and ancient sediments. *Earth. Sci. Rev.*, 23, 75–174.

Pickering, K.T., Hiscott, R.N. & Hein, F.J. (1989) *Deep Marine Environments: Clastic Sedimentation and Tectonics*. Unwin Hyman, London, 416 pp.

Piper, D.J.W., Kontopoulos, N., Chronis, G. & Panagos, A.G. (1990) Modern fan deltas in the Western Gulf of Corinth, Greece. *Geo-Marine Letters*, 10, 51–12.

Pirson, S.J. (1977) *Geological Well Analysis*. Gulf Publishing, Houston.

Pitman, W.C. (1978) Relationship between eustasy and stratigraphic sequences on passive margins. *Geol. Soc. Am. Bull.*, **89**, 1389–1403.

Plint, A.G. (1988) Sharp based shoreface sequences and "offshore bars" in the Cardium Formation of Alberta; their relationship to relative changes in sea level. In: *Sea Level Changes: An Integrated Approach* (ed. by C.K. Wilgus, B.S. Hastings, H.S. Posamentier, J. Van Wagoner, C.A. Ross & C.G. Kendall). Special Publication, Society of Economic Paleontologists and Mineralogists, Tulsa, **42**, 357–370.

Plint, A.G. & Walker, R.G. (1987) Cardium Formation 8. Facies and environments of the Cardium shoreline and coastal plain in the Kakwa field and adjacent areas, northwestern Alberta. *Bull. Can. Petrol. Geol.*, **35**, 48–64.

Plint, A.G., Walker, R.G. & Bergman, K.M. (1986) Cardium Formation 6. Stratigraphic framework of the Cardium Formation in subsurface. *Bull. Can. Petrol. Geol.*, **34**, 213–225.

Plint, A.G., Walker, R.G. & Duke, W.L. (1988) An outcrop to subsurface correlation of the Cardium Formation in Alberta. In: *Sequences, Stratigraphy, Sedimentology: Surface and Subsurface* (ed. by D.P. James & D.A. Leckie). Memoir of the Canadian Society of Petroleum Geologists, Calgary, **15**, 167–184.

Pomar, L. (1991) Reef geometries, erosion surfaces and high-frequency sea level changes, Upper Miocene Reef Complex, Mallorca, Spain. *Sedimentology*, **38**, 243–269.

Posamentier, H.W. (1988) Fluvial deposition in a sequence stratigraphic framework. In: *Sequences, Stratigraphy, Sedimentology: Surface and Subsurface* (ed. by D.P. James & D.A. Leckie). Memoir of the Canadian Society of Petroleum Geologists, Calgary, **15**, 582–583.

Posamentier, H.W. & Allen, G.P. (1993a) Siliciclastic sequence stratigraphic patterns in foreland ramp type basins. *Geology*, **21**, 455–458.

Posamentier, H.W. & Allen, G.P. (1993b) Recent advances in sequence stratigraphy: the lowstand and transgressive systems tracts (abstract). *Am. Assoc. Petrol. Geol. Bull.*, **77**, 1655.

Posamentier, H.W. & Chamberlain, C.J. (1993) Sequence stratigraphic analysis of Viking Formation lowstand beach deposits in the Joarcam field, Alberta. In: *Sequence Stratigraphy and Facies Associations* (ed. by H.W. Posamentier, C.P. Summerhayes, B.U. Haq & G.P. Allen). Special Publication, International Association of Sedimentologists, **18**, 469–486. Blackwell Scientific Publications, Oxford.

Posamentier, H.W. & Erskine, R.D. (1991) Seismic expression and recognition criteria of ancient submarine fans. In: *Seismic Facies and Sedimentary Processes of Modern and Ancient Submarine Fans and Turbidite Systems* (ed. by P. Weimer & M.H. Link). Springer-Verlag, New York, pp. 197–222.

Posamentier, H.W. & James, D.P. (1993) An overview of sequence stratigraphic concepts: uses and abuses. In: *Sequence Stratigraphy and Facies Associations* (ed. by H.W. Posamentier, C.P. Summerhayes, B.U. Haq & G.P. Allen). Special Publications, International Association of Sedimentologists, **18**, 3–18. Blackwell Scientific Publications, Oxford.

Posamentier, H.W. & Vail, P.R. (1988) Eustatic controls on clastic deposition I — sequence and systems tract models. In: *Sea-level Changes: An Integrated Approach* (ed. by C.K. Wilgus, B.S. Hastings, C.G. St. C. Kendall, H.W. Posamentier, C.A. Ross & J.C. Van Wagoner). Special Publication, Society of Economic Paleontologists and Mineralogists, Tulsa, USA, **42**, 109–124.

Posamentier, H.W. & Weimer, P. (1993) Siliciclastic sequence stratigraphy and petroleum geology — where to from here? *Am. Assoc. Petrol. Geol. Bull.*, **77**, 731–742.

Posamentier, H.W., Jervey, M.T. & Vail, P.R. (1988) Eustatic controls on clastic deposition II — conceptual framework. In: *Sea-level changes: An integrated Approach* (ed. by C.K. Wilgus, B.S. Hastings, C.G. St. C. Kendall, H.W. Posamentier, C.A. Ross & J.C. Van Wagoner). Special Publication, Society of Economic Paleontologists and Mineralogists, **42**, 125–154.

Posamentier, H.W., Allen, H.W., James, D.P. & Tesson, M. (1992) Forced regressions in a sequence stratigraphic framework: concepts, examples and sequence stratigraphic significance. *Am. Assoc. Petrol Geol. Bull.*, **76**, 1687–1709.

Postma, G. (1986) Classification of sediment gravity flow deposits based on flow conditions during sedimentation. *Geology*, **14**, 291–294.

Pratt, B.R. & James, N.P. (1986) The St George Group (Lower Ordovician) of western Newfoundland: tidal flat island model for carbonate sedimentation in shallow epeiric seas. *Sedimentology*, **33**, 313–343.

Prior, D.B. & Bornhold, B.D. (1989) Submarine sedimentation on a developing Holocene fan delta. *Sedimentology*, **36**, 1053–1076.

Puigdefabregas, C. (1973) Miocene point bar deposits in the Ebro Basin, northern Spain. *Sedimentology*, **20**, 133–144.

Puigdefabregas, C. & Van Vliet, A. (1978) Meandering stream deposits from the Tertiary of the southern Pyrenees. In: *Fluvial Sedimentology* (ed. by A.D. Miall). Memoir of the Canadian Society of Petroleum Geologists, Calgary, **5**, 469–485.

Pulham, A.J. (1989) Controls on internal structure and architecture of sandstone bodies within Upper Carboniferous fluvial-dominated deltas, County Clare, western Ireland. In: *Deltas: Sites and Traps for Fossil Fuels* (ed. by M.K. Whateley & K.T. Pickering). Special Publication, Geological Society of London, **41**, 179–203.

Purdy, E.G. (1974) Karst-determined facies patterns in British Honduras: Holocene carbonate sedimentation model. *Am. Assoc. Petrol. Geol. Bull.*, **58**, 825–855.

Purdy, E.G. & Bertram, G.T. (1993) Carbonate Concepts from the Maldives, Indian Ocean. *Am. Assoc. Petrol. Geol. Studies in Geology 34*. American Association of Petroleum Geologists, Tulsa, USA.

Purser, B.H. (1973) *The Persian Gulf: Holocene Carbonate Sedimentation in a Shallow Epicontinental Sea.* Springer-Verlag, Berlin, p. 471.

Putman, P.E. (1983) Fluvial deposits and hydrocarbon accumulations: examples from the Lloydminster area, Canada. In: *Modern and Ancient Fluvial Systems* (ed. by J.D. Collinson & J. Lewin). Special Publication, International Association of Sedimentologists, **6**, 517–532. Blackwell Scientific Publications, Oxford.

Ramasayer, G.R. (1979) Seismic stratigraphy; a fundamental exploration tool. *Offshore Technology Conference, Houston Texas, April–May 1979, Proceedings*, **3**, 1859–1867.

Ramsay, A.T.S. (1987) Depositional environments in the Dinantian Limestones of Gower, South Wales. In: *European Dinantian Environments* (ed. by J. Miller, A.E. Adams & V.P. Wright). John Wiley, Chichester, pp. 265–308.

Read, J.F. (1982) Carbonate margins of passive (extensional) continental margins: types, characteristics and evolution. *Tectonophysics*, **81**, 195–212.

Read, J.F. (1985) Carbonate platform facies models. *Am. Assoc.*

Petrol. Geol. Bull., 69, 1–21.

Read, J.F. & Goldhammer, R.K. (1988) Use of Fischer plots to define third-order sea-level curves in Ordovician peritidal cyclic carbonates, Appalachians. *Geology*, 16, 895–899.

Read, J.F. & Horbury, A.D. (1993) Eustatic and tectonic controls on porosity evolution beneath sequence-bounding unconformaties and parasequence disconformities on carbonate platforms. In: Diagenesis and Basin Development. *Am. Assoc. Petrol. Geol. Studies in Geology 36* (ed. by A.D. Horbury & A.G. Robinson) American Association of Petroleum Geologists, Tulsa, USA, pp. 155–197.

Read, J.F., Elrick, M.E., Horbury, A.D. & Osleger, D.A. (1992) High frequency sea level fluctuations through time: their imprint on sequence boundaries and their diagenesis. In: *Carbonate Stratigraphic Sequences: Sequence Boundaries and Associated Facies, Conference Abstracts* (ed. by A. Simo, E. Franseen & M. Harris). La Seu, Spain, pp. 82–83.

Reading, H.G. (1986) *Sedimentary Environments and Facies.* Blackwell Scientific Publications, Oxford.

Reading, H.G. (1991) The classification of deep-sea depositional systems by sediment calibre and feeder system. *J. Geol. Soc. Lond.*, 148, 427–430.

Reading, H.G. & Orton, G.J. (1991) Sediment calibre: a control on facies models with special reference to deep sea depositional systems. In: *Controversies in Modern Geology* (ed. by D.W. Muller, J.A. McKenzie & H. Weissert). Academic Press, London, pp. 85–111.

Reading, H.G. & Richards, M.T. (1994) Turbidite systems in deep water basin margins classified by grain-size and feeder system. *Am. Assoc. Petrol. Geol. Bull.*, 78, 792–822.

Reijmer, J.J., Schlager, W. & Droxler, A.W. (1988) Site 632: Pliocene–Pleistocene sedimentation in a Bahamian Basin. *Proc. Ocean Drill. Proj.*, 101, 213–220.

Reijmer, J.J., Ten Kate, W.G.H.Z., Sprenger, A. & Schlager, W. (1991) Calciturbidite composition related to the exposure and flooding of a carbonate platform (Triassic, Eastern Alps). *Sedimentology*, 38, 1049–1074.

Reinson, G.E., Clarke, J.E. & Foscolos, A.E. (1988) Reservoir geology of Crystal Viking field, Lower Cretaceous estuarine tidal channel-bay complex, South-central Alberta. *Am. Assoc. Petrol. Geol. Bull.*, 72, 1270–1294.

Reynolds, A.D. (1994a) Sequence stratigraphy from core and wireline log data: Viking Formation, Albian South Central Alberta. *Marine Petrol. Geol.*, 11, 258–282.

Reynolds, A.D. (1994b) Sequence stratigraphy and the dimensions of paralic sandstone bodies. In: *High Resolution Sequence Stratigraphy: Innovations and Applications* (ed. by S.D. Johnson). University of Liverpool, pp. 69–72.

Reynolds, A.D. (1995) Sedimentology and sequence stratigraphy of the Thistle field, northern North Sea. In: *Sequence stratigraphy on the Northwest European Margin* (ed. by R.J. Steel, V.L. Felt, E.P. Johannessen, C. Matthieu). Norwegian Petroleum Society, Spec. Publ. 5, pp. 257–271.

Ricci-Lucchi, F. & Valmori, E. (1980) Basin-wide turbidites in a Miocene, over supplied deep-sea plain: a geometrical analysis. *Sedimentology*, 27, 241–270.

Ricci-Lucchi, F., Collella, A., Gabbianella, G., Rossi, S. & Normark, W.R. (1984) Crati fan Mediterranean. In: *Submarine Fans and Related Turbidite Systems* (ed. by A.H. Bouma, W.R. Normark & N.E. Barnes). Springer-Verlag, Berlin, pp. 51–57.

Richards, M.T. (1994) Transgression of an estuarine channel-tidal flat complex: the Lower Trias of Barles, Alpes de Haute Provence. *Sedimentology*, 41, 1–27.

Richards, M.T. & Bowman, M.B.J. (in press a) Submarine fan systems II: Variability in reservoir architecture. *Mar. Petrol. Geol.*

Richards, M.T., Corwin, L.C., Knock, D.G., Puls, D.P. & Deacon, M.W. (1994) Application of sequence stratigraphy to the description of a fluvio-deltaic reservoir: Zone 1/2A Prudhoe Bay Field, North Slope Alaska. *Proceedings of the 1994 Hedberg Research Conference on Sequence Stratigraphy of Petroleum Reservoirs and Applications to Development Geology*, Paris.

Richards, M.T., Reading, H.G. & Bowman, M.B.J. (in press) Submarine fan systems I: characterisation and stratigraphic prediction. *Mar. Petrol. Geol.*

Rider, M. (1986) *The Geological Interpretation of Well Logs.* Blackie, London.

Rine, J.M., Tillman, R.W., Stubblefield, W.L. & Swift, D.J.P. (1986) Lithostratigraphy of Holocene sand ridges from the nearshore and middle continental shelf of New Jersey. In: *Modern and Ancient Shelf Clastics: a Core Workshop* (ed. by T.F. Moslow & E.G. Rhodes). Society of Economic Paleontologists and Mineralogists, Tulsa, OK, Core Workshop 9.

Roberts, H.H. & Phipps, C.V. (1988) Proposed oceanographic controls on modern Indonesian Reefs: a turn-off/turn-on mechanism in a monsoonal setting. *Proceedings of the Sixth International Coral Reef Symposium, Townsville*, 3, 529–534.

Rossi, M.E. & Rogledi, S. (1988) Relative sea-level changes, local tectonic settings and basin margin sedimentation in the interference zone between two orogenic belts: seismic stratigraphic examples from the Padan foreland basin, northern Italy. In: *Fan Deltas* (ed. by W. Nemec and R.J. Steel). Blackie, Glasgow, pp. 368–384.

Rudolph, K.W. & Lehmann, P.J. (1989) Platform evolution and sequence stratigraphy of the Natuna Platform, South China Sea. In: *Controls on Carbonate Platform and Basin Development* (ed. by P.D. Crevello, J.L. Wilson, J.F. Sarg & J.F. Read). Special Publication, Society of Economic Paleontologists and Mineralogists, Tulsa, 44, 353–361.

Rust, B.R. (1978) A classification of alluvial channel systems. In: *Fluvial Sedimentology* (ed. by A.D. Miall). Memoir of the Canadian Society of Petroleum Geologists, Calgary, 5, 187–198.

Rust, B.R. (1981) Sedimentation in an arid zone anastomosing fluvial system: Cooper's Creek, Central Australia. *J. Sediment Petrol.*, 51, 745–755.

Rust, B.R. & Legun, A.S. (1983) Modern anastomosing fluvial deposits in arid Central Australia and a Carboniferous anologue in New Brunswick Canada. In: *Modern and Ancient Fluvial Systems* (ed. by J.D. Collinson & J. Lewin). Special Publication, International Association of Sedimentologists, 6, 385–392. Blackwell Scientific Publications, Oxford.

Ryer, T.A. (1981) Deltaic coals of the Ferron Sandstone Member of Mancos Shale: predictive model for Cretaceous coal-bearing strata of Western Interior. *Am. Assoc. Petrol. Geol. Bull.*, 65, 2323–2340.

Ryer, T.A. (1984) Transgressive–regressive cycles and the occurrence of coal in some Cretaceous strata of Utah, U.S.A. In: *Sedimentology of Coal and Coal Bearing Strata* (ed. by R.R. Rahmani & R.M. Flores). Special Publication, International Association of Sedimentologists, 7, 217–227. Blackwell Scientific Publications, Oxford.

Ryer, T.A., Phillips, R.E., Bohor, B.F. & Pollastro, R.M. (1980) Use of altered volcanic ash falls in stratigraphic studies of coal bearing sequences: an example from the Upper Cretaceous Ferron Sandstone Member of the Mancos Shale in central Utah. *Bull. Geol. Soc. Am.*, **91**, 579–586.

Ryseth, A. (1989) Correlation of depositional patterns in the Ness Formation, Oseberg area. In: *Correlation in Hydrocarbon Exploration* (ed. by J.D. Collinson). Graham & Trotman, London, pp. 313–326.

Sadler, P.M., Osleger, D.A. & Montanez, I.P. (1993) On the labeling, length and objective basis of Fischer plots. *J. Sediment Petrol.*, **63**, 360–368.

Saller, A.H., Barton, J.W. & Barton, R.E. (1989) Slope sedimentation associated with a vertically building shelf, Bone Spring Formation, Mescalero Escarpe Field, southeastern New Mexico. In: *Controls on Carbonate Platform and Basin Development* (ed. by P.D. Crevello, J.L. Wilson, J.F. Sarg & J.F. Read). Special Publication, Society of Economic Paleontologists and Sedimentologists, Tulsa, **44**, 275–288.

Salter, T. (1993) Fluvial scour and incision; models for their influence on the development of realistic reservoir geometries. In: *Characterization of Fluvial and Aeolian Reservoirs* (ed. by C.P. North & D.J. Prosser). Special Publication, Geological Society of London, **73**, 33–51.

Sanfilippo, A. & Riedel, W.R. (1985) Cretaceous Radiolaria. In: *Plankton Stratigraphy* (ed. by H.M. Bolli, J.B. Saunders & K. Perch-Nielsen). Cambridge Earth Science Series, Cambridge University Press, Cambridge, UK, pp. 573–630.

Sanfilippo, A., Westberg, M.J. & Riedel, W.R. (1981) Cenozoic radiolarians at Site 462, Deep Sea Drilling Project Leg 61, Western Tropical Pacific. *Initial Rep. Deep Sea Drill. Proj.*, **61**, 495–505.

Sangree, J.B. & Widmier, J.M. (1977) Seismic stratigraphy and global changes of sea-level, part 9: seismic stratigraphic interpretation of clastic depositional facies. In: *Seismic Stratigraphy – Applications to Hydrocarbon Exploration* (ed. by C.E. Payton). Memoir of the American Association of Petroleum Geologists, Tulsa, **26**, 1165–1184.

Sangree, J.B., Vail, P.R. & Mitchum, R.M. (1991) Summary of exploration applications of sequence stratigraphy. In: *Sequence Stratigraphy as an Exploration Tool: Concepts and Practices from the Gulf Coast* (ed. by J.M. Armentrout & B.F. Perkins). Eleventh Annual Research Conference Gulf Coast Section. Society of Economic Paleontologists and Mineralogists, pp. 321–327.

Sarg, J.F. (1988) Carbonate sequence stratigraphy. In: *Sea Level Changes: an Integrated Approach* (ed. by C.K. Wilgus, B.S. Hastings, C.G. St. C. Kendall, H.W. Posamentier, C.A. Ross & J.C. van Wagoner). Special Publication, Society of Economic Paleontologists and Mineralogists, Tulsa, **42**, 155–181.

Sarg, J.F. & Skjold, (1982) Stratigraphic traps in Palaeocene sands in the Balder Area, North Sea. In: *The Deliberate Search for the Subtle Trap* (ed. by M.T. Halbouty). Memoir of the American Association of Petroleum Geologists, Tulsa, **32**, 197–206.

Sassen, R. (1988) Geochemical and carbon isotopic studies of crude oil destruction, bitumen precipitation and sulfate reduction in the deep Smackover Formation. *Org. Geochem.*, **30**, 79–91.

Scaturo, D.M., Kendall, C.G. St. C. & Wendte, J.C. (1987) Simulation of variables controlling carbonate geometry and facies; Judy Creek Reef, West Canada; an example. *Am. Assoc. Petrol. Geol. Bull.*, **71**, 610.

Scaturo, D.M., Strobel, J.S., Kendall, C. St. G., Wendte, J.C., Biswas, G., Bezdek, J. & Cannon, R. (1989) Judy Creek: a case study for a twodimensional sediment deposition simulation. In: *Controls on Carbonate Platform and Basin Development* (ed. by P.D. Crevello, J.L. Wilson, J.F. Sarg & J.F. Read). Special Publication, Society of Economic Paleontologists and Mineralogists, Tulsa, **44**, 3–76.

Schlager, W. (1981) The paradox of drowned reefs and carbonate platforms. *Bull. Geol. Soc. Am.*, **92**, 197–211.

Schlager, W. (1989) Drowning unconformities on carbonate platforms. In: *Controls on Carbonate Platform and Basin Development* (ed. by P.D. Crevello, J.L. Wilson, J.F. Sarg & J.F. Read). Special Publication, Society of Economic Paleontologists and Mineralogists, Tulsa, **44**, 15–25.

Schlager, W. (1992) *Sedimentology and Sequence Stratigraphy of Reefs and Carbonate Platforms.* American Association of Petroleum Geologists, Tulsa, Continuing Education Course Notes Series 34.

Schlager, W. & Camber, O. (1986) Submarine slope angles, drowning unconformities and self-erosion of limestone escarpments. *Geology*, **14**, 762–765.

Schlager, W. & Philip, J. (1990) Cretaceous carbonate platforms. In: *Cretaceous Resources, Events and Rhythms* (ed. by R.N. Ginsburg & B. Beaudoin). NATO ASI Series C **304**, 173–195.

Schlanger, S.O. (1981) Shallow water limestone in oceanic basins as tectonic and paleoceanographic indicators. In: *The Deep Sea Drilling Project: A Decade of Progress* (ed. by J.E. Warme, R.G. Douglas & E.L. Winterer). Special Publication, Society of Economic Paleontologists and Mineralogists, Tulsa, **32**, 209–226.

Schlesinger, W.H. & Mellack, J.M. (1981) Transport of organic carbon in the world's rivers. *Tellus*, **33**, 172–187.

Scholle, P.A., Bebout, D.G. & Moore, C.H. (1983) *Carbonate Depositional Environments.* Memoir of the American Association of Petroleum Geologists, Tulsa, **33**, 708 pp.

Schultz, E.H. (1982) The chronosome and supersome: terms proposed for low-rank chronostratigraphic units. *Bull. Can. Petrol. Geol.*, **30**, 29–33.

Schumm, S.A. (1968) Speculation concerning palaeohydrologic controls on terrestrial sedimentation. *Bull. Geol. Soc. Am.*, **79**, 1573–1588.

Schumm, S.A. (1971) Fluvial geomorphology; channel adjustment and river metamorphosis. In: *River Mechanics* (ed. and published by H.W. Shen). Fort Collins, Colorado, Ch. 5, p. 22.

Schumm, S.A. (1977) *The Fluvial System.* John Wiley & Sons, New York, 338 pp.

Schumm, S.A. (1981) Evolution and response of the fluvial system, sedimentologic implications. In: *Recent and Ancient Nonmarine Depositional Environments* (ed. by F.G. Ethridge & R.M. Flores). Special Publication, Society of Economic Paleontologists and Mineralogists, Tulsa, **31**, 19–29.

Schumm, S.A. (1993) River response to base level change: implications for sequence stratigraphy. *J. Geol.*, **101**, 279–294.

Schumm, S.A. & Brakenridge, G.R. (1987) River responses. In: *The Geology of North America*, Vol. K-3, *North America and the Adjacent Oceans during the Last Deglaciation* (ed. by W.F. Ruddiman & H.E. Wright, Jr.). Geological Society of America, Boulder, Colorado, pp. 221–240.

Schumm, S.A. & Ethridge, F.G. (1991) The effect of baselevel change on the fluvial system (abs). *Geol. Soc. Am. Abstr. Programs*, **23**, A170.

Schuppers, J.D. (1992) Quantification of turbidite facies in a reservoir-analogous submarine fan channel sandbody, south-central Pyrenees, Spain. In: *The Geological Modelling of Hydro-*

carbon Reservoirs and Outcrop Analogues. Special Publication, International Association of Sedimentologists, **15**, 99–112. Blackwell Scientific Publications, Oxford.

Schwarzacher, W. (1975) Sedimentation models and quantitative stratigraphy. *Developments in Sedimentology*, **19**. Elsevier, Amsterdam, 382 pp.

Schwarzkopf, T.A. (1993) Model for prediction of organic content in possible source rocks. *Mar. petrol. Geol.*, **10**, 478–492.

Scoffin, T.P. (1987) *An Introduction to Carbonate Sediments and Rocks*. Blackie, London, 274 pp.

Sedgwick, A. & Murchison, R.I. (1839) On the classification of the older rocks of Devon and Cornwall. *Proc. Geol. Soc. Lond.*, **3**, 121–123.

Seguret, M. (1972) Etude tectonique des nappes et series decollees de la partie centrale due versant sud des Pyrenees. *Publ. USTELA, Ser. Geol. Struct.*, **2**, 155 pp.

Shaffer, B.L. (1987) The potential of calcareous nannofossils for recognizing Plio-Pleistocene climatic cycles and sequence boundaries on the shelf. In: *Innovative Biostratigraphic Approaches to Sequence Analysis: New Exploration Opportunities*. 8th Annual Research Conference, Gulf Coast Section, Society of Economic Paleontologists and Mineralogists, pp. 142–145.

Shanley, K.W. (1991) Sequence stratigraphic relationships and facies architecture of Turonian–Campanian strata, Kaiparowits Plateau, south-central Utah, PhD dissertation, Department of Geology and Geological Engineering, Colorado School of Mines, Golden, Colorado, 390 pp.

Shanley, K.W. & McCabe, P.J. (1989) Sequence stratigraphic relationships and facies architecture of Turonian–Campanian strata, Kaiparowits Plateau, south-central Utah. *Am. Assoc. Petrol. Geol. Bull.*, **73**, 410–411.

Shanley, K.W. & McCabe, P.J. (1990) Tidal influence in fluvial strata — a key element in high resolution sequence stratigraphic correlation. *Am. Assoc. Petrol. Geol. Bull.*, **74**, 762.

Shanley, K.W. & McCabe, P.J. (1991) Predicting facies architecture through sequence stratigraphy — an example from the Kaiparowits Plateau, Utah. *Geology*, **19**, 742–745.

Shanley, K.W. & McCabe, P.J. (1994) Perspectives on the sequence stratigraphy of continental strata. *Am. Assoc. Petrol. Geol. Bull.*, **78**, 544–568.

Shanley, K.W., McCabe, P.J. & Hettinger, R.D. (1993) Tidal influence in Cretaceous fluvial strata from Utah, USA: a key to sequence stratigraphic interpretation. *Sedimentology*, **39**, 905–930.

Shanmugam, G. & Moiola, R.J. (1982) Prediction of deep sea reservoir facies. *Trans. Gulf Cst Assoc. Geol. Soc.*, **32**, 275–281.

Shanmugam, G. & Moiola., R.J. (1985) Submarine fan models: problems and solutions. In: *Submarine Fans and Related Turbidite Systems* (ed. by A.H. Bouma, W.R. Normark & N.E. Barnes). Springer-Verlag, Berlin, Chap. 6, pp. 29–34.

Shanmugam, G., Moiola, R.J. & Damuth, J.E. (1985) Eustatic control on submarine fan development. In: *Submarine Fans and Related Turbidite Systems* (ed. by A.H. Bouma, W.R. Normark & N.E. Barnes). Springer-Verlag, Chap. 5, pp. 23–28.

Shanmugam, G., Spalding, T.D. & Rofheart, D.H. (1993) Process sedimentology and reservoir quality of deep-marine bottom-current reworked sands (sandy contourites): an example from the Gulf of Mexico. *Am. Assoc. Petrol. Geol. Bull.*, **77**, 1241–1259.

Shepard, F.P. (1960) Gulf Coast barriers. In: *Recent Sediments, Northwest Gulf of Mexico* (ed. by F.P. Shepard, F.B. Phleger & Tj. H. van Andel). American Association of Petroleum Geologists, Tulsa, pp. 197–220.

Sherrif, R.E. (1977) Limitations on resolution of seismic reflections and geologic detail derivable from them. In: *Seismic Stratigraphy — Applications to Hydrocarbon Exploration* (ed. by C.E. Payton). Memoir of the American Association of Petroleum Geologists, Tulsa, 26 pp.

Sheriff, R.E. (1985) Aspects of seismic resolution. In: *Seismic Stratigraphy, II* (ed. by O.R. Berg & D.G. Woolverton). American Association of Petroleum Geologists, Tulsa, OK, Memoir 39, 1–10.

Shinn, E.A. (1983) Tidal flat environment. In: *Carbonate Depositional Environments* (ed. by P.A. Scholle, D.G. Bebout & C.H. Moore). Memoir of the American Association of Petroleum Geologists, Tulsa, **33**, 173–210.

Simmons, M.D. & Williams, C.L. (1992) Sequence stratigraphy and eustatic sea-level change: the role of micropalaeontology. *J. Micropalaeontol.*, **11**, 112.

Sissingh, W. (1977) Biostratigraphy of Cretaceous nanoplankton. *Geol. Mijnbouw*, **56**, 37–65.

Sloss, L.L. (1962) Stratigraphic models in exploration. *J. sediment. Petrol.*, **32**, 415–422.

Sloss, L.L. (1963) Sequences in the cratonic interior of North America. *Geol. Soc. Am. Bull.*, **74**, 93–114.

Sloss, L.L., Krumbein, W.C. & Dapples, E.C. (1949) Integrated facies analysis. *Geol. Soc. Am. Mem.*, **39**, 91–124.

Smith, D.G. (1983) Anastomosed fluvial deposits: modern examples from Western Canada. In: *Modern and Ancient Fluvial Systems* (ed. by J.D. Collinson & J. Lewin). Special Publication, International Association of Sedimentologists, **6**, 155–168. Blackwell Scientific Publications, Oxford.

Smith, A.E. (1966) Modern deltas: comparison maps. In: *Deltas and their geologic framework* (ed. M.L. Shirley). Houston Geol. Soc., 234–251.

Smith, D.G. (1986) Anastomosing river deposits: sedimentation rates and basin subsidence Magdalena River, northwestern Colombia, Southern America. *Sediment. Geol.*, **46**, 177–196.

Smith, D.G. & Putman, P.E. (1980) Anastomosed river deposits: modern and ancient examples in Alberta, Canada. *Can. J. Earth Sci.*, **17**, 1396–1406.

Smith, D.G. & Smith, N.D. (1980) Sedimentation in anastomosed river systems: examples from alluvial valleys near Banff, Alberta. *J. Sediment. Petrol.*, **50**, 157–164.

Smith, N.D. (1974) Sedimentology and bar formation in the upper Kicking Horse River, a braided outwash stream. *J. Geol.*, **82**, 205–223.

Sonnenfeld, M. (1991) Anatomy of offlap in a shelf-margin depositional sequence: Upper San Andres Formation (Permian, Guadalupian), Last Chance Canyon, Guadalupe Mountains, New Mexico, USA. In: *Dolomieu Conference on Carbonate Platforms and Dolomitization*, Abstract Volume (ed. by A. Bosellini, R. Brandner, E. Flugel, B. Purser, W. Schlager, M.E. Tucker & B. Zenger). Ortisei, Italy, pp. 254–255.

Sonneland, L., Barkved, O., Olsen, M. and Snyder, G. (1989) Application of seismic wave field attributes in reservoir characterization. In: *Abstracts of the 59th Annual Meeting of the Society of Exploration Geophysicists*, Vol. 2, pp. 813–817.

Stamp, L.D. (1921) On cycles of sedimentation in the Eocene strata of the Anglo-France–Belgium Basin. *Geol. Mag.*, **58**, 108–114.

Steel, R.J. (1974) New Red Sandstone Floodplain and piedmont sedimentation in the Hebridean Province. *J. Sediment Petrol.*,

Stewart, D.J. (1983) Possible suspended load channel deposits from the Wealden Group (Lower Cretaceous) of southern England. In: *Modern and Ancient Fluvial Systems* (ed. by J.D. Collinson & J. Lewin). Special Publication, International Association of Sedimentologists, 6, 369–384. Blackwell Scientific Publications, Oxford.

Stewart, I.J. (1987) A revised stratigraphic interpretation of the Early Palaeogene of the Central North Sea. In: *Petroleum Geology of North West Europe* (ed. by J. Brooks & K.W. Glennie). Graham & Trotman, London, pp. 557–576.

Stille, H. (1924) Grundfragen der vergleichenden Tektonik. Borntraeger, Berlin, 443 pp.

Stoakes, F.A. (1980) Nature and control of shale basin fill and its effect on reef growth and termination: Upper Devonian Douvernay and Ireton Formations of Alberta, Canada. *Bull. Can. Petrol. Geol.*, 28, 345–410.

Stoakes, F.A. & Wendte, J.C. (1988) The Woodbend Group. In: *Devonian Lithofacies and Reservoir Styles in Alberta* (ed. by F.F. Krause & O.G. Burrows). Canadian Society of Petroleum Geology 13th Core Conference, Calgary, pp. 153–170.

Stow, D.A.V. & Bowen, A.J. (1980) A physical model for the transport and sorting of fine grained sediments by turbidity currents. *Sedimentology*, 27, 31–46.

Stow, D.A.V. & Holbrock, J.A. (1984) North Atlantic contourites: an overview. In: *Fine Grained Sediments: Processes and Facies* (ed. by D.A.V. Stow & D.J.W. Piper). Special Publication, Geological Society of London, 15, 245–256.

Stow, D.A.V., Bishop, C.B. & Mills, S.J. (1982) Sedimentology of the Brae Oilfield, North Sea: fan models and controls. *J. Petrol. Geol.*, 5, 129–148.

Stride, A.H. (1982) *Offshore Tidal Sands*. London, Chapman & Hall, 222 pp.

Stride, A.H., Belderson, R.H., Kenyon, N.H. & Johnson, M.A. (1992) Offshore tidal deposits: sand sheet and sand bank facies. In: *Offshore Tidal Sands* (ed. by A.H. Stride). Chapman & Hall, London, pp. 95–125.

Suess, E. (1888) *Das Antlitz der Erde*. Tempsky, Vienna, 778 pp.

Suess, E. (1980) Particulate organic carbon flux in the oceans — surface productivity and oxygen utilization. *Nature*, 288, 260–262.

Summerfield, M.A. (1991) *Global Geomorphology*. Longman, London, 537 pp.

Surlyk, F. (1978) Submarine fan sedimentation along fault scarps on tilted fault blocks (Jurassic–Cretaceous boundary, East Greenland). *Bull. Grnl. Geol. Unders.*, 128, 108 pp.

Suter, J.R. & Berryhill, H.L. (1985) Late Quaternary shelf-margin deltas, north-west Gulf of Mexico. *Am. Assoc. petrol. Geol. Bull.*, 41, 741–746.

Swift, D.J.P. (1968) Coastal erosion and transgressive stratigraphy. *J. Geol.*, 76, 444–456.

Tanner, M.T. & Sherrif, R.E. (1977) Application of amplitude, frequency and other attributes to stratigraphic and hydrocarbon interpretation. In; *Seismic Stratigraphy — Applications to Hydrocarbon Exploration* (ed. by C.E. Payton). Memoir of the American Association of Petroleum Geologists, Tulsa, 26, 301–327.

Taylor, A.M. & Gawthorpe, R.L. (1993) Application of sequence stratigraphy and trace fossil analysis to reservoir description: examples from the Jurassic of the North Sea. In: *Petroleum Geology of Northwest Europe* (ed. by J.R. Parker). Geol. Soc. Lond., 317–336.

Tetzlaff, D.M. & Harbaugh, J.W. (1989) Simulating clastic sedimentation. In: *Computer Methods in the Geosciences* (ed. D.F. Merriam). Van Nostrand Reinhold, New York.

Tipper, J.C. (1993) Do reflections necessarily have chronostratigraphic significance? *Geol. Mag.*, 1, 47–55.

Tissot, B.P., Durand, B., Espitalie, J. & Combaz, (1974) Influence and nature of diagenesis of organic matter in formation of petroleum. *Am. Assoc. Petrol. Geol. Bull.*, 58, 499–506.

Tornqvist, T.E. (1993) Holocene alternation of meandering and anastomosing fluvial systems in the Rhine–Meuse delta (Central Netherlands) controlled by sea level rise.

Trudgill, S. (1985) *Limestone Geomorphology*, Longman, London, p. 196.

Tucker, M.E. (1985) Shallow marine carbonate facies and facies models. In: *Sedimentology: Recent Development and Applied Aspects* (ed. by P.J. Brenchley & B.P.J. Williams). Special Publication, Geological Society of London, 18, 139–161.

Tucker, M.E. (1991) Sequence stratigraphy of carbonate-evaporite basins: models and application to the Upper Permian (Zechstein) of northeast England and adjoining North Sea. *J. Geol. Soc. Lond.*, 148, 1019–1036.

Tucker, M.E. & Wright, V. (1990) *Carbonate Sedimentology*. Blackwell Scientific Publications, Oxford. 496 pp.

Turner, J.P. (1992) Evolving alluvial stratigraphy and thrust front development in the West Jaca piggy-back basin, Spanish Pyrenees. *J. Geol. Soc. Lond.*, 149, 51–63.

Twitchell, D.C., Schwab, W.C., Nelson, H.C., Kenyon, N.H. & Lee, H.J. (1992) Characteristics of a sandy depositional lobe on the outer Mississippi fan from Sea Marc A sidescan sonar images. *Geology*, 20, 289–692.

Tye, R.S., Bhattacharya, J.P., Lorson, J.A. & Sindelar, S.T. (1994) Stratigraphic controls on reservoir performance and development in deltaic deposits of the Ivishak Sandstone, Prudhoe Bay Field, Alaska. *Abstract of the American Association of Petroleum Geologists Regional Meeting Denver*, Colorado.

Tyler, N. & Gholston, J.C. (1988) *Heterogeneous Deep Sea Fan Reservoirs, Shackeford and Preston Waterflood Units, Spraeberry Trend, West Texas*. Bureau of Economic Geology, The University of Texas at Austin, Report of Investigations, 171, 38 pp.

Tyler, N., Galloway, W.E., Garett, C.M. Jr. and Ewing, T.E. (1984) *Oil Accumulation, production characteristics, and targets for additionaly recovery in major oil reservoirs in Texas*. Texas University *Bureau of Economic Geology, Geologic Circular*, 84-2, 31 pp.

Tyler, N., Barton, M., Fisher, R.S. & Gardner, M.H. (1991) *Architecture and Permeability Structure of Fluvial-Delatic Sandstones. A Field Guide to Selected Outcrops of the Ferron Sandstone, East-Central Utah*. Bureau of Economic Geology, The University of Texas at Austin, 103 pp.

Underhill, J. & Partington, M. (1993) Jurassic thermal doming and deflation in the North Sea: implications of the sequence stratigraphic evidence. In: *Petroleum Geology of Northwest Europe: Proceedings of the 4th conference* (ed. by Parker). Geological Society of London, Bath, pp. 337–346.

Vail, P. & Todd, R.G. (1981) North Sea Jurassic unconformities, chronostratigraphy and global sea level changes from seismic stratigraphy. In: *Petroleum Geology of the NW Continental Shelf, Proceedings*, pp. 216–235.

Vail, P.R. & Wilbur, R.O. (1966) Onlap, key to worldwide unconformities and depositional cycles. *Am. Assoc. Petrol. Geol.*

Bull. (abs.), **50**, 638–639.

Vail, P.R. & Wornardt, W.W. (1990) Well log seismic sequence stratigraphy: an integrated tool for the 90's. In: *Sequence Stratigraphy as an Exploration Tool: Concepts and Practices from the Gulf Coast* (ed. by J.M. Armentrout & B.F. Perkins), Eleventh Annual Research Conference, Gulf Coast Section, Society of Economic Paleontologists and Mineralogists, pp. 379–388.

Vail, P.R., Mitchum, R.M., Jr, Todd, R.G., Widmier, J.M., Thompson, S., III, Sangree, J.B., Bubb, J.N. & Hatleid, W.G. (1977a) Seismic stratigraphy and global changes in sea level. In: *Seismic Stratigraphy — Applications to Hydrocarbon Exploration* (ed. by C.E. Payton). Memoir of the American Association of Petroleum Geologists, Tulsa, **26**, 49–62.

Vail, P.R., Mitchum, R.M., Jr & Thompson, S., III (1977b) Seismic stratigraphy and global changes of sea-level, part 3: relative changes of sea level from coastal onlap. In: *Seismic Stratigraphy — Applications to Hydrocarbon Exploration* (ed. by C.E. Payton). Memoir of the American Association of Petroleum Geologists, Tulsa, **26**, 63–82.

Vail, P.R., Mitchum, R.M., Jr & Thompson, S., III (1977c) Seismic stratigraphy and global changes of sea level, part 4: global cycles of relative changes in sea level. In: *Seismic Stratigraphy — Applications to Hydrocarbon Exploration* (ed. by C.E. Payton). Memoir of the American Association of Petroleum Geologists, Tulsa, **26**, 83–97.

Vail, P.R., Hardenbol, J. & Todd, R.G. (1984) Jurassic unconformities, chronostratigraphy and sea level changes from seismic stratigraphy. In: *Interregional Unconformities and Hydrocarbon Exploration* (ed. by J.S. Schlee). Memoir of the American Association of Petroleum Geologists, Tulsa, **33**, 129–144.

Vail, P.R., Audemart, F., Bowman, S.A., Eisner, P.N. & Perez-Cruz, G. (1991) The stratigraphic signatures of tectonics, eustasy and sedimentation — an overview. In: *Cyclic Stratigraphy* (ed. by G. Einsele, W. Ricken & A. Seilacher). Springer-Verlag, New York, pp. 617–659.

Van Gorsel, J.T. (1988) Biostratigraphy in Indonesia: methods, pitfalls and new directions. In: *Proceedings of the Indonesian Petroleum Association Seventeenth Annual Convention*, October 1988, pp. 275–300.

Van Vliet, A. & Schwander, M.M. (1989) Stratigraphic interpretation of a regional seismic section across the Labuan syncline and its flank structures, Sabah, North Borneo. In: *Atlas of Seismic Stratigraphy*, Vol. 3 (ed. by A.W. Bally), pp. 163–167.

Van Wagoner, J.C. (1992) Non marine sequence stratigraphy and facies architecture of the Updip Desert and Castlegate Sandstones in the Book Cliffs of western Colorado and eastern Utah Sections 3 & 4. In: *Sequence Stratigraphy. Applications to Shelf Sandstone Reservoirs: Outcrop to Subsurface Examples* (ed. by J.C. Van Wagoner, C.R. Jones, D.R. Taylor, D. Nummedal, D.C. Jennette & G.W. Riley), American Association of Petroleum Geologists Fieldtrip Conference, September 1991.

Van Wagoner, J.C., Posamentier, H.W., Mitchum, R.M., Vail, P.R., Sarg, J.F., Loutit, T.S. & Hardenbol, J. (1988) An overview of the fundamentals of sequence stratigraphy and key definitions. In: *Sea-Level Changes: An Integrated Approach* (ed. by C.K. Wilgus, B.S. Hastings, C.G. St C. Kendall, H.W. Posamentier, C.A. Ross & J.C. Van Wagoner). Special Publication, Society of Economic Paleontologists and Mineralogists, Tulsa, **42**, 39–45.

Van Wagoner, J.C., Mitchum, R.M., Jr, Campion, K.M. & Rahmanian, V.D. (1990) Siliciclastic Sequence Stratigraphy in Well Logs, Cores and Outcrop: Concepts for High Resolution Correlation of Time and Facies. American Association of Petroleum Geologists Methods in Exploration Series, Tulsa, **7**, 55 pp.

Van Wagoner, J.C., Jones, C.R., Taylor, D.R., Nummedal, D., Jenette, D.C. & Riley, G.W. (1991) *Sequence Stratigraphy: Applications to Shelf Sandstone Reservoirs Outcrop and Subsurface Examples*. American Association of Petroleum Geologists, Tulsa, Field Conference.

Verdier, A.C., Oki, T. & Suardy, A. (1980) Geology of the Handil field (east Kalimantan, Indonesia). In: *Giant Oil and Gas Fields of the Decade 1968–1978* (ed. by M.T. Halbouty). American Association of Petroleum Geologists, Memoir, Tulsa, OK, **30**, 399–422.

Visser, M.J. (1980) Neap-spring cycles reflected in Holocene subtidal large-scale bedform deposits: a preliminary note. *Geology*, **8**, 543–546.

Walkden, G. & Davies, L.J. (1983) Polyphase erosion of subaerial omission surfaces in the Late Omentian of Anglesey, North Wales. *Sedimentology*, **30**, 861–878.

Walker, R.G. (1978) Deep water sandstone facies and ancient submarine fans: models for exploration for stratigraphic traps. *Am. Assoc. Petrol Geol. Bull.*, **62**, 932–966.

Walker, R.G. (1985) Mudstone & thin-bedded turbidites associated with the Upper Cretaceous Wheeler Gorge Conglomerates, California: a possible channel-levee complex. *J. Sediment. Petrol.*, **55**, 279–290.

Walker, R.G. (1990) Facies modelling and sequence stratigraphy. *J. Sediment. Petrol.*, **60**, 777–786.

Walker, R.G. (1992) Turbidites and submarine fans. In: *Facies Models: Response to Sea Level Change* (ed. by R.G. Walker & N.P. James). Geoscience Canada, pp. 239–263.

Walker, R.G. & Eyles, C.H. (1991) Topography and significance of basin-wide sequence bounding erosion surfaces in the Cretaceous Cardium Formation, Alberta, Canada. *J. Sediment. Pet.*, 473–498.

Walker, R.G. & James, N.P. (1992) *Facies Models: Response to Sea Level Change*. Geoscience Canada.

Wanless, H.R. & Shepard, F.P. (1935) Permo-Carboniferous coal series related to Southern Hemisphere glaciation. *Science*, **81**, 521–522.

Warson, H.J. (1982) Casablanca Field, offshore Spain, a paleogeomorphic trap. *Am. Assoc. Petrol. Geol. Bull.*, **32**, 237–250.

Webb, G.W. (1981) Stevens and Earlier Miocene turbidite sandstones, southern San Joaquin Valley, California. *Am. Assoc. Petrol. Geol. Bull.*, **65**, 438–465.

Weimer, P. (1990) Sequence stratigraphy, facies geometry and depositional history of the Mississippi fan, Gulf of Mexico. *Am. Assoc. Petrol. Geol. Bull.*, **74**, 425–453.

Weise, B.R. (1980) *Wave Dominated Delta Systems of the Upper Cretaceous San Miguel Formation, Maverick Basin, South Texas*. Report of Investigations, 107, Bureau of Economic Geology, University of Texas, Austin, 33 pp.

Wells, N.A. & Dorr, J.A., Jr (1987) Shifting of the Kosi River, northern India. *Geology*, **15**, 204–207.

Westcott, W.A. (1993) Geomorphic thresholds and complex response of fluvial systems — some implications for sequence stratigraphy. *Am. Assoc. Petrol. Geol. Bull.*, **77**, 208–218.

Weuller, D.E. & James, W.C. (1989) Braided and meandering submarine fan channel deposits, Tesnus Formation, Marathon Basin, West Texas. *Sediment Geol.*, **62**, 27–45.

Wheeler, D.M., Scott, A.J., Devine, P.E. & Home, J.C. (1990) A spectrum of transgressive valley fill sequences: implications for sequence stratigraphy and petroleum exploration. *Am. Assoc. Petrol. Geol. Bull.* (abs.), **74**, 789.

Wheeler, H.E. (1958) Time stratigraphy. *Am. Assoc. Petrol. Geol. Bull.*, **42**, 1047–1063.

Wignall, P.B. (1991a) A model for transgressive black shales. *Geology*, **19**, 167–170.

Wignall, P.B. (1991b) Test of the concepts of sequence stratigraphy in the Kimmeridgian of England and northern France. *Mar. Petrol. Geol.*, **8**, 430–441.

Wilcox, D.N. (1967) Coarse bedload as a factor in determining bedslope. *Publ. Intl. Assoc. Sci. Hydrol.*, **75**, 143–150.

Wilgus, C.K., Hastings, B.S., Kendall, C.G. St. C., Posamentier, H.W., Ross, C.A. & Van Wagoner, J.C. (eds) (1988) *Sea-level Changes: An Integrated Approach*. Special Publication, Society of Economic Paleontologists and Mineralogists, Tulsa, **42**, 407 pp.

Wilkinson, B.H. (1979) Biomineralization, paleoceanography and the evolution of calcareous marine organisms. *Geology*, **7**, 524–527.

Williams, G.L. (1977) Dinocysts: their paleontology, biostratigraphy and paleoecology. In: *Oceanic Micropalaeontology* (ed. by A.T.S. Ramsey). Academic Press, London, pp. 1231–1325.

Wilson, J.L. (1975) *Carbonate Facies in Geologic History*, Springer-Verlag, Berlin, 471 pp.

Winn, R.D., Jr. & Dott, R.H., Jr. (1979) Deep-water fan-channel conglomerates of Late Cretaceous age, southern Chile. *Sedimentology*, **26**, 203–228.

Wood, G.V. & Woolfe, M.J. (1969) Sabkha cycles in the Arab/Darb Formation off the Trucial coast of Arabia. *Sedimentology*, **12**, 165–191.

Wood, J.M. & Hopkins, J.C. (1989) Reservoir sandstone bodies in estuarine valley fill: Lower Cretaceous Glauconitic Member, Little Bow Field, Alberta, Canada. *Am. Assoc. Petrol. Geol. Bull.*, **73**, 1361–1382.

Wood, L.J. (1991) Effects of basin type on coastal plain-shelf slope systems during base level fluctuations: an experimental approach. *Am. Assoc. Petrol. Geol. Bull* (abst), **75**, 969.

Wood, L.J., Ethridge, F.G. & Schumm, S.A. (1991) Influence of subaqueous shelf angle on coastal plain-shelf slope deposits resulting from a rise or fall in base level. *Am. Assoc. Petrol. Geol. Bull.* (abst), **75**, 969.

Wright, L.D. (1977) Sediment transport and deposition at river mouths: a synthesis. *Bull. Geol. Soc. Am.*, **88**, 857–868.

Wright, V.P. (1986) Facies sequences on a carbonate ramp: the Carboniferous Limestone of South Wales. *Sedimentology*, **33**, 221–241.

Zaitlin, B.A. & Schultz, B.C. (1990) Wave-influenced estuarine sandbody, Senlac heavy oil pool, Saskatchewan, Canada. In: *Sandstone Petroleum Reservoirs* (ed. by J.H. Barwis, J.G. McPherson & J.R.J. Studlick). Springer-Verlag, New York, pp. 363–387.

Zezza, F. (1975) Significance of the subsidence collapse phenomena in the carbonate areas of southern Italy. *Mem. Geol. Soc. Italy*, **14**, 9–34.

Index

Page numbers in **bold** type refer to tables, those in *italics* refer to figures.

AAPG Memoir 26, 6, 45
abundance events, planktonic 101
accommodation (space/volume) 6, 74, 104, **120**, 138–9, 153
   controlled by eustasy and subsidence 16
   exceeded (carbonate production) 214
   filling of 23–4
   foreland-basin margins 15
   increased 28, 118, 176, 193
   and law of sigmoidal growth 213
   parasequences in paralic successions 145, *148*, 148, *149*
   results of changes in *124*, 130–3
   shelfal 24, 193
   through time 16–17, *20*, 133
   topset 24, 27, 29, *29*, 32
accumulation/relative sea-level/facies relationship 23, 23–4
aerial photographic data 87
aggradation 24, 62, 72, 75, *132*
   by carbonate 227
   and coal formation 242
   coastal plains 242
   flood-plain 133
   fluvial 122
Akhdar Group supercycle 261
algae 212
allocyclic controls 128, 181
allocyclic events 117
alluvial sand-body architecture 127–8
alluvial strata
   highstand systems tracts in 125
   sequence boundaries and lowstand systems tracts in 122–3
   transgressive systems and flooding surfaces in 123–4
alluvial successions, problem of sequence boundary recognition 122
anoxia 243, 244, 248
   conditions for development 239, 241, 245, 247, 250–1
anoxic bottom waters 239, 241, 244, *249*, 257
apparent conformity 54, 81
apparent extinctions 104
apparent sequence boundaries 237
apparent toplap surface 53
apparent truncation 54, *54*, 58, 60, 81, 83
Arab Formation (Abu Dhabi) 234
Arabian Platform, hierarchical cyclicity 267
atolls 220
autocyclic controls/mechanisms 117, 181, 235, 235–6
autocyclic sedimentary processes 11, 18, 36, 75, 154
automatic gain control (AGC) 50
avulsion 36

Bahamas, development of different platform margin types 231, *234*
Bahamas platform, an overproducing system 214

base level 14, 16, *17*, **120**, 191
   fall and rise cycle 26
   fluvial system response to falls in *121*
   rapidly rising 122
      preference for anastomosing channels 112
baselap 53
basin architecture 24, *25*
basin fill
   computer modelling of 258–69
      forward model types 259–61
      practical forward modelling 263–9
   wedge-shaped 13
basin-floor fan systems, vs. slope fan systems 191, 193–7
basin-floor fans 27–8, 206, 266
   seismic expression of 193–4, *195*, 197
   on wireline logs 197
basin-margin concepts 13–15
basin-margin types 15
basin-margin units, prograding 54
basinal environments, log responses from 75
basinal fan unit 58
basins
   extensional 11–13
   intracratonic and back-arc 241
   nutrient starved 98
   permanently anoxic 244, *247*
   silled 252
'bay line' concept 30
beach 137
Beatrice Field, interpretation tools, comparison of resolution 46, 61
bedload geology 118
Belize, erosional shelf-margins off 224
Belize lagoon 212, 227
bell trend *see* dirtying-up trend
benthic marker–lithofacies relationship 89
benthos
   environmental controls on distribution of 93
   epifaunal 98, 104
   and palynofacies 91–3
   and water depth 94
bias 49
biochronozones 90–1
biofacies 93, 95–6, *96*
   correlations based on 106
   diachronous, retrogradational 101
biofacies indicators, actual and apparent 95
biofacies trends 107
biostratigraphy 89–107
   and sequence stratigraphy 97–104
Blackhawk Formation (Utah) 123, *125*
Bone Spring Formation (New Mexico) 220
bottom currents 180
bottomsets *14*, 14
bow trend 69, 70, 71, 73
boxcar trend 69, 71, 72, 75, 197
braid deltas 136
braided channel deposits **120**, 123
braided channel systems 113–15, 118, 127
   incision by 122

Brent Group
   log suite and stratigraphic interpretation of 68, 76
   sequence boundaries in *78*, 78
Brown Limestone Formation (Egypt) 251–2, *257*
build-up
   intercarbonate 249–50, *252*
   'pinnacle' type 231, *232*
   'platform' type 231, *233*

calcisponges 212, 221
calciturbidites 214, 231
calcrete 226
caliche soils 127
Cap Blanc, Mallorca, observed and simulated reef 261, *265*
carbonate 216
   effects of exposure 219
   pelagic 104
   peritidal 222
   redeposition into deep water 214
   Shublik Formation 128, *130*, 130, 131, 133
'CARBONATE', carbonate modelling program 261, *265*
carbonate cycles 237
   in arid settings 232, 234
   in more humid settings 234
   origins 235–6
   and parasequences — examples 232, 234, *235*
   stacking patterns 234–5
carbonate factory 211, 214
carbonate margins
   backstepping 218
   erosion, bypass and accretionary 220–1, *221*
   sediment composition vs. slope angle 221, *222*
carbonate ooze sedimentation 216
carbonate platforms 211
   classification of 220–1
   cyclicity and parasequences on 232, 234–6
   differential growth 212
   drowning of 217–18, **218**, *219*, 225, 226–7, 227, 231
   effects of geometry of 220–1
   exposure of 218–20
   facies belts on 221–4
   importance of slope angle 220–1
   isolated 216, 220, 224
      symmetrical and asymmetrical 231, *232, 233, 234*
   ramp-to-rimmed shelf progression 227
   sequence stratigraphic models for 224–31, *232*, 236–7
   subaerial exposure 218–20
carbonate precipitation 220
carbonate production 215, 237
   lagoonal 228
   organic and inorganic 211–14

291

carbonate ramps 211, 220
  development into rimmed shelves 222
  drowning of 226–7
  facies belts 222, 222–3
  petroleum source-rock accumulation during transgression 225
  sequence stratigraphy of 225, 225–7
carbonate reef, Mallorca, observed and simulated 261, 265
carbonate sedimentation, controls on 211–20
carbonate systems 7, 211–36
  inorganically dominated 211–12
  low-productivity 251
  organic, sigmoidal growth pattern 213
carbonate-evaporite cycles 232, 234
Cardium Formation (Alberta) 162, 162, 163
  parasequence-set signature 153
Castell Coch Limestone, oolitic facies 226
Castissent Formation (Spain), fluvial/deltaic deposition, small thrust-top basin 126–8, 129
Castlegate Member lowstand systems tract 123
Catalan Basin (Spain), ramp drowning 226–7
cementation, marine 231
channel belts, increased lateral migration 118
channel cycles, and parasequences 62
channel facies
  high-sinuosity systems 113
  stacked 62
channel fills, fining-upwards 153
channel sand bodies 30, 120, 185
channel sands, fluvial 71
channel sandstone 64, 120
channel stacking patterns, change in 120, 123
channel systems
  anastomosing 111–12, 112, 118, 120
  high-sinuosity 112–13
  low-sinuosity 113–15, 118
channel-levee complexes/systems 28, 181, 188, 191, 197, 206
chronosomes 80, 266, 267
chronostratigraphic charts 80–8
  construction of
    from seismic data 80–3
    from other data 87–8
  filling in 82–3
  from well data 87–8
  interpretation of 83–4
  purpose of 80
  scaling to absolute time 83
cleaning-up units 77
cleaning-upwards trend (motif)/signature 69, 69, 72, 74, 153, 158–9, 167
  Kakwa Member 162
  in shallow marine settings 70, 70
climate 11, 20, 118, 227
  and carbonate cycles 232, 234
  effect on exposed carbonate surfaces 219–20, 224, 226, 237
  influence of 148, 151, 224
  and kerogen production/preservation 242–3
  past, from fossil assemblages 96
climate modelling, and sediment input 263–4
clinoforms 13–14, 14, 24, 54, 55, 84, 173
  log response of 72, 74
  low-angle 226
  prograding 83–4
  seismic and cleaning-up motifs 70
  shelf deltas 171
  shingled 225
  slope angle 14
clipping (of seismic data) 49
coal 64, 101, 123, 124, 238, 248, 252
  autochthonous 241–2
  controls on formation of 242
  in lowstand systems tracts 243
  sequence stratigraphic significance of 241–2
  thick, regionally extensive 243
coal beds 36, 74, 84
coarsening-up signature 62, 62, 64, 153, 158–9, 162
coarsening-upwards units 36, 70
coastal onlap 13, 26, 30, 53, 81, 81–2, 83, 122
  asymmetrical cycles 85
  downward shift in 30, 34, 39, 56, 57, 78, 193
  lateral movement of 84
  significance of 60
coastal onlap charts 80, 85
coastal onlap curves 84–5, 265
coastal onlap point 74, 225
coastal plain facies 23, 24
coastal plain to shoreline-shelf systems 137, 156
  stratigraphy of 152–3
coastal upwelling 257
  of nutrient-rich water 239, 241
cohesive flows 178, 179
complex attributes 48, 50, 50
composite sequences 18, 24, 33, 33, 144
composite systems tracts 33, 60
computer modelling, of basin fill 258–69
  examples of stratigraphic use 261–3
  forward or inverse 260–61
  forward model types 259–61
  practical forward modelling 263–69
condensed facies 29, 34, 244, 247, 252
  anoxic 241, 243
condensed intervals/units 24, 35
  hemipelagic, abundant fossil assemblages 101
  marine 53, 72, 75, 77, 78
  shelfal 77
  see also marine condensation
condensed sections 29, 36, 40, 67, 176, 218
  associated with maximum flooding surfaces 101, 104
  chronostratigraphic expression of 82, 83
  fossil and seismic ties in 106
  as source-rock intervals 242
conglomerate 133, 163, 167, 185
Cooking Lake platform 250
corals 212
Crati fan 206
crevasse-splay deposits/sandstones 64, 138, 139
cross-bedding, dune(-scale) 63, 146–7, 153, 156, 166
cross-beds 152
cross-cutting surfaces 45
cross-lamination 167
  ripple-scale 62
cross-stratification 114–15
  hummocky 152, 158–9
crustal shortening 81
current activity, effect on carbonate platforms 221
cyclicity 261
  Hanifa Formation 251
  orders of, and global correlation 17–20
  and parasequences on carbonate platforms 232, 234–6
  see also periodicity
cyclothems, Pennsylvanian 6
cylindrical trend see boxcar trend

Dampier Basin (Australia), prograding rimmed shelf systems 227, 230
debris flows 28, 104, 179, 185, 188, 191, 223
deep-marine clastic systems 178–210
  classification of 180–91
  controls on development of 181, 181
  depositional processes 178–80

principal architectural elements 182
  sedimentological charateristics of 183
  subsurface characteristics of 184
deformation, syn-sedimentary 173
delta lobe switching 36, 74, 154, 160, 161–2
delta lobes, and coal formation 242, 243
delta plains 136, 154
  maximum flooding zones 144–5
deltaic systems 131, 234
  stratigraphy off 153–4
deltas 134–5, 136
  delta physiography 135–6
  fluvially-dominated 136
  grain size in 148
  gravel shelf 158–9
  and sedimentary processes 136–7
  terminology 135
  tide-dominated 137, 156, 158–9, 176
  wave- and storm-dominated 136–7, 158–9, 176
density (FDC) log 67
density-neutron suites 75
  linking lithology and depositional trend 67, 68
denudation 20, 22, 23
depositional environments
  correlation procedure 155–7
  paralic, and seismic facies 174
depositional packages see systems tracts
depositional processes, deep-marine clastic systems 178–80
depositional shoreline break 14
depositional styles, chronostratigraphic expression of 83–4
depositional systems
  biofacies boundaries and movement in 95
  fluvial 115, 120
  paralic 134–7
Desert Member lowstand systems tract 123
diagenesis, meteoric, effects of 219–20
diffusion models 261
diluvial theory 4–5
dirtying-up trend 62, 69, 70–1, 71, 73
dirtying-up units 77
disconformities 130
dissolved oxygen minimum 242
distributary channels 154, 157, 160, 161
Dolomites, Italian
  prograding Triassic carbonates 64, 66
  ramp-to-rimmed-shelf transition 227
  slope angles, carbonate platforms 221, 222
  'pentacycle' in 235
Dornoch Formation 58
Douvernay Formation 250, 253–4
downlap 53, 64, 75, 83, 225
downlap point 225
downlap surfaces 28, 66, 74, 77, 101, 227, 244
drainage basins 20
drowning unconformities 217–18, 219, 225, 237
dunes, aeolian 137, 151
Dunvegan Formation 162

E/T surfaces 39, 122, 162
El Paso Group (West Texas), use of Fischer plots 235, 236
environmental factors, and carbonate platform drowning 217–18, 218
equilibrium point 16
erosion 84, 140
erosion surfaces 83, 163
erosional truncation 53–4, 81
erosional-aggradational cycles 127–8
escarpment margins 211, 220, 227–8, 230, 231
  drowning of 231
  facies belts 224

estuaries 29, 122, 137
  tide-dominated 137, *139*
  wave-dominated 137, *138*
estuarine systems *163*
  stratigraphy of 155
Ettrick Field, North Sea 71, *73*
  topset and clinoform, controls on 84
  see also seismic facies; shoal facies
eustasy 11, 17, 235
  glacial 5, 18
  global 5, 15–16
  tectono-eustasy 17
eustatic control on deposition, theory of 18
eustatic cycles 17, 33
eustatic sea-level 16
  fall in and fan development 191, *192*
eustatic signal, global 19–20
evaporites 71, 220, 232
  and evaporitic structures 222
  as intraformational seals 234
  lowstand 227
  sabkha 226
event beds 167
'expanding puddle' model 245, *248*
extension 11, 81, 197, 203, 241
extinctions, platform carbonate 214

facies
  aeolian 226
  basinward shift in 98, **120**, 122, 139, 142, 167
  changes in represented by seismic surfaces 81
  channel-fill 188
  estuarine 167
  marine, mud-prone 23
  oolitic 226
  rapid variations, gravel-rich systems 185
  sand-rich 185
facies associations, and parasequence geology 62
facies belts
  on carbonate platforms 221–4
  controls on 20
  on a rimmed shelf *223*
  stacking of 242
facies dislocation 26, 39, 64, 66, 77, 176
fan deltas 136
fan formation 27–8
fan lobes 54
  channelized 185
  constructional 188, *189*, 197
  down-dip 188
fan systems
  basin-floor vs. slope 191, 193–7
  development during highstand and transgression 206, 210
fast AGC 50
fault truncation 54
fault-plane surfaces 81
faults
  extensional 15, 204
  see also growth faults
faunal abundances 77, 96, 101
  minimum 97
fining-up trend 71
fining-up units, large-scale 71
fining-upwards succession 133, 152, 154
finite-difference numerical solutions 260–1
firmgrounds 167
first order (continental encroachment) cycles 17, 241
first/last appearance data 90
Fischer plots 234–5, 237
flood plains 120
flooding surfaces 64, 123–4, 153, 155
  generated by eustatic sea-level rise 161
  Viking Formation 62, *63*, *146–7*, 166

fluidized/liquidized flows 178, **179**
fluvial architecture 117–25
  controls on 117–22
  reconstruction of 125–33
fluvial avulsion cycles 36
fluvial channel styles 124, 127
  evolution of *129*, *131*, *133*
  and fluvial processes 111–15
  vertical variation in 123
fluvial sand bodies 123, 126, 131, 133
fluvial systems
  anastomosing 111–12, *112*, 118, **120**
  classification of 115
  downstream controls 119, 121–2
  geographical division of 117, 117–18
  high-sinuosity 112–13, *113*, 123
  low-sinuosity 113–15, *114*
  midstream controls 118–19
  responses
    to base level rise 122
    to relative sea-level fall 121–2, *124*
    to relative sea-level rise *126*
    to uplift and subsidence *119*
  upstream controls 118
foraminifera
  benthic 93, 95
  planktonic 93
forced regressions 141–3
  producing stratigraphic traps 168
forced regressive systems tract 32–3
forced regressive wedge systems tract 31
forced regressive wedges 31
foreland basins 13
fossil assemblages
  highstand systems tracts 104, *105*
  lowstand systems tracts 98
  shelfal 104
  transgressive systems tracts 101, *102*
fossil groups 89–90, 106
fossil signature
  lowstand fan 99
  lowstand wedge 98, *100*
fossil zonation schemes, and biochronostratigraphic resolution 90–1
fossils
  inception/extinction events, usefulness of 91
  preservation of 90
  reworked 97–8, 98
  terrestrially-derived, indicating fluvial influx 96
fourth order 'parasequence' cycles 18, 242
Fresnel Zone 47, *48*
fringing reefs 216, *217*, 220, 231

gain (seismic data) 50
gamma-ray logs 61, 67
  see also wireline logs
genetic stratigraphic sequences/units 7, 34–5
geological periods 4
geometrical models 260, 269
geophysical workstation technology, use of 55
Gilbert deltas 136, *136*, 153
glacial theories, Lyell and Agassiz 5
global coastal onlap chart, Haq 85
global eustatic signal 19–20
Glossifungites ichnofacies 168, *169*
Golden Meadow field, Louisiana 261, *262*
graded stream profile
  changes in 118
  concept of 115–17, *117*
  effect of coastal and marginal marine processes 122
  gradient changes imposed on lower reaches 119, 121
grain flows 178, **179**
grain size 148, 176
  in deep-marine clastic systems *182*, 188

and fluvial depositional systems 115, 115, *116*
  increase in, and facies dislocations 39
  and seismic expression of fan systems 182, 194, 197
  shelf deltas 158–9
Grantham Formation 226
gravity flow deposits 104
gravity flow processes 98, 136, 178, **179**
Great Bahama Bank 231
Gristhorpe Member 64
growth faults 154, *160*, 197, 203–4
growth-fault margins 15, *15*, 195, *205*
'gull-wing' seismic geometry 188, 191, 197
gullies, erosional 191

Hanifa Formation (Saudi Arabia) 250, 251, *255*
Haq global sea-level cycle chart 6, 18
  criticised 19
hardgrounds 36, 101
'healing phase' component, transgressive systems tracts 29
heavy minerals, influencing logs 67
hiatus(es) 83, 98
  biostratigraphic 97
  erosional and non-depositional 98
  marine 24, 81
  non-marine 84
high-resolution sequence stratigraphy 176–7
  correlations within paralic successions 155–67
  and parasequences 35–41
  problems and pitfalls 41
highstand, margin failure during 231
highstand fan systems 206
highstand prograding wedges 39, 74, 231
highstand sequence set 33
highstand shedding 214–16, 227, 231, 237
highstand systems tracts 30, 64, 75, 104, *105*, 176
  in alluvial strata 125, *127*
  carbonate ramps 225, 225–6
  fluvial deposition patterns and relative sea-level change **120**
  potential for source rock development 248, *251*
  as a prograding basin-margin unit 79
  on a ramp margin 31–2
  recognition of on seismic data 60
hydraulic process–response models 259–60, 261–69
hydrocarbon migration 168, 169

incised valley-fill deposits 31, 64, 78, 122, 123, 126, 127, 168
  in paralic successions 138, 139–40
  up-dip 197
incised valleys 27, 28, 39, 41, 63, 77, **120**, 122, 193
  further reasons for 123
  recognition of 137–40
  as stratigraphic traps 168, *170*, *171*
  variable patterns 140
incision 27, 31, 60, 118, 123, 128, 137
  by braided river systems 122
  and falling base levels 121
  fluvial 231
intercarbonate build-up, source rocks 249–50, *252*
interfan lobes 98
interfluve sequence boundaries 63, 123, 140, 146–7, 167
interfluves, raised 119
intraplate stress 7, 19
intraplatform depression, and source rocks 250–1, *254*

Ireton Formation 250, *253—4*
irregular trends 69, *71—2*
isotope stratigraphy *106—7*
Ivishak Formation (Alaska) 128, *130—3*

Jebel Bou Dahar (Moroccan Atlas) 227
Judy Creek carbonate platform complex, simulation of 261, 263, 266

Kakwa Member *162, 163*
karst/karstification 226, 227, 231
　in carbonate reservoirs 220
　see also palaeokarst
Kavik Formation (Alaska), shelf mudstones 128
kerogen, oil- and gas-prone *242—3*
Kimmeridge Clay Formation (North Sea), wireline log response 249

La Luna Formation (S. America) 251
Labuan syncline (offshore Sabah), seismic facies of paralic successions 175
lacustrine flooding surfaces 154
lagoonal successions 145, *152—3*, 223
lags, 36, 62, 63, *146—7*
　basal 71
　coarse 104
　transgressive 39, 140, 167
lapout 53
Latemar Limestone (Italy) 233
Leduc Formation, pinnacle reefs 250, *253—4*
light intensity, and reef distribution *212*, 212
limestone–dolomite pairs 233
Lincolnshire Limestone Formation, ramp transgressive system *225*, 226
Lisburne Group (Alaska) 128
lithospheric stretching 11, 13
Liuhua carbonate platform, Pearl River Mouth Basin *217—18*, *220*, 231
Llanos Basin, Colombia, tectonic subsidence curve *13*, 13
log markers 75
logs *see* wireline logs
low-nutrient environment, and carbonate platform drowning **218**
lowstand
　chemical processes acting on carbonates 220
　delta-fluvial braid plain complex 131
　escarpment margins 231
　fan development during *191—206*
　rimmed shelves 227
lowstand deposits
　conglomeratic *163*
　fans, exotic fossil assemblages in 98
　fluvial, complexities of 123
　wedge deposits 98
lowstand evaporite wedge 226, 227
lowstand fans *27—8*, 33, 58, 98, 99, 197, 206
　carbonate fans 216
　mud-prone 98
　recognition of 78
　sand-rich 98
lowstand prograding wedge *28—9*, 39, 57, 74, 75, 78, 176, 197, 243
lowstand systems tracts *26—9*, 98, 176
　and deep-marine clastic systems 197, *203—4*
　effect of sediment calibre on evolution of *205—6*, *207, 208, 209*
　fluvial depositional patterns and relative sea-level change **120**
　potential for organic-rich facies development *243—4, 246*
　on a ramp margin *31—2*, *225*, 226
　recognition of on seismic data *57—8, 58*
　and sequence boundaries in alluvial strata *122—3*

lowstand wedge 31, *58, 58*, 98
Luconia Province (offshore Sarawak)
　carbonate platform drowning **218**, *218*, 218
　platform-type build-up 231, *233*

magneto-stratigraphy 125
Maldive Atolls 211, *212*
margins 211
　foreland-basin *15*, 15
　oolitic 212
　see also growth-fault margins; passive continental margins; ramp margins
marine condensation 74, 81, 83
marine flooding surfaces 36, 37, 40, *74*, 74, 101, 123, *162*, 169
　recognition of 62
　as subregional correlation markers 37
marine microfaunas, clear-water 101
marine onlap 53, 60, 81
　boundaries 83
marine onlap surfaces 57
marine processes 174
mass flow processes.185, 188, 206
mass wasting 104, 185
mathematical inversion 267, 270
maximum flooding surfaces 29, 32, 33, 40, 77, 79, 107, 176, *235*, 244
　at basin margins 104
　as cycle boundary (genetic stratigraphic sequences) 34, *34—5*
　development of widespread condensed sections 101
　fluvial deposition patterns associated with relative sea-level change **120**
　forming regional seals 169
　in paralic successions *144—5*
　and planktonic foraminiferal abundance 101, *103*
　rimmed shelves 227
　on seismic data 57, 58, 60
　shown in well data 40
maximum flooding zones *144—5*
maximum progradation surface 28, 29, 33, 34, 35, 77
megasequences 11
Mexico, Gulf of
　biozonation of 91
　time-span of growth-fault history *203—4*
　turbidity currents in ramp settings 197
microfloral assemblages 96
microfossils 89, 92
　stratigraphically useful 90
Midland Basin, Texas, ramp-to-rimmed-shelf transition 227, *229*
midstand systems tract *32—3*
Milankovitch theory of orbital forcing 7, *233*
Miller Field, North Sea, turbidite sandstone reservoir *71, 72*
Mississippi delta, a ramp margin 14
Mississippi fan *205—6*, 210
Moor Grit, an incised-valley fill 64
mouth bars 136, 145
mudstone
　basinal 223
　hemipelagic 185, 193
　marine 140, 168
　overbank 36
　shelf 128
Muschelkalk, oolitic facies 226

nannofossils 89, 101, 104
Navier–Stokes equations 259
Neptunian theory 4
Ness Formation, stacked topset parasequences 74
neutron log (CNL) 67
Nisku Formation (Alberta) 226

non-deposition 81, 82, 84
'non-uniqueness', phenomena of 260
North Sea, central
　Palaeocene/Eocene
　　coastal onlap curve 86, *87*
　　relative sea-level curve 86, *88*
　systems tracts on seismic data *57, 58, 59, 60*

oblique offlap 45, *55*, 86
oceanic basins, restricted 240
oceanic circulation 242
offlap, aggradational *55*, 86
offlap break *14*, 14, 54, 74, 78, 82
　lateral movement of 84, *86*
　vertical movement of 86
oil and gas fields, paralic *155—6, 161*, 167
omission surfaces 36
onlap 53, 75, *163*, 168, *170*
　see also coastal onlap; marine onlap
onlap/offlap patterns, and global sea-level change 5
ooid sedimentation 223
ooids 212
oolites, ramp, South Wales 226
orbital forcing 5, 7, 235
organic matter 252
　degradation and recycling of 239
　enhanced preservation of 241, 244
　preservation of 238, *239*, 239, 241
　terrestrial *238—39*
organic productivity 238, 239
organic-rich facies
　controls on deposition of *238—41*
　delta plain, geochemistry of *242—3*
　and source rocks, delta/coastal plain *242—4*
　and systems tracts in clastic systems *244—9*
organic-rich intervals, source potential varies 249, *250*
orogenesis, as a continuous process 6
Outer Moray Firth
　chronostratigraphic expression of regional seismic line 84, *85*
　seismic data, interpreted/uninterpreted 51, *52*
overbank deposits 36, 113, 193
oxygen, resupply to bottom water restricted 240
oxygen deficiency/depletion (ODD) *see* anoxia
oxygen minimum, most intense areas 242
oxygenation, and carbonate platform drowning **218**

palaeoenvironmental analysis *91—6*
palaeokarst *222—3*, 226
palaeosols 140
　poorly drained **120**, 124
　value of as correlation tool 125
　wet 36, 145
palynofacies analysis 93, *94*
palynomorphs 89
Paradox Basin 235
paralic successions *134—77*
　at a seismic scale 171, *173—4*
　correlation procedures *155—66*
　environments and subenvironments in **134**
　parasequences in *145—51*
　reservoirs in *167—9, 170*
　sequences in *137—45*
　variations in within a sea-level cycle 174, 176
parasequence boundaries 36
parasequence sets 24, 37, *37, 38*, 144, 145, 162
　aggradational 243
　forced regressive, storm-dominated 153
　progradational 31, *146—7, 150—1,* 153, *163*, 166, 197

and ravinement surfaces 41
retrogradational 79, 101, *151*, 153
in storm-dominated shoreline-shelf
  successions *150*–2, 153
transgressive *152*, 153
parasequence stacking geometry/patterns 37,
  38, 75, 77
  change in 28, 30, 193
  relationships
    with basin processes 174
    with stratigraphic position of coals 242,
      244
shelf-margin systems tracts 98
and systems tracts 62–4
parasequences 6, 24, 62, 139
  back-stepping 153
  carbonate 233
  and continental equivalents 36–7
  correlation procedure 161–2
  and cyclicity on carbonate shelves 232,
    234–6
  and high-resolution sequence
    stratigraphy 35–41
  local *161*, 161–2
  log response of 74–5
  in lowstand wedges 98
  in outcrops and cores 61–2
  in paralic successions 145–51
  retrograding topset 58
  stacked 75
  storm-dominated 152–3
  thickness trends 37, 39
  tide-dominated 153
  topset 74, *74*, 75, 78, 79
  wave-dominated 153, *154*–5
Paris Basin, actual and simulated stratigraphy
  261, *263*–4
passive continental margins 14, 197, 203–4,
  247
peat accumulation, modern 238
  rates of *242*, 242
pedogenesis 140
pedogenic horizons 64, 123, 131
periodicity
  in the stratigraphic record 5–6
  see also cyclicity; Milankovitch theory of
    orbital forcing
permeability contrasts 169, *172*
petroleum exploration, and sequence boundaries
  34
Philippines, offshore carbonate platform 231,
  232
photic zone 213–14, 239
phytoplankton 93, 239
pinchouts 168, *170*
plankton 93, 104
  main environmental controls on distribution
    of 95
planktonic markers 101, 107
point-bar deposits 123
polar water, oxygen-rich 242
post-rift megasequence model 11, 13
potential gradient models 261
predation, and carbonate platform
  drowning **218**
progradation 62, 75
  oblique 104
progradational wedges 31, 33, 60
Prudhoe Bay Oilfield 128, 130
pulsation theory 6

Quaternary, evidence of eustatic control on fan
  development 191

raised mires 243
ramp margins 14, *15*
  lowstand systems tracts on 31–2, 197, *205*

ramps
  low-angle clastic *173*, 173
  see also carbonate ramps; submarine ramp
    system
ravinement 153
ravinement surfaces 40–1, 78, 140
red algal interval 218, *220*
red beds 127
reefs 211, 212, 223
  major framework builders 212, *213*
  see also fringing reefs; Leduc Formation
reflection terminations 81, 82
reflections, transference to a time-scale 82
reflectors, geological age of 267
regression, of the shoreline 23–4
regressive systems tracts 33, *33*
relative sea-level 16, 30, 64
  changes in 80, **120**
  fall in 193, 206, 237
    at offlap break of shelf-break margin 14,
      27
    and forced regressions 141–3
    response of fluvial system 121, *124*
    a major control 262
    and offlap break movement 86
  rise in 29, 33, 75, 125, 140, 176, 193
    changes in rate of 37
    rapid 101
    responses of fluvial system 123, *126*, 131
  variations in 36, 128
    can be read from coastal onlap curve 84
relative sea-level curves 84, *85*–6, 265
relative sea-level cycles 26
  and fan deposition 205–6, *207*, *208*, *209*
reservoir facies 107
reservoirs 58
  basinal 34
  division into flow units 169
  karstic carbonate 227
  in paralic successions 167–9, *170*
  reservoir studies and biostratigraphy 91
  stacked 167
  turbidite sandstone 71, *72*
resistivity logs 67, *68*, 69
resolution
  chronostratigraphic, of fossil groups **91**, *91*
  of fossils, actual and apparent 106
  of marker fossils 97
  of seismic data 46–7, *49*
  of well data 61
  see also seismic resolution
retrogradation 62, 75
retrogradational geometries 24
rhythmic successions 7
rift margins 14–15, *15*
rimmed shelves 211, 220, 227
  cf. carbonate ramps 223
  drowning of 227
  facies belts 223
  sequence stratigraphic models for *228*
rivers 28
  not fully equilibrated with lowered base levels
    121–2
  straight and anastomosing 111–12
  see also incised valleys; incision
rockfall *179*, 185
Rocknest Formation, shallowing upwards cycles
  232
rudists 212

sabkhas, supratidal 232
sacred theories 3–4
Sadlerochit Group 128, *130*
  sequences in 130–3
Sag River Formation *130*, 131
salinity variation, and carbonate platform
  drowning **218**
sand : mud ratio, prediction of 265

sand : shale ratio 131, 188
sand bodies 113
  fluvial 123, 126, 131, 133
  lenticular 114–15
  sand-body dimensions 145, 148, 169
  storm-dominated 145
  see also alluvial sand-body architecture;
    channel sand bodies; fluvial sand bodies;
    sandstone bodies
sand sheets 153, *156*
  tidal 62, 166
    prograding *163*, 167, *168*
sand/shale distribution, empirical and sine-wave
  RSL curves 266, 269
sands
  aeolian 71
  sheet 123, 191
sandstone *163*, 185, 193
  channel-fill 191
  paralic **140**, 174
  sheet 133
  valley-fill 127
  see also channel sandstone
sandstone bodies
  closely superposed 174
  highly connected 128
  scale and geometry 155–7, *161*
  shore-parallel 145
Scalby Formation 172
scale, and resolution 263
Scarborough Formation, retrogradational
  parasequence architecture 64, *65*
Sea Level Changes — an Integrated Approach,
  SEPM 42–6
  questioned 6–7
sea-level
  definitions of 15–16, *16*
  see also eustatic sea-level; relative sea-level
sea-level change, eustatic vs. tectonic controls
  5, 6
seals
  intraformational 169, 234
  regional 169
  to topset reservoirs 29
  up-dip 168, *171*
second order cycles 18
sediment bypass 31, 84, 98, 104
  on carbonate escarpment margin 221, *224*,
    224
  of shelf system 193
  to nutrient starved basins 98
sediment calibre
  effect on evolution of lowstand systems tracts
    205–6, *207*, *208*, *209*
  increase in 193
  and sedimentary facies 20
sediment input 265–6
sediment load 20, 22
sediment starvation 252
sediment supply 20–4, 29, 98
  in deltaic systems 153
  for fans during transgression 206
  and fluvial deposits of transgressive systems
    tract 124
  and marine condensed intervals 75, 101
  principles of 20, *22*, 23
  to carbonate slopes 221
  see also accommodation (space/volume)
sediment supply-accommodation volume
  balance 36
sedimentary processes 145, 153, 176
sedimentary sequences, control of
  biostratigraphic character 104
sedimentation, carbonate, controls on 211–20
sedimentation rates 62, 71
  basinal 215–16, *217*
  and organic preservation 242
SEDPAK program (University of South
  Carolina) 260

simulation of Judy Creek carbonate
    platform 261, 263, 266
SEDSIM3 model (Stanford University) 259
    application to Golden Meadow field,
        Louisiana 261, 262
seismic amplitude 55
seismic data
    chronostratigraphic analysis of 86–7
    construction of chronostratigraphic charts
        from 80–3
    display of 47–8
    migration of 47, 174
    recognition of systems tracts on 57–60
    resolution of 46–7
seismic facies 174, 175
    and attribute analysis 54–5
seismic facies mapping
    A, B, C technique 54–5, 55
    and construction of geological models 55
    and geological maps 55, 56
seismic geometries 265, 268
seismic interpretation 47
seismic packages
    and component reflections, numbering of 82
    fan/ramp systems as single package 185,
        188
    mixed sand-mud systems 188
'seismic' reef 211
seismic reflections
    following gross bedding 45, 45
    termination patterns 51–7
    as time lines 45
seismic resolution 173–4
seismic sections 45
    marking up of 51–2
seismic stratigraphy 6, 45–60
    seismic stratigraphic interpretation 45–51
        pitfalls 60
    seismic processing and display for 47–51
seismic surfaces 51–2, 57
    identification of 81–2
    marine 81
    non-marine 81, 81–2
seismic-scale models, paralic systems 171, 173
    deltaic models 171, 173
    low-angle clastic ramps 173
    wide-shelf model 173
Sele Formation 58
SEPM Special Publication 42, 61
sequence boundaries 24, 26, 39–41, 57, 74,
        77–8, 78, 107, 163, 166, 197
    chronostratigraphic expression of 83
    coinciding with flooding surfaces 78
    and correlative conformities 97–8
    and downward shift in coastal onlap 30, 34,
        39, 56
    fluvial deposition patterns associated with
        relative sea-level change 120
    of genetic stratigraphic sequences 34–5
    and lowstand systems tracts in alluvial strata
        122–3
    recognition from seismic data 56, 60
    Sadlerochit Group 130, 130–1, 133
    type 1 27, 28, 32, 56, 77, 78, 82, 83, 98,
        121, 176, 227
        increased benthic foraminiferal extinction
            rates at 97
    type 2 56, 78, 79, 82, 83, 97, 121, 227
        and shelf-margin systems tracts 30–1, 31,
            98
        and valley incision 137–40
sequence sets 145, 162
sequence stratigraphy 253
    and biostratigraphy 97–104
    definitions 33
    of distinct paralic systems 151–5
    evolution of 3–7
    framework combining shelf width and
        sedimentary processes 176

high resolution 7
    interpretation of well-log database, check-list
        79
    log suites used in 64, 67–8
    models
        for carbonate platforms 224–30, 232
        lowstand models 191
    of outcrops and cores 61–4
    and reservoirs in paralic successions 167–9,
        170
    of wireline logs 64, 67–79
sequences 11, 17, 35
    clinoform units in 74
    defined 24, 26
    fourth order 60
    in paralic successions 137–45
    second and third order 33
    and systems tracts 24–35
    type 1 26, 26, 29, 29
        variation of coastal onlap with time 84,
            86
    type 2 26, 29, 29, 31
    see also genetic stratigraphic sequences
Shaftesbury Formation (Canada) 247
shale complexes, floodplain and lacustrine 131
shales 74, 75, 101
    condensed 98
    hemipelagic 185, 188, 197
    transgressive 23, 98, 168
        black 238, 244, 245
shallowing-up cycles 18, 232, 234
shallowing-up signature 62, 98
    in benthic fossil assemblages 104
sheet sands 123, 191
sheet sandstone 133
shelf break 13, 14, 211
shelf deltas 29, 104, 136, 136, 153, 171, 173
    classified 158–9
shelf edge 14, 14
shelf processes 174, 176
shelf width
    lowstand systems tracts 98
    relationship with sedimentary processes 176
shelf-break margins 14, 15, 26, 30
    midstand systems tracts 32–3
shelf-edge deltas 104, 135–6, 136, 153, 171,
    173, 173, 176, 239
shelf-margin systems tracts 79, 98
    and type 2 sequence boundaries 30–1
shelf-margin wedge systems tracts 227
shelf-slope break 135, 197
shelf-slope systems 173, 205, 206
    relict 185
shoal facies 232, 234
    carbonate ramps 225, 226
shoreface 137
    barred and non-barred 152
    erosion 155
    lowstand 142–3, 144
    retreat 40, 41
    upper 152, 153
shoreface erosion plane see ravinement surfaces
shoreface stacking pattern-fluvial architecture
    relationship 125, 128
shoreline-shelf depositional systems
    accommodation/relative sea-level/water
        depth
        relationship 16, 18
        tide-dominated 153
shorelines 145
    basinwards movement of 141–2
    lowstand, types of 142–3, 144
    wave-dominated 154–5
Shublik Formation carbonate 128, 130, 130,
    131, 133
sigmoidal offlap 55, 86
siliciclastic input 235
    and carbonate platform drowning 218, 226
simulation 260

input parameters for 263–5, 267, 268
    of seismic clinoform geometries 265–6, 268
slide gullies 191
slope aprons 181, 182, 206, 224
    gravel-rich 185, 186
    mixed sand-mud 188, 189
    mud-dominated 188, 190, 191
    sand-rich 185, 187
slope failure 27, 206
slope fan systems, vs. basin-floor fan systems
    191, 193–7
slope fans 27, 28
    seismic expression of 193–4, 196, 197
    on wireline logs 197
'slope front fill' 27, 197
slope systems, shelf-edge delta fronts 171, 173
slump packages 188
slumps/slumping 104, 178, 179
    fault-induced 136
    rotational 185, 191
Smackover Formation (USA) 225, 251
    ramp system 225
soils 222–3, 226
sonic logs 67, 68
sonic transit time 67, 69
source rocks 34, 241, 243
    and delta/coastal plain organic-rich facies
        241–3
        prediction of oil-prone source rocks 242–
            3, 245
    distribution of through geological time 240
    marine carbonate 248–52
        deep ocean basin 252
        intercarbonate build-up 241–50, 252
        intraplatform depression 250–1, 254
        unrestricted basin margin 251–2, 256
        models for development of 244–5, 247,
            247–8, 249, 250
South Viking Graben, tectonic subsidence
    curve 13, 13
SP (spontaneous potential) logs 67, 68
spectral gamma logs 67
squash plots 51
storm beds 223
strandplain ramp system 225, 225
stratigraphic correlation, in fluvial deposits
    125–6
stratigraphic cycles, hierarchy of 17–18, 21
stratigraphic surfaces
    key surfaces 64, 75, 77
    recognition of 55–7
stratigraphic traps 29, 31
    in paralic successions 168, 170, 171
stratigraphy, control by world-wide sea level
    change 5–6
stream/river rejuvenation 26, 118, 123, 140
strike-slip basins 13
stromatoporoids 212
subaerial erosion surfaces 24
subaerial exposure 122, 216, 232, 237
subaerial exposure surfaces 123
submarine canyons 60, 206, 210, 239
submarine fans 14, 27, 75, 101, 181–2
    deposition and the relative sea-level cycle
        205–6
    development during transgressions 206, 210
    gravel-rich 185, 186
    hydrocarbon-bearing, relating to periods of
        relative sea-level fall 191, 194
    mixed sand-mud 188, 189, 194, 197,
        200–1, 204, 206
    mud-rich 179, 180, 191, 197, 202–3, 206
    sand-rich 185, 187, 188, 194, 198–9
    waxing/waning of sedimentation 71, 73
    see also lowstand fans
submarine ramp systems 181, 181–2
    gravel-rich 185, 186
    mixed sand-mud 206
    mud-rich 191

sand-rich 185, *187*, 188
submarine slides 178, **179**, 191
subsidence 15
 differential 11, 227, 250
 see also tectonic subsidence
'super sequences' 6
surface mixing layer, and anoxia 241, 244
swamp peats 120
swamps/marsh areas 238
swing (of seismic data) 49
symmetrical trend *see* bow trend
syn-rift megasequence model 11
systems tract boundaries 32, 41
systems tracts 11, 82
 bounded by seismic surfaces 81–2
 defined 26
 idealized development of *205*
 identification of from log response 78–9
 and parasequence stacking patterns 62–4
 partitioning of sand-body types into **145**, 145
 and preservation of terrestrial organic matter 238–9
 recognition of on seismic data 57–60
 and sequences 24–35

talus cones 231
Tarbert Formation *70*, 70
 dirtying-up unit *71*
tectonic cycles 17
tectonic subsidence 11–13, 15, 31, 161
tectonic tilting 118
tectonics 11, 241
 intrabasinal, and midstream fluvial systems 118
 in upstream fluvial systems 118
temperature 212
thin bed effect 46
third-order cycles 19, 242
 foundation of sequence stratigraphy 18
tidal deposits, transgressive 153
tidal flat progradational model, and carbonate rhythms 235–6
tidal flats 153, *156*, 222–3
 prograding 153
tidal island model 236
tidal ranges 174, 176
tidal scour 153, 155, 167
tidal successions, progradational 153, *156*
tides 174, 176
time resolution 263
time-stratigraphy 6
TOC (total organic carbonate) 245
Top Beauly Formation Coal 58
Top Dornoch Formation Coal 58
top lowstand fan surface 57
top lowstand surface 29
toplap 53, 81, 82, 83, 145
topseal 34
topset accommodation volume 24, 29, 32
topset highstand deposits 104
topset parasequences 74, *74*
 in a highstand wedge 79
 in a lowstand wedge 8
 retrogradational stacking of 75

topset reflectors 84
topset-clinoform profile *14*, 14
topset-clinoform system 28, 54
 prograding 27, 30, 60
topsets 13, *14*, 53, 84, 225
 and construction of relative sea-level curves 85–6
 deposits, Ivishak Formation 131
 transgressive system 29, *29*
 type 1 sequence boundary in 78
trace equalization 47–8, *50*
trace shape 47, 48–50
trace summing 51
transfer zones, rift margin 15
transgression 23, 24, 74, 84
 fan development during 206, 210
 regional, and enhanced organic preservation 241
 and rimmed shelves 227
transgressive surfaces 29, 35, 56–7, 58, 98, 101, 155, *163*, 167, 197, 257
 in paralic successions 139–40
 represents a retrogradational biofacies boundary 101
transgressive systems tracts 29–30, 64, 75, 101, 176, 252
 carbonate ramps *225*, 225
 deposition of organic-rich carbonates 249–50
 and flooding surfaces in alluvial strata 123–4
 fluvial depositional patterns associated with relative sea-level change **120**
 general features 244
 important for development of marine oil-prone source rocks 244–5, *247*, 247–8, *249*, 250
 on a ramp margin 31–2
 recognition of on seismic data 58, *59*
 recognized as a retrogradational parasequence set 79
transport efficiency 179, *180*
truncation 53, 83, 166, *170*
tunnel trend *see* cleaning-upwards trend (motif)/signature
turbidite boxcar units 71, *72*
turbidite facies 178
turbidite systems, deep-marine 206
 mud-rich 188, *190*, 191
 sandstone packages 191
 variability of 181
turbidites 28, 104, 223
 delta-front 14
 high-density 185
 thin-bedded 71
turbidity currents 14, 178, **179**, 206
 high-density 180
 high-efficiency 179
 in ramp settings 197

unconformities 4, 5, 24, 30
 at tectonically enhanced sequence boundaries 97
 boundaries of 83
 subaerial 35, 39

see also drowning unconformities
unrestricted basin margin source rocks 251–2, 256
Upper San Andres Formation (New Mexico), mixed carbonate-siliciclastic cycles 234
upward-shoaling cycles 74
upward-shoaling facies 36
upwards-deepening facies 36
uranium 101
 affecting gamma ray logs 67, 249, *257*

'variable area' format 48, *50*
'variable area and wiggle trace' display 48, *50*
'variable intensity (density)' display 48
Viking Formation (Alberta) *164*–7, 166–7, *168*
 erosion surfaces in 143, *144*
 parasequences deposited in tidal sand-sheet setting 146–7
 progradational parasequences 62–3, *63*
 stratigraphy of *163*
visual enhancement (of seismic data) 48
 techniques 50–1

water balance, negative 242
water column stratification 94, 241, 244
water depth, and supply of marine organic matter to bottom sediment 239
water tables, rising 124
waves 176
wells
 addition to chronostratigraphic chart 83
 core data from 61
West Florida Platform, steep-walled 228, 230, *231*
Western Interior Seaway
 coal positions related to parasequence stacking patterns 242, *244*
 maximum organic richness 248
Wheeler diagrams 80, 265, 266, *268*
'wiggle' trace 48, *50*
Wilmington Platform (USA), drowning unconformity *219*, 227
Wilson Cycles 241
wind regime 148, 151
wireline logs
 basin-floor and slope fans on *196*, 197, *204*
 Brown Limestone *257*
 calibration by core data 61
 Douvernay Formation 253–4
 Hanifa Formation *255*
 log suites, in sequence stratigraphy 64, 67–9
 log trends 69–72
 organic-rich Kimmeridge Clay Formation *249*
 pitfalls/ambiguities in sequence analysis of data 79
 sequence stratigraphy of 64, 67–79
 use in characterising source rocks 245

Yoredale Series, carbonate cycles 234